GeoJournal Library

Volume 116

More information about this series at http://www.springer.com/series/6007

Ying Long • Zhenjiang Shen

Geospatial Analysis to Support Urban Planning in Beijing

Springer

Ying Long
Beijing Key Lab of Capital Spatial
Planning and Studies
Beijing Institute of City Planning
Beijing, China

Zhenjiang Shen
2C718
Kanazawa University Natural Science
and Technology Hall
Kanazawa, Japan

Tsinghua University
Beijing, China

Fuzhou University
Fuzhou, China

ISSN 0924-5499 ISSN 2215-0072 (electronic)
GeoJournal Library
ISBN 978-3-319-19341-0 ISBN 978-3-319-19342-7 (eBook)
DOI 10.1007/978-3-319-19342-7

Library of Congress Control Number: 2015942626

Springer Cham Heidelberg New York Dordrecht London

Printed on acid-free paper

Springer International Publishing AG Switzerland is part of Springer Science+Business Media (www.springer.com)

Foreword

Geospatial analysis as one of key techniques for supporting urban and regional planning has been developed for decades and used widely in the world. Authors of this book focus on planning support for the planning compilation and evaluation procedure of planners in the Beijing Institute of City Planning. Such experiences enrich advanced geospatial analysis techniques for planning support in various aspects. The book, consisting 13 chapters, elaborates diverse planning support efforts, which primarily introduced how geospatial analysis could be applied to the planning practice to facing multi-issues in the mega capital city.

The applications of geospatial analysis in Beijing are described as two essential aspects of urban system: urban space and urban activity. In the book, urban activity refers to human behaviour taking place in urban space, and their spatial pattern can reflect how human beings behave relative to the urban form. Authors have tested urban activity using large-scale personal traffic surveys and their moving tracks recorded in mobile devices. The analysis was intended to identify the conflicts between urban activity and urban spaces, therefore providing solutions to the problems of urban planning and design.

While reading these manuscripts, we are witnessing a dramatically changing world in the twenty-first century, especially our data environment. In such a background, the book includes several chapters which adopt open/big data as the main data source, e.g. public transit smart card records. And they also propose a new research diagram called "Big Model". We expect there would be more and more new data sources for feeding geospatial analysis in the coming works of new urbanization and promoting urban planning support to a new stage.

School of Architecture
Tsinghua University
Beijing, China

Qizhi Mao

Foreword

Humanity is facing the inevitable fact that most of the global population will be concentrated in urban areas in the not too distant future. Human population movement and displacement are quickly becoming important contemporary issues. The resulting urban growth from population movement is a key factor contributing to the reduction of valuable agricultural and forested lands, increase in environmental degradation, and loss of natural ecosystems. Population displacement arising from economic growth and uncertainty, climatic hazards, landslides and earthquakes impose unplanned relocations that have consequences on the evolving local and global dynamics of cites around the world. The overall dynamic is that some cities decline while others expand at seemingly uncontrolled rates.

Consequently, there is a vital need to monitor, analyze, and model the spatial dynamics of land use change and urban growth in order to better understand the changes from a human-environmental perspective so that we can improve planning for a sustainable future. One approach is to integrate the suite of available geospatial methods and tools to bridge the gap between the complex technical models of geocomputation and geosimulation with the needs of urban planners in order to make decisions that have minimal unintended consequences. This new book provides a valuable step in that direction.

Beijing is one of the world's megacites with a rich history making it a vibrant spot on the earth that has attracted people for centuries. Today, the city is experiencing increased economic growth and faster than ever physical expansion, thus the urgent need for advanced urban planning support systems. The timing for this book could not have been more relevant as it provides a comprehensive list of the cutting-edge approaches that the geospatial community can currently offer to better understand and forecast urban change patterns to manage their ecological impacts and assist in robust planning.

The book is organized at both a conceptual level and a technical level. At the conceptual level, the important issues related to the causes and effects of urban form changes are emphasized. At the technical level, the analysis and simulation of spatial patterns of land use change caused by human interactions, stakeholder desires, and constant urban development are emphasized. Cellular automata and agent-based modeling are presented as useful planning support tools that can represent and handle the complexity of changes in the urban fabric from local levels to larger city levels. Issues such as human mobility and transportation are tackled with care as they are key components in the massive daily interactions between commuters. Economic elements are also considered in the simulation approaches as land and rent prices together with economic constrains that impact labor demand and supply all having an impact on the future urban landscape. Using a foundation of geosimulation and other geospatial tools in an era when big data is becoming increasingly accessible, this book offers a unique approach to urban planning by using geospatial models to support and shape future development in Beijing.

Spatial Analysis and Modeling Laboratory
Simon Fraser University, Burnaby, Canada

Suzana Dragicevic

Acknowledgements

We would like to express our deep appreciation to the series editors of *GeoJournal Library*, Professor Daniel Sui of the Ohio State University for his kind support and encouragement, Stefan Einarson, Urmila Baidjoe and Andre Tournois of Springer for their kind editorial work and help to have such diverse topics on *Geospatial Analysis to Support Urban Planning in Beijing* published as a monograph. Special thanks to Bernadette Deelen-Mans and Mireille le Grand-van Kan for their efforts on helping us to make this book possible in the initial step of preparing the proposal of this title. This book is partially funded by the National Natural Science Foundation of China (51408039 and 51078213).

We are deeply indebted to Professor Qizhi Mao from the School of Architecture, Tsinghua University, whose help, stimulating suggestions and encouragement helped us in all the time of research. We want to express our thanks to the contributors listed below. We enjoyed the academic collaborations with them. Without them, the monograph cannot be fulfilled and reach the current quality.

Contents

Contributors

Sangni Gao Department of City and Regional Planning, Cornell University, Ithaca, NY, USA

Song Gao UCSB Center for Spatial Studies and Space and Time Knowledge Organization Lab (STKO), University of California Santa Barbara, Santa Barbara, CA, USA

Haoying Han Department of Urban and Regional Planning, College of Civil Engineering and Architecture, Zhejiang University, Zhejiang, China

Cuiling Liu Ministry of Housing and Urban-Rural Development of the People's Republic of China, Beijing, China

Xingjian Liu Faculty of Architecture, The University of Hong Kong, Hong Kong, China

Yao Shen Bartlett School, University College London, London, UK

Zhenjiang Shen 2C718, Kanazawa University Natural Science and Technology Hall, Kanazawa, Japan

Tsinghua University, Beijing, China

Fuzhou University, Fuzhou, China

Jianghao Wang Institute of Geographic Sciences & Natural Resources Research, Chinese Academy of Sciences, Beijing, China

Kang Wu Department of Urban Economics, Capital University of Economics and Business, Beijing, China

Xiaobai Yao Department of Geography, University of Georgia, Athens, GA, USA

Xiang Yu College of Public Affairs, Zhejiang University, Hangzhou, China

Yongping Zhang School of Planning and Geography, Cardiff University, Cardiff, UK

Nijmegen School of Management, Radboud University Nijmegen, Nijmegen, The Netherlands

Jiangping Zhou School of Geography, Planning and Environmental Management, The University of Queensland, Brisbane, Australia

About the Authors

Dr. Ying Long, the associate professor in the Beijing Institute of City Planning and a senior researcher in Beijing Key Lab of Capital Spatial Planning and Studies is an interdisciplinary scholar with a global vision and substantive planning experiences in China. He received his PhD from School of Architecture, Tsinghua University. In the past few years, his research focuses on urban planning, quantitative urban studies and applied urban modelling. Familiar with planning practices in China and versed in the international literature, Dr. Long's academic studies creatively integrate international methods and experiences with local planning practices. Dr. Long is also the founder of Beijing City Lab (BCL, www.beijingcitylab.com), an open research network for quantitative urban studies.

Prof. Zhenjiang Shen, the corresponding author of this book, a professor in Kanazawa University, is a visiting professor in Fuzhou University, Chongqing University and Fudan University, and he also served as an associate instructor for the Ph.D. education programme in Tsinghua University, China. His research interests include policy-making support system for planning and design using GIS and VR. He is now serving as an academic commissioner of the City Planning Institute of Japan and the planning advisor in local cities such as Nanao City and Kanazawa City in Japan and developed online design tools for enhancing public participation. Prof. Shen also participated in the spatial strategic planning of local cities in China and cooperated with the Beijing Municipal Commission of Urban Planning for metropolitan growth simulation. Prof. Shen is now the editor in chief of *IRSPSD International*, Japanese editor of *IJSSS* and organizing an International Community on Spatial Planning and Sustainable Development from 2010.

Biography of Contributors

Sangni Gao obtained dual degrees of bachelor in economics and environmental engineering from Tsinghua University in 2012 and a master of regional and city planning from Cornell University in 2014. In 2012, she worked as a real estate planning assistant in Xin-Tai-Cang Historic Courtyard Hotel in Beijing. In 2013, she worked for Hudson River Park Trust in New York as a land use and zoning consultant. From January to May 2014, she worked as a master planner for R.F.K. Memorial Stadium Campus in Washington, DC. From November 2014 to January 2015 she worked as a city planner in New York City Department of City Planning. From February 2015 to present, she has been working as an urban planner/real estate project manager in Brooklyn, NY.

Song Gao is a Ph.D. student majoring in GIScience and a research assistant in the Department of Geography at UC Santa Barbara, co-supervised by Dr. Krzysztof Janowicz and Dr. Michael F. Goodchild. He obtained his M.S. degree from the Institute of RS and GIS, Peking University, and B.S. from Beijing Normal University. His research interests include Big Geo-Data Analytics and spatio-temporal analysis on urban informatics. He has published about twenty peer-reviewed articles and serves as a peer reviewer for several prominent journals including *Proceedings of the National Academy of Sciences*, *Transactions in GIS*, *Computers, Environment and Urban Systems*, *GIScience & Remote Sensing*, *International Journal of Digital Earth* and *International Journal of Geographical Information Science*.

Haoying Han is a professor in the Department of Urban and Regional Planning and director of the Institute of Planning Theories and Technologies at Zhejiang University. He received his B. Arch. and M. E. degrees from Tsinghua University and Ph. D. degree from the University of Tokyo. His research interests are urban complexity and planning theories. He is the associate editor of *Journal of Urban Management* (Elsevier journal) and author of several academic papers published in peer-reviewed journals, including *Environment and Planning B*, *Landscape and Urban Planning*, *Land Use Policy*, etc.

Cuiling Liu master of science in Cartography and Geographic Information System, Capital Normal University, Beijing, China. Research interests include urban expansion simulation and analysis, geo-simulation and urban micro-simulation, big date, geo-spatial analysis, data mining, spatial data structures, spatial databases, urban planning evaluation.

Xingjian Liu is an assistant professor in the Department of Urban Planning and Design at the University of Hong Kong. His research interests are in urban and regional development, globalization and world cities as well as spatial/network analytics. Xingjian received his degrees from Wuhan, Texas State and Cambridge. Xingjian has authored over 30 scholarly articles, and his research has appeared in peer-reviewed journals such as *Environment and Planning A/B*, *Urban Studies*, *Urban Geography* and *Regional Studies*. Xingjian's most recent research efforts focus on understanding urban form and function through the analytical lens of emerging data sources.

Yao Shen is a trained architect, urban planner and designer. Currently, he is a Ph.D. candidate in Space Syntax Laboratory in Bartlett School of Architecture, University College London (UCL). Yao received his B. Eng. and M. Eng. from the School of Architecture in Tianjin University with honours. His main research interests focus on urban morphology, big data-based spatial analysis, generative urban modelling and data-informed urban design. Professionally, Yao has been involved in a wide range of projects in China and Southeast Asia, wining several national and international prizes. In parallel with his practice works, Yao has published 10 peer-reviewed papers until now; simultaneously, he works as the reviewer in several academic conferences and journals (e.g. *IJGIS*). He is also the founder and funding chief designer of CityMatrix, a data-based design and research studio in China.

Jianghao Wang is an assistant professor in the State Key Laboratory of Resources and Environmental Information System, Institute of Geographic Sciences and Natural Resources Research, Chinese Academy of Sciences. His research interests are in geographic information system, spatial statistics, spatio-temporal data mining and remote sensing of the environment. Jianghao Wang has authored over 20 scholarly articles, and his research has appeared in peer-reviewed journals such as *International Journal of Geographic Information Science*, *IEEE Transactions on Geoscience and Remote Sensing* and *IEEE Geoscience and Remote Sensing Letters*. Jianghao's most recent research efforts focus on earth surface processes modelling with spatio-temporal geostatistics and spatio-temporal data mining through volunteer geographic information.

Kang Wu received his Ph.D. in geography from the Chinese Academy of Sciences and then joined the department of Urban Economics at Capital University of Economics and Business as an assistant professor in July 2013. His research interests are urban economic geography and regional planning and recent works focus on understanding the spatial organization of polycentric city regions and urban networks in China.

Xiaobai Yao is an associate professor of geography at the University of Georgia (UGA). She received her B.S. degree from Wuhan Technical University of Surveying and Mapping in China, MSc degree from the International Institute of Aerospace and Earth Science in the Netherlands and Ph.D. in geography from the State University of New York at Buffalo. Her research interests centre on geospatial analysis and modelling, transportation and urban/regional analysis, network analysis as well as health. Among other honours and awards, Xiaobai received the ESRI-UCGIS Young Scholarship Award in 2005. She is the UGA delegate to the University Consortium of Geographical Information Science (UCGIS) and the incoming chair of the Geographical Analysis and Modeling Commission of International Cartographic Association (ICA).

Xiang Yu is a master student in College of Public Affairs, Zhejiang University. His research interest is applying big data for understanding urban structure of Chinese cities.

Yongping Zhang is now a master's student at Cardiff University (Wales) and Radboud University Nijmegen (the Netherlands), simultaneously, and expected to obtain his master's degrees in European spatial planning and environmental policy from Cardiff and in social and political science of environment from Nijmegen in August 2015. Before these, he has obtained a master's degree in urban and regional planning from Peking University in 2013 and bachelor's degree in geographical information system from Wuhan University in 2010. He aims to be an urban scientist who can understand urban issues comprehensively and can provide workable—usually quantitative and GIS-related—solutions to support sustainable development of cities. His recent interests are urban modelling, spatial analysis using urban big data and Chinese urbanization.

Jiangping Zhou is an assistant professor in community and regional planning, transportation and sustainable environments at Iowa State University. Before that, he was China policy director, Institute for Transportation and Development, and UCLA's principal transportation planner, respectively. He has degrees in urban planning from both China and the USA. His interdisciplinary Ph.D. degree in policy, planning and development is from the University of Southern California (USC). He completed graduate certificates in sustainable cities and GIS at USC. His interests are sustainable urban/transportation systems, urban form and related policies and planning. He is elected member of the congestion pricing subcommittee of the US Transportation Research Board and elected board member of the International Association for China Planning. He also served on editorial boards of *China City Planning Review* and *Urban Planning International*.

Chapter 1
Geospatial Analysis and Application: A Comprehensive View of Planning Support Issues in the Beijing Metropolitan Area

In planning practice, geospatial analysis and modelling serves as a planning support tool to search for solutions of a better urban form by ascertaining the patterns of activities in urban space (Neutens et al. 2010; Yuan et al. 2012), because it is usual that planners start to consider the future plan of a city based on the findings of geospatial analysis. Conventionally, planners' visions for the future can be divided as demand-oriented or solution-oriented scenarios. Demand-oriented scenario is designed for allocating the defined economic demand and supply in urban spaces, and solution-oriented scenario is for solving some particular planning issues that take place in urban areas. Geospatial analysis usually serves as the starting point of planning process and the geospatial analysis results are the key inputs to set up those scenarios for urban simulation. For example, geospatial analysis researcher (e.g. Xie 1996) took the initiative to simulate the evolution of urban forms and to capture interactions among various dynamics driving the development of cities using cellular automata (CA) models and GIS methods. Along this line, researchers further explored the use of geospatial methods in simulating and predicting urban forms for planning sustainable cities (Batty et al. 1999; Cheng and Masser 2004; Li and Yeh 2000, 2002; Oguz et al. 2007; Torrens 2006). Thus, geospatial analysis and modelling are serving as important tools to support urban plan.

Since China adopted reform and opening up policies in the late 1970s, the Beijing Metropolitan Area (BMA) has rapidly developed into one of the largest metropolitan areas in the world. This development has been at the expense of environmental quality and ecological sustainability. The entire area has long suffered from citywide traffic congestion, severe air pollution, and environmental deterioration. These problems have drawn considerable attentions from both academia and practice. Some studies have investigated the urbanization issues in Beijing from various perspectives, ranging from land-use changes (Wu et al. 2006), traffic impact (Huang and Pan 2007; Wang 2005; Zhang and Gao 2008), air pollutions (Gualtieri and Tartaglia 1998), to housing issues (He et al. 2011; Wang and Chai 2009; Wang and Li 2004). In this book, we primarily introduced how geospatial analysis can be used

Y. Long, Z. Shen, *Geospatial Analysis to Support Urban Planning in Beijing*,
GeoJournal Library 116, DOI 10.1007/978-3-319-19342-7_1

to the planning practice to solve these planning issues in the Beijing Metropolitan Area (BMA).

The application of geospatial analysis would focus on two essential aspects of urban system: urban space and urban activity. In this book, urban activity refers to human behaviour taking place in urban space, and their spatial pattern can reflect how human beings behave relative to the urban form. We tested urban activity using large-scale personal traffic surveys and their moving tracks recorded in mobile devices. The analysis was intended to identify the conflicts between urban activity and urban spaces, therefore providing solutions to the planning and design issues.

1.1 How Geospatial Analysis Help Planners

There are many researches on geospatial analysis and modelling using Geographic Information System (GIS) (Maguire et al. 2005). The applications of GIS for spatial analysis and modelling range from simulation to prediction and monitoring (Bernard and Kruger 2000; Carrara et al. 2000; Huang and Cai 2007), from human-related issues to natural hazards modelling (Al-Sabhan et al. 2003; Perry et al. 1999; Thornton et al. 2011; Xu et al. 2014), from real-world problems to virtual-world activities modelling (Croitoru et al. 2014; Turgeon 2013), from using data collected by digitally instrumented devices (e.g. remote sensing images, GPS coordinates, sensor network data) to using data collected by field survey and mobile devices (e.g. survey data, geo-social media data, mobile phone data) (Blaschke et al. 2011; Liu et al. 2014; Patino et al. 2014; Widener and Li 2014). Thus, there are many types of urban data, which are important sources for planners to find out planning issues and solutions in urban areas using geospatial analysis technologies.

1.1.1 Geospatial Analysis: Spatial Patterns and Urban Development

Geospatial analysis involves GIS analytical techniques and advanced techniques such as geocomputational methods. The basic components of geospatial analysis include surface analysis, network and locational analysis, distance and directional analysis, and geometrical processing (e.g. exploratory spatial data analysis and spatial statistics). Geocomputational methods usually include Cellular automata and agent based modelling. Geospatial analysis is built upon GeoDatabase, which usually come from a range of national surveys. Urban form is a result of the interactions between stakeholders under spatial planning and relevant planning policies while considering economic, social, and ecological aspects (Kawakami et al. 2013). For instance, driver forces from private development are always competing for low cost and high profit; thus, the allocation of land demand of such private sectors (e.g.

industries, shopping malls, and housing development projects) tends to locate at the urban fringes, thereby resulting in urban sprawl. Simulating spatial patterns of land use development using geocomputation methods can represent the urban form and it is possible to identify the roles of driving forces for achieving a better urban form of Beijing.

Geospatial analysis is utilized for investigating current situation of urban areas in BMA and making evaluation to planning outcome in its planning practice. Integrating urban theories and GIS methods allows researchers to understand how planning policies influence urban form by simulating activities of stakeholders in urban spaces (Shen 2012). In this book, geospatial analysis is a tool to understand the spatial patterns of land use development. urban boundary area is simulated in order to investigate the policy impacts on land use development as driving forces. An approach is labeled as "form scenario analysis" (FSA) in order to analyze the consistency between a predefined urban form and the required spatial policies, through simulating the impacts of different spatial policies. The simulated urban form, actually is divided as developed and undeveloped areas according to the methodology of cellular automatic simulation and the boundaries of the metropolitan are presented based on different urban growth policies.

1.1.2 Better Urban Form: Human Behaviour and Their Spatial Patterns

Many studies examined the capacity of geospatial analysis and modelling in aid in understandings of household behaviours, such as housing location choices (Ettema 2011; Gaube and Remesch 2013; Huang et al. 2013) for investigating residential demand (Fontaine and Rounsevell 2009), and relative land-use changes (Jokar Arsanjani et al.2013; Kocabas and Dragicevic 2013). The analysis target is the human behaviours and their spatial patterns of activities. It is possible to make analysis at the level of macro and micro urban space and to investigate the spatial patterns reflecting the behaviour patterns of collective population between zones or of individual person in urban spaces. In the work of analysis, human attributes such as age, sex, affiliation and professional skills can be classified into different groups to examine how the population move from their houses to work and shopping in urban space. The conventional national surveys are limited in tracking urban activities. Big data collected from mobile devices and traffic cards store rich real-time individual dataset and therefore can help planners to understand the diverse urban activities and their interaction pattern over time and space (Nabian et al. 2013).

In this book, we argue that it is possible to identify problems hidden in urban structure using geospatial analysis on human behaviours happening in real or simulated urban space. For problem-oriented analysis using GeoDatabase of diverse surveys, it is important to discuss how to figure out the solution-oriented scenario for considering effective urban policies regarding urban form, such as urban growth boundary, transportation network and land use pattern through human behaviours in the process of spatial planning implementation. Thus, "urban activities" in this book refers broadly to human activities and their spatial patterns on urban spaces under planning and policy implementation. Thereby, the solution-oriented scenario is possible to be the outcome of geospatial analysis on human behaviors and their spatial patterns in urban space. We believe that urban data provide new possibilities for geospatial analysis to solve planning issues in BMA.

1.1.3 Planning Support: Developing Tools for Planning and Design

In this book, we have considered how to simulate the planners' work and to generate alternatives to land use planning using agent-based models based on the results of geospatial analysis. We also described the framework of planning support tools developed in Beijing and try to connect our thinking of "big model" for our future work. In general, most of planning support tools are developed based on GIS technologies, such as Index, WhatIf?, CommunityVIZ, Cube and other software. Longley et al. (2005) described basic GIS concepts and addressed the radical shifts that have taken place in GIS technologies and its applications. GIS data collection, data transformation and linkage, implementing GIS and using it as a management tool are key technologies for urban planning management.

Planning support tools are developed for supporting planning and design during the planning process, which includes establishing planning objectives within a planning framework, defining scenarios, simulating urban activities in urban space, visualising simulated results, and supporting decision making. Geospatial analysis and modelling are widely discussed in the context of planning support in the world. For example, simulation can be used for predicting the future of urban spaces based on different planning alternatives, planning scenarios and strategies through geosimulation and geovisualisation (Shen 2012). With the help of urban models that are designed for finding optimal solution and simulating the change of urban space, geospatial analysis can be a useful tool for planners and decision makers to understand the current situation based on survey data and evaluate the possible future based on simulated results. Examples in this strand include studying cities' social areas using census data (Gu et al. 2005; Li and Shanmuganathan 2007; Yao and Zhang 2014), measuring life qualities and neighbourhood disparities (Li and Weng 2007; Neckerman et al. 2009), and predicting transportation needs based on land-use patterns and demographic characteristics (Yao 2007). With respect to geospatial analysis and application for planning support, issues such as space and time, problems with data quality, and uncertainty and error detection are important for reaching correct and reasonable results from analysis. Yeh and Li (2006), for example, pointed out that errors in GIS source can propagate through modelling processes.

As mentioned above, geospatial analysis is conducted for finding problems in urban areas or evaluating the outcome of planning in practice. Therefore, a planning solution for solving the current issues can be designed as solution-based scenario for planners to conduct a simulation and investigate the possibility of the suggested planning solution. In this book, in order to understand how people use the urban space under different plan alternatives, we show how urban policy impact on urban boundary during urban development period and how human behaviors take place in the planned land use patterns of employment, shopping locations and housing locations in BMA. Furthermore, we try to setup solution-based scenario through geospatial analysis on geodatabase of conventional survey data and big data in order to simulate the change of urban boundary, transportation network and land use pattern in urban space of BMA.

1.2 Urban Form: Spatial Patterns and Land Use Development

We argue that spatial patterns of land use development can represent the urban form and it is possible to identify an optimal or to simulate a better urban form of Beijing using geocomputation methods. For this, geospatial analysis is a tool to understand the spatial patterns of land use development. There are some chapters on simulating and visualising land use changes using cellular automata (CA) for urban growth simulation (UG). This can be used to represent the urban boundary of Beijing, simulated by the Beijing urban development model (BUDEM).

1.2.1 Planning Targets and Raster Dataset for Simulating Urban Form

In Chap. 2, the authors propose an alternative process to set up an urban form as a planning scenario, instead of conventional urban growth scenario analysis (SA) by urban growth models using CA. In most of urban growth models using CA, development policies are designed as scenario conditions for reflecting impact of driving forces on future uncertainties of urban developments (Landis 1994, 1995; Klosterman 1999). Couclelis (2005) argued that the land use models have done little to investigate desired or unwanted urban form. In support of Couclelis's argument, we use the expected urban form as the scenario condition to identify the spatial policies required to realize the predefined urban form. Because the favor of artistic planning rather than strategic planning is still predominant in China, the lack of advanced planning theories and technical tools largely limited the professional scope and skills of planners. However, the unprecedented progress of urbanization and resulted environmental crises in Chinese cities call for a revolutionary transform of the existing planning procedure. We propose an approach labeled as "form scenario analysis" (FSA) in order to analyze the consistency between a predefined urban form and the required spatial policies, by simulating the impacts of different spatial policies. FSA can potentially be applied to identify policies required to create user-defined sustainable urban forms.

For predicting the urban form of BMA, the Beijing urban development model (BUDEM) is developed. As a case study, the authors explain the simulated urban form of Beijing-Tianjin-Hebei Area (BTH) using cellular automaton in Chap. 3. Based on prevalent urban growth theory using constrained cellular automatic, BUDEM has been developed in 2008 for analysing and simulating urban growth for the BMA. It has been proved that the model is capable of analysing historical urban growth mechanisms and predicting future urban growth for BMA. In the case study of BUDEM-BTH, the authors identify urban growth mechanisms in two historical phases from 2000 to 2005 and from 2005 to 2010. The heterogeneity of driving force as model parameters were investigated for fulfilling accurate simulation in Beijing-Tianjin-Hebei area. The simulated urban form, actually is divided as developed and undeveloped areas according to the methodology of cellular automatic simulation and the boundaries of the metropolitan are presented based on different urban growth scenarios for 2049 from 2020.

All the work of the two chapters, we use the raster data set such as cell-based land use patterns and town center location data using rasterized map for calculating the development potential using CA.

1.2.2 *Vector Database for Measuring and Simulating the Urban Form*

In the BUDEM work, spatial dataset such as road network, town centers and rivers are reorganized as raster data, and slope data, land use data regenerated from remote sensing data are used for urban form simulation. Nowadays, most of spatial database in the world are prepared as vector database, such as roads, building footprints, and parcels together with their attributes. In the practice of Beijing, parcel direction (PD) is concerned with planners of Beijing because the historical urban form is important to be preserved. Actually, most or roads in Beijing are north–south or east–west direction. Thus, the parcel directions that can be recognized as the same direction of road network are possible to be used to measure the change of the historical urban form in Beijing. In Chap. 4, the authors tried to develop a novel indicator for measuring urban form quantitatively. In order to quantitate the polygon-based urban form, the approaches for calculating the PD based on GIS and for measuring urban form spatiotemporally are explained in this chapter. Urban form of Beijing is measured at three spatial scales, namely the parcel, zone, and region scales. It is also examined over time horizon from 1949 using PD. The extent to which the planned urban form inherits the historical urban form therefore can be identified by comparing the PD index of the two forms in various stages.

For investigating the interaction between parcels for establishing the land use pattern, cellular automata is used as a methodology for simulating the formulation of land use. BUDEM is a raster-based CA model for finding urban boundary driven by urban policies and natural constrains in Beijing. We developed a vector-based version of BUDEM (V-BUDEM). In Chap. 5, urban space consists of irregular parcels, and the neighborhoods are defined as all parcels surrounding the cell within a certain distance. Additionally, a framework of parcel subdivision was adopted to subdivide existing parcels for facilitating future development. After describing the conceptual model of V-BUDEM, we tested it in Beijing's Yanqing Town for simulating urban growth from 2010 to 2020. Results show the V-BUDEM can be used to simulate urban growth scenarios. The main contributions of this study are as follows: (1) the model adopts a vector-based CA method using land parcels to represent urban space, and can simulate urban growth in a way closer to real world situations; (2) the model integrates a process of parcel subdivision. While considering the impacts of existing parcel boundaries and planned parcel boundaries, we developed a straight-forward and automatic parcel subdivision tool to partition existing large parcels; (3) compared with other urban models, V-BUDEM is developed specifically to identify policies required for implementing the planned urban form desired by planners and decision makers.

In our work of the first part, we present some case studies for setting planning targets (scenario) and simulating urban form via raster dataset and vector dataset in order to understand urban boundary, land use pattern within urban space and driving forces of urban growth. However, even though we can try to understand the mechanism of planning issues regarding land use pattern and urban boundary via simulating interactions in urban space, the reasons of current planning issues taking place in urban space are not easy to be verified via simulation.

1.3 Urban Form: Human Behaviour and Their Spatial Patterns

Geospatial analysis is a powerful tool for understanding the reasons about why and how there are planning issues taking place in Beijing. In the developing period, planners focus on how to allocate economic demand on urban space; after development, planners in Chinese cities have to pay more attention to transportation modes carrying people between their houses, working and shopping places because of over heavy traffic jam in Beijing. Thus, researchers have to understand how people move and try to find out solutions for solving the planning issues. For this, how to analyze the increasingly available big/open data which store detailed human mobility information becomes a hot topic in Beijing and it is important to understand the methodology of how to check the human mobility in urban space for their daily life. Some initiatives in this area include studying Beijing's function zones using human mobility and POI data (Yuan et al. 2012) and evaluating the effectiveness of a future road plan based on taxi trajectories (Zheng et al. 2011b). Through adjusting urban structures and providing more effective public transportation services based on the findings of data research, planners feel confident to improve the local plan.

The second part of this book explains human behaviour in urban space using geometric patterns for geospatial analysis. These patterns, such as points, networks, and polygons, contain urban activities in Beijing, and help to explain spatial patterns of human behaviour in urban spaces. Recently, spatial data mining has become a popular research topic. Collecting, distributing, and managing large amounts of data using GIS have been widely discussed. Specifically, researchers have been interested in making use of so-called big geo-data to understand how people use, perceive, and interact with city places (Feick and Robertson 2014; Girardin et al. 2009; Hollenstein and Purves 2010), investigating human movement patterns (Jankowski et al. 2010; Kisilevich et al. 2010; Liu et al. 2014), creating intelligent traffic systems (Noei et al. 2014; Sakaki and Matsuo 2012), and detecting connections between cyber and physical communities (Croitoru et al. 2014).

Here, we try to answer how and why geometric spatial patterns are necessary to understand urban forms in Beijing. We examine human behaviour in urban space, visualising human behaviour and determining the main planning problems that exist in the current form of Beijing. Recently, many reports attempt to identify, characterise, and develop geospatial frameworks and algorithms for individual behavioural models for spatiotemporal knowledge discovery. In fact, the study of individual behaviours has a long history in the GIS field. Survey data, such as the survey of people's daily activities, is initially exploited as the main data source (Kwan 1998). Geospatial frameworks were developed to analyse people's activities patterns over time and geography (Kwan 2004). Geo-visualization methods were introduced to aid in understanding of those patterns by delivering interactive visual outputs (Kwan and Lee 2003). More recently, the increasing use of volunteered geographic information (VGI) and open data has further spurred the study of individual behaviours. In this thread, researchers built upon the previous work of spatio-temporal data mining, and developed methods to discover spatio-temporal patterns of human activities from geotagged photos and mobile phone data (Andrienko et al. 2010; Sagl et al. 2014; Zheng et al. 2011a).

1.3.1 Open Data and Survey for Investigating Mechanism in Urban Space

With the open-government initiatives in China and the emerging Web 2.0 techniques, open data are becoming increasingly available. Dataset with locational and individual attributes has been applied as an essential input for micro-level urban modelling. However, population distribution at a fine scale, as the input for population synthesis, is not universally available. In Chap. 6, we propose an automatic process using open data for population spatialization and synthesis. Specifically, the road network in OpenStreetMap is used to identify and delineate parcel geometries, while crowd-sourced points of interest (POIs) are gathered to infer urban parcels with a vector cellular automata model. Housing-related online Check-in records are then applied to distinguish residential parcels from all of the identified urban parcels. Finally the published census data, in which the sub-district level of attributes' statistical distributions and relationships are available, is used for synthesizing population attributes with a previously developed tool: Agenter (Long and Shen 2013). The results are validated with ground truth manually-prepared dataset by planners from Beijing Institute of City Planning and it is possible to use open data to analysis human mobility on urban space through population spatialization and synthesis.

In Chap. 7, the empirical effort helps us to understand that people's mobility behaviours can be influenced by spatially exploring relevant variables of urban form. In this study, the Beijing Household Travel Survey in 2005 has been used. Most of the existing studies neglected the spatial heterogeneity of the impact of urban form on human mobility. This study performs an examination of the spatially varying impacts of urban form and socioeconomic attributes on human mobility (average travel distance in this chapter) in the Beijing Area. By using mixed Geographically Weighted Regression (MGWR), we incorporate the spatial stationary and non-stationary in one model to estimate the influence surfaces of elements that maters in residential mobility at the Traffic Analysis Zone (TAZ) level. Compared with the results produced by Ordinary Least Square (OLS) and Spatial Auto-Regression (SAR), the outputs of MGWR indicate that semi-parametric model has better performance in presenting the spatial variation of predictive factors' impacts. The maps further show that urban form features do impact people's mobility to various extents, both positively and negatively. This study yields that the analysis on the human mobility at the TAZ level reveals the spatial variability for understanding the urban structure.

However, at the level of TAZ, the individual behaviors are aggregated as the average characteristics in different zones, it is not possible to check the individuals' characteristics. Thus, big data have been paid more attention from young researchers for understanding the planning issues in Beijing. In the case of Big data, the spatial units of urban space can be checked in more complicated units than the TAZ zonings. More useful information for understanding the planning issues is expected via big data.

1.3.2 Big Data and Findings of the Human Mobility in Urban Space

Public transportation is a crucial part of urban transportation system. Nowadays, smartcard for transportation are widely used in Chinese cities. Many planning authorities try to use smart card to visualize the human mobility over time and space. The collective information of transit usage can also reflect the land use pattern and be a valuable source for identifying possible solutions to planning issues.

The availability of large-scale smart card data (SCD) offers new opportunities to understand intra-urban transportation structure and spatial interaction dynamics between communities. In Chap. 8, we examined the dynamic spatial interaction structures of public transportation communities in the Beijing Metropolitan Area using SCD. We found that the daily community detection results using SCD are different from that using household travel surveys. The SCD results match better with the planned urban area boundary, which means that the actual operation data of publication transportation might be a good source to understand the relationship between the land use patterns and public transportation. For instance, it is helpful to identify the ground-truth community structure of strongly connected TAZs by SCD of public transportation, which can provide insights for urban planners on and transportation engineers on traffic congestion and land use patterns. Thus, exploring the spatiotemporal patterns of public trips can help us understand dynamic transportation patterns and the complex urban systems thus supporting better urban planning.

For understanding the change of residential locations during long- or mid-term urban dynamics using real-time SCD, namely "big data", we attempt to profile different types of residents in Beijing at a fine scale. Mobility of economically underprivileged residents in China has seldom been well profiled due to privacy issue and the characteristics of Chinese people over poverty. In Chap. 9, we identify and characterize underprivileged residents in Beijing using ubiquitous public transport smartcard transactions in 2008 and 2010, respectively. We regard these frequent bus/metro riders (FRs) in China, especially in Beijing, as economically underprivileged residents. Our argument is tested against (1) the household travel survey in 2010, (2) a small-scale survey in 2012, as well as (3) our interviews with local residents in Beijing. Cardholders' job and residence locations are identified using SCD in 2008 and 2010, respectively. Our analysis is restricted to cardholders that appear in both years. We then classify all identified FRs into 20 groups by residence changes (change, no change), workplace changes (change, no change, finding a job, losing a job, and all-time employed) during 2008–2010 and housing place in 2010 (within the fourth ring road or not).

It is important for city planners to identify different functional zones and understand their spatial structure within the city in order to make better urban plans. In Chap. 10, we used 77,976,010 bus smart card records of Beijing in one week in April 2008 and converted them into two-dimensional time series data of each bus platform, Then, through data mining in the big database system and previous studies on citizens' trip behavior, we established the DZoF (discovering zones of different functions) model based on SCD and POIs, and pooled the results at the TAZ (traffic analysis zone) level. Cities comprise various functional zones, including residential, educational, commercial zones, etc. The results suggested that DzoF model and cluster analysis based on dimensionality reduction and EM (expectation-maximization) algorithm can identify functional zones that well match the actual land uses in Beijing. The methodology in the present research can help urban planners and the public understand the complex urban spatial structure and contribute to the academia of urban geography and urban planning.

In this part, we use SCD to detect traffic activities in urban area. We also profile different types of residents for understanding how they move in Beijing. We attempt to identify different functional zones and understand their spatial structure within the city via visualising the traffic activities of household in order to improve the land use planning and transportation network.

1.4 Planning Support and Its Future in Beijing

Planning support systems (PSSs) have attracted extensive attentions from scholars and decision makers for decades. Most of the existing research on PSSs is related to system design, implementation, application as well as evaluation of a stand-alone system in one area, e.g. What if?, CommunityViz and INDEX. There is no existing research on an entire framework of PSSs for various types of plans.

In Chap. 11, we propose a PSS framework for various types of plans in Beijing, China, e.g. master plan, detailed plan, municipal infrastructure plan and transport plan. The framework has been applied in the Beijing Institute of City Planning (BICP) for over two years, and has attracted hundreds of application requests from planners. Based on an extensive literature review and multiple rounds of planner and decision maker surveys, the framework focuses on two aspects. On one hand, we itemize the types of planning (termed as "plan elements into various steps for each type of plan, e.g. population forecasting and establishing urban growth boundaries in a master plan. In our research, PSSs have been discussed from three viewpoints, which are different models of PSS to be developed or already developed, existing PSS software (e.g. What if? and INDEX) as well as quantitative methods and modules that can be used in PPS (e.g. scenario analysis, systems analysis, and logistic regression). The two dimensional framework provides a full picture of PSS applications in various types of plans.

In Chap. 12, we propose a systematic framework to help planners consider the compilation of land use plan. A land use pattern, also known as land use layout, is composed of different land use types in all parcels of an urban area. Planner Agents are classified into three types: Non-spatial Planner Agent (NPA), Spatial Planner Agent (SPA) and Chief Planner Agent (CPA), who work together in order to establish the land use pattern, based on the negotiation with the government agent (GA) and residential agents (RA). We simulate the behaviour of planner agents (PA) for planner to consider possible land use patterns in the process of planning compilation according to negotiation between planners and stakeholders. Government, planners and residents are the agents participating in the establishment of land use planning in Chinese cities. Among them, planners of a land use plan play a role in negotiating with related agents and then establishing land use patterns. In this work, we show the possibility of simulating the planned urban form via planner agents' interactions in the virtual planning process using Geocomputation technology. After proposing the framework, it is tested in Beijing.

Finally, as a future perspective instead of summary to this book, the Chap. 13, proposes the concept of big model as a novel research paradigm for regional and urban studies. Big models are fine-scale regional/urban analysis and simulation models for a large geographical area. With the widespread use of big/open data, the increased computation capacity, as well as the advanced regional and urban modelling methodologies, big models make it possible to overcome the trade-off between simulated scale and spatial unit. In this chapter, the concept, characteristics, and potential applications of big models have been elaborated. We present several case studies to illustrate the progress of our research and the application of big models. They include mapping urban areas for all Chinese cities, performing parcel-level urban expansion simulation, and several ongoing research projects. Most of these applications can be adopted across the country, and all of them are focusing at a fine-scale level, such as a parcel, a block, or a township (sub-district). It is expected that big models will mark a promising new era for the urban and regional studies in the age of big data.

References

Al-Sabhan, W., Mulligan, M., & Blackburn, G. A. (2003). A real-time hydrological model for flood prediction using GIS and the WWW. *Computers, Environment and Urban Systems, 27*, 9–32. doi:10.1016/S0198-9715(01)00010-2.

Andrienko, G., Andrienko, N., Mladenov, M., Mock, M., & Pölitz, C. (2010). Discovering bits of place histories from people's activity traces. 2010 IEEE Symposium on Visual Analytics Science and Technology VAST, 59–66. Retrieved from http://geoanalytics.net/and/papers/vast10.pdf

Batty, M., Xie, Y., & Sun, Z. (1999). Modeling urban dynamics through GIS-based cellular automata. *Computers, Environment and Urban Systems, 23*(3), 205–233. doi:10.1016/S0198-9715(99)00015-0.

Bernard, L., & Kruger, T. (2000). Integration of GIS and spatio-temporal simulation models: Interoperable components for different simulation strategies. *Transactions in GIS, 4*, 197–215.

Blaschke, T., Hay, G. J., Weng, Q., & Resch, B. (2011). Collective sensing: Integrating geospatial technologies to understand urban systems – an overview. *Remote Sensing, 3*(8), 1743–1776. doi:10.3390/rs3081743.

Carrara, A., Guzzetti, F., Cardinali, M., & Reichenbach, P. (2000). Use of GIS technology in the prediction and monitoring of landslide hazard. *Natural Hazards, 20*, 117–135. doi:10.1023/A:1008097111310.

Cheng, J., & Masser, I. (2004). Understanding spatial and temporal processes of urban growth: Cellular automata modelling. *Environment and Planning B: Planning and Design, 31*(2), 167–194. doi:10.1068/b2975.

Couclelis, H. (2005). "Where has the future gone?" Rethinking the role of integrated land-use models in spatial planning. *Environment and Planning A, 37*(8), 1353–1371.

Croitoru, A., Wayant, N., Crooks, A., Radzikowski, J., & Stefanidis, A. (2014). Linking cyber and physical spaces through community detection and clustering in social media feeds. *Computers, Environment and Urban Systems*. doi:10.1016/j.compenvurbsys.2014.11.002.

Ettema, D. (2011). A multi-agent model of urban processes: Modelling relocation processes and price setting in housing markets. *Computers, Environment and Urban Systems, 35*, 1–11. doi:10.1016/j.compenvurbsys.2010.06.005.

Feick, R., & Robertson, C. (2014). A multi-scale approach to exploring urban places in geotagged photographs. *Computers, Environment and Urban Systems*. doi:10.1016/j.compenvurbsys.2013.11.006.

Fontaine, C. M., & Rounsevell, M. D. A. (2009). An agent-based approach to model future residential pressure on a regional landscape. *Landscape Ecology, 24*, 1237–1254. doi:10.1007/s10980-009-9378-0.

Gaube, V., & Remesch, A. (2013). Impact of urban planning on household's residential decisions: An agent-based simulation model for Vienna. *Environmental Modelling and Software, 45*, 92–103. doi:10.1016/j.envsoft.2012.11.012.

Girardin, F., Vaccari, A., Gerber, A., Biderman, A., & Ratti, C. (2009). Quantifying urban attractiveness from the distribution and density of digital footprints. *International Journal of Spatial Data Infrastructures Research, 4*, 175–200. doi:10.2902/1725-0463.2009.04.art10.

Gu, C., Wang, F., & Liu, G. (2005). The structure of social space in Beijing in 1998: A socialist city in transition. *Urban Geography*. doi:10.2747/0272-3638.26.2.167.

Gualtieri, G., & Tartaglia, M. (1998). Predicting urban traffic air pollution: A GIS framework. *Transportation Research Part D: Transport and Environment*. doi:10.1016/S1361-9209(98)00011-X.

He, C., Tian, J., Shi, P., & Hu, D. (2011). Simulation of the spatial stress due to urban expansion on the wetlands in Beijing, China using a GIS-based assessment model. *Landscape and Urban Planning, 101*, 269–277. doi:10.1016/j.landurbplan.2011.02.032.

Hollenstein, L., & Purves, R. (2010). Exploring place through user-generated content: Using Flickr to describe city cores. *Journal of Spatial Information Science, 1*(1), 21–48. doi:10.5311/JOSIS.2010.1.3.

Huang, Q. H., & Cai, Y. L. (2007). Simulation of land use change using GIS-based stochastic model: The case study of Shiqian County, Southwestern China. *Stochastic Environmental Research and Risk Assessment, 21*, 419–426. doi:10.1007/s00477-006-0074-1.

Huang, B., & Pan, X. (2007). GIS coupled with traffic simulation and optimization for incident response. *Computers, Environment and Urban Systems, 31*, 116–132. doi:10.1016/j.compenvurbsys.2006.06.001.

Huang, Q., Parker, D. C., Sun, S., & Filatova, T. (2013). Effects of agent heterogeneity in the presence of a land-market: A systematic test in an agent-based laboratory. *Computers, Environment and Urban Systems, 41*, 188–203. doi:10.1016/j.compenvurbsys.2013.06.004.

Jankowski, P., Andrienko, N., Andrienko, G., & Kisilevich, S. (2010). Discovering landmark preferences and movement patterns from photo postings. *Transactions in GIS, 14*(6), 833–852. doi:10.1111/j.1467-9671.2010.01235.x.

Jokar Arsanjani, J., Helbich, M., & de Noronha Vaz, E. (2013). Spatiotemporal simulation of urban growth patterns using agent-based modeling: The case of Tehran. *Cities, 32*, 33–42. doi:10.1016/j.cities.2013.01.005.

Kawakami, M., Shen, Z., Pai, J., Gao, X., & Zhang, M. (2013). Spatial planning and sustainable development: Approaches for achieving sustainable Urban form in Asian cities. Dordrecht: Springer.

Kisilevich, S., Keim, D., & Rokach, L. (2010). A novel approach to mining travel sequences using collections of geotagged photos. In M. Painho (Ed.), *Geospatial thinking* (pp. 163–182). Berlin/Heidelberg: Springer. Retrieved from http://link.springer.com/chapter/10.1007/978-3-642-12326-9_9.

Klosterman, R. E. (1999). The what if? Collaborative planning support system. *Environment and Planning B: Planning and Design, 26*(3), 393–408.

Kocabas, V., & Dragicevic, S. (2013). Bayesian networks and agent-based modeling approach for urban land-use and population density change: A BNAS model. *Journal of Geographical Systems, 15*, 403–426. doi:10.1007/s10109-012-0171-2.

Kwan, M. P. (1998). Space-time and integral measures of individual accessibility: A comparative analysis using a point-based framework. *Geographical analysis, 30*(3), 191–216.

Kwan, M.-P. (2004). GIS methods in time-geographic research: Geocomputation and geovisualization of human activity patterns. *Geografiska Annaler: Series B, Human Geography, 86*, 267–280. doi:10.1111/j.0435-3684.2004.00167.x.

Kwan, M., & Lee, J. (2003). Geovisualization of human activity patterns using 3D GIS: A time-geographic approach. In M. Goodchild & D. G. Janelle (Eds.), *Spatially integrated social science: Examples in best practices*. Oxford: Oxford University Press.

Landis, J. D. (1994). The California urban future model: A new generation of metropolitan simulation models. *Environment and Planning B: Planning and Design, 21*(4), 399–420.

Landis, J. D. (1995). Imaging land use futures: Applying the California urban future model. *Journal of American Planning Association, 61*(4), 438–457.

Li, Y., & Shanmuganathan, S. (2007). *Social area analysis using SOM and GIS: a preliminary research*. Ritsumeikan Center for Asia Pacific Studies (RCAPS) Working Paper, Ritsumeikan Asia Pacific University.

Li, G., & Weng, Q. (2007). Measuring the quality of life in city of Indianapolis by integration of remote sensing and census data. *International Journal of Remote Sensing, 28*(2), 249–267. doi:10.1080/01431160600735624.

Li, X., & Yeh, A. G.-O. (2000). Modelling sustainable urban development by the integration of constrained cellular automata and GIS. *International Journal of Geographical Information Science, 14*(2), 131–152. doi:10.1080/136588100240886.

Li, X., & Yeh, A. G.-O. (2002). Neural-network-based cellular automata for simulating multiple land use changes using GIS. *International Journal of Geographical Information Science, 16*(4), 323–343. doi:10.1080/13658810210137004.

Liu, Y., Sui, Z., Kang, C., & Gao, Y. (2014). Uncovering patterns of inter-urban trip and spatial interaction from social media check-in data. *PloS One, 9*(1), e86026. doi:10.1371/journal.pone.0086026.

Long, Y., & Shen, Z. (2013). Disaggregating heterogeneous agent attributes and location from aggregated data, small-scale surveys and empirical researches. *Computers, Environment and Urban Systems, 42*, 14–25.

Longley, P. A., Goodchild, M. F., Maguire, D. J., & Rhind, D. W. (2005). *Geographic information systems: Principles, techniques, management and applications*. Hoboken: Wiley. Abridged edition.

Maguire, D. J., Batty, M., & Goodchild, M. F. (2005). *GIS, spatial analysis and modelling*. Redlands: ESRI Press.

Nabian, N., Offenhuber, D., Vanky, A., & Ratti, C. (2013). Data dimension : Accessing urban data and making it accessible. *Urban Design and Planning, 166*(DP1), 60–75.

Neckerman, K. M., Lovasi, G. S., Davies, S., Purciel, M., Quinn, J., Feder, E., & Rundle, A. (2009). Disparities in urban neighborhood conditions: Evidence from GIS measures and field observation in New York City. *Journal of Public Health Policy, 30*(1), S264–S285. doi:10.1057/jphp.2008.47.

Neutens, T., Versichele, M., & Schwanen, T. (2010). Arranging place and time: A GIS toolkit to assess person-based accessibility of urban opportunities. *Applied Geography, 30*(4), 561–575. doi:10.1016/j.apgeog.2010.05.006.

Noei, S., Santana, H., Sargolzaei, A., & Noei, M. (2014). Reducing traffic congestion using geo-fence technology: Application for emergency car. In EMASC'14 (pp. 15–20). Retrieved from http://dl.acm.org/citation.cfm?id=2661709

Oguz, H., Klein, A., & Srinivasan, R. (2007). Using the SLEUTH urban growth model to simulate the impacts of future policy scenarios on urban land use in the Houston-Galveston-Brazoria CMSA. *Research Journal of Social Sciences, 2*, 72–82. Retrieved from http://www.ncgia.ucsb.edu/projects/gig/Repository/references/Houston_TX/10.1.1.118.7250.pdf.

Patino, J. E., Duque, J. C., Pardo-Pascual, J. E., & Ruiz, L. A. (2014). Using remote sensing to assess the relationship between crime and the urban layout. *Applied Geography, 55*, 48–60. doi:10.1016/j.apgeog.2014.08.016.

Perry, G. L. W., Sparrow, A. D., & Owens, I. F. (1999). A GIS-supported model for the simulation of the spatial structure of wildland fire, Cass Basin, New Zealand. *Journal of Applied Ecology, 36*, 502–518. doi:10.1046/j.1365-2664.1999.00416.x.

Sagl, G., Delmelle, E., & Delmelle, E. (2014). Mapping collective human activity in an urban environment based on mobile phone data. *Cartography and Geographic Information Science, 41*(3), 272–285. doi:10.1080/15230406.2014.888958.

Sakaki, T., & Matsuo, Y. (2012). *Real-time event extraction for driving information from social sensors*. In Proceedings of 2012 IEEE International Conference on Cyber Technology in Automation, Control and Intelligent Systems (pp. 221–226). Retrieved from http://ieeexplore.ieee.org/xpls/abs_all.jsp?arnumber=6392557

Shen, Z. J. (2012). *Geospatial techniques in urban planning, in series advances in geographic information science*. Berlin/Heidelberg: Springer-Verlag/GmbH.

Thornton, L. E., Pearce, J. R., & Kavanagh, A. M. (2011). Using geographic information systems (GIS) to assess the role of the built environment in influencing obesity: A glossary. *The International Journal of Behavioral Nutrition and Physical Activity, 8*, 71. doi:10.1186/1479-5868-8-71.

Torrens, P. M. (2006). Simulating sprawl. *Annals of the Association of American Geographers, 96*, 248–275.

Turgeon, M. F. (2013). *The pizza map: Using social media, crowd-sourcing, & web-based mapping to analyze regional pizza preferences in the United States*. Northridge: California State University.

Wang, X. (2005). Integrating GIS, simulation models, and visualization in traffic impact analysis. *Computers, Environment and Urban Systems, 29*, 471–496. doi:10.1016/j.compenvurbsys.2004.01.002.

Wang, D., & Chai, Y. (2009). The jobs-housing relationship and commuting in Beijing, China: The legacy of Danwei. *Journal of Transport Geography, 17*, 30–38. doi:10.1016/j.jtrangeo.2008.04.005.

Wang, D., & Li, S. M. (2004). Housing preferences in a transitional housing system: The case of Beijing, China. *Environment and Planning A, 36*, 69–87. doi:10.1068/a35263.

Widener, M. J., & Li, W. (2014). Using geolocated Twitter data to monitor the prevalence of healthy and unhealthy food references across the US. *Applied Geography, 54*, 189–197. doi:10.1016/j.apgeog.2014.07.017.

Wu, Q., Li, H., Wang, R., Paulussen, J., He, Y., Wang, M., & Wang, Z. (2006). Monitoring and predicting land use change in Beijing using remote sensing and GIS. *Landscape and Urban Planning*. doi:10.1016/j.landurbplan.2005.10.002.

Xie, Y. (1996). A generalized model for cellular urban dynamics. *Geographical Analysis, 28*(4), 350–373. doi:10.1111/j.1538-4632.1996.tb00940.x.

Xu, J., Nyerges, T. L., & Nie, G. (2014). Modeling and representation for earthquake emergency response knowledge: Perspective for working with geo-ontology. *International Journal of Geographical Information Science, 28*(1), 185–205. doi:10.1080/13658816.2013.845893.

Yao, X. (2007). Where are public transit needed – Examining potential demand for public transit for commuting trips. *Computers, Environment and Urban Systems, 31*, 535–550. doi:10.1016/j.compenvurbsys.2007.08.005.

Yao, X., & Zhang, S. (2014). Social-spatial structure of Beijing: A spatial-temporal analysis. *International Journal of Society Systems Science, 6*(1), 18. doi:10.1504/IJSSS.2014.059923.

Yeh, A. G. O., & Li, X. (2006). Errors and uncertainties in urban cellular automata. *Computers, Environment and Urban Systems, 30*, 10–28.

Yuan, J., Zheng, Y., & Xie, X. (2012). Discovering regions of different functions in a city using human mobility and POIs. In ACM KDD'12. Beijing, China.

Zhang, W., & Gao, X. (2008). Spatial differentiations of traffic satisfaction and its policy implications in Beijing. *Habitat International, 32*, 437–451. doi:10.1016/j.habitatint.2008.01.005.

Zheng, Y., Li, Y., Zha, Z., & Chua, T. (2011a). Mining travel patterns from GPS-tagged photos. In Advances in multimedia modeling (pp. 262–272). Retrieved from http://link.springer.com/chapter/10.1007/978-3-642-17832-0_25

Zheng, Y., Liu, Y., Yuan, J., & Xie, X. (2011b). Urban computing with taxicabs. In Proceedings of the 13th international conference on ubiquitous computing – UbiComp '11 (p. 89). New York: ACM Press. doi:10.1145/2030112.2030126.

Part I
Urban Form: Spatial Patterns and Land Use Development

Chapter 2
Target or Dream? Examining the Possibility of Implementing Planned Urban Forms Using a Constrained Cellular Automata Model

2.1 Introduction

This chapter proposes an alternative process to conventional urban growth scenario analysis (SA), which has been extensively studied in urban growth models. Conventional urban growth models regard development policies as scenario conditions for reflecting future uncertainties of urban developments (for instance, Landis 1994, 1995; Klosterman 1999). Couclelis (2005) argued that the standard land use models have done little to investigate desired or unwanted outcomes in future plans. In support of Couclelis' argument, we use the future urban form as the scenario condition to identify the spatial policies required to realize the predefined urban form. We propose an approach which we have labeled "form scenario analysis" (FSA). This analyzes the consistency between the predefined urban form and the corresponding spatial policies, together with the effects of the required spatial policies on various planning alternatives. Urban form is associated with many issues, and inappropriate urban forms (e.g. sprawling cities) may create various negative impacts such as over-consumption of land resources, increased commuting distances and traffic jams, decreased provision of affordable housing, increased urban infrastructure construction costs, reduced water supplies, poor neighborhood interactions, and poor public health (Kahn 2000; Ewing et al. 2003). FSA can potentially be applied in identifying policies required to create sustainable urban forms.

The driving force for FSA in China is threefold. First, unlike in the West, most urban planners in China work in state-owned planning institutions. They tend to retain the approaches of the planned economy before the 1978 reform and opening-up policy. This has resulted in widespread ignorance of market factors (such as location of commercial centers and roads). Second, senior planners in China generally have a tradition of hand drawing plans, which originates from limited computer

Y. Long, Z. Shen, *Geospatial Analysis to Support Urban Planning in Beijing*, GeoJournal Library 116, DOI 10.1007/978-3-319-19342-7_2

resources decades ago. Junior planners and students have partially inherited this tradition. Third, decision makers prefer a good plan based on an artistic approach (such as concepts like axis and cluster), rather than considering its implementation. These three points have driven planners to create the desired planned form first, and then consider the spatial policies required to realize the desired urban planning form. Consequently, the actual urban growth often differs from the planned form. For instance, the planned form evaluation results indicate that more than 35 % of urban developments in Beijing and Guangzhou exceed the original planned form (Han et al. 2009; Tian and Shen 2011). Planning departments, lacking appropriate policy guidance, have little knowledge of the policies required to create the planned form and the differences between the policies needed and the current policies. From this point of view, the government is commonly concerned with the urban policies required to create the planned form. Therefore, FSA has great practical potential for solving this problem in China.

The term "scenario" is defined as "the assumption of a reasonable event with uncertainty happening in the next period of time" (see Kahn and Wiener 1967; Pearman 1988). The term "scenario analysis", defined as "the entire process of predicting and analyzing the possible influences of the scenario" (see Ratcliffe 1999), acknowledges that the future development is diversified and has a wide range of possible trends (Schoemaker 1995; Ringland 1998). Scenario analysis is widely applied in urban growth models since accurate prediction of the future urban form is often difficult. Various sets of policy parameters can be employed to generate the corresponding urban form (Landis 1994, 1995; Klosterman 1999). Recently, constrained cellular automata (CA) models have also been extensively applied for urban growth scenario analysis (Wu 1998; Li and Yeh 2000; White et al. 2004; Long and Shen 2011).

In an alternative approach, this chapter proposes FSA with constrained CA, which analyzes the development policies required for form scenarios in order to present institutional implications for urban planning practices. This novel exploration of FSA can identify the existence of the required policies as well as the policy variations among planning alternatives (namely planned urban forms). This approach differs from traditional applications of constrained CA. This chapter is organized as follows. In Sect. 2.2, we explain in detail the methodology for FSA using the constrained CA. We use four planning alternatives from the latest urban master plan of the Beijing Metropolitan Area as the case study for the FSA approach. In Sect. 2.3, the research materials including the study area, location constraints and planning alternatives are described. The form scenario analysis results are provided in Sect. 2.4. Finally, we discuss the findings and conclusions and then consider the next steps for the FSA research in the last two sections of this chapter.

2.2 Method

2.2.1 Form Scenario Analysis

Urban growth SA, as a vision of future urban form (namely urban layout) in essence, is an allocation process for imaging future urban form based on initial urban form at the base year and the total land to be developed in future, while considering constraints and a model for combining these constraints. Constraints and the combination model are for prioritizing locations for future urban development and are the key elements in the allocation process for urban growth SA. In detail, urban growth SA can then be expressed as follows (x and a are all spatial explicit variables which should contain ij subscript to represent spatial location. The subscript is omitted in order to simplify the equation):

$$\begin{aligned} Y &= \Psi(P) \\ P &= f(X,A) \\ X &= \{x_k^{\,t} \mid k = 1,2,3,\ldots,n; t = 1,2,3,\ldots,p\} \\ A &= \{a_k^{\,t} \mid k = 1,2,3,\ldots,n; t = 1,2,3,\ldots,q\} \\ T^p &= \{T_t^{\,p} \mid t = 1,2,3,\ldots,p; T_t^{\,p} \in [T_s, T_e]\} \\ T^q &= \{T_t^{\,q} \mid t = 1,2,3,\ldots,q; T_t^{\,q} \in [T_s, T_e]\} \end{aligned} \tag{2.1}$$

where A as the term "policy", stands for the urban development policy, which is a temporal dynamic variable. X as the term "policy parameter" and a temporal dynamic variable, stands for the implementation intensity of the corresponding policy A. X can be regarded as the acceptable degree for the corresponding urban form of A. P is the development probability based on A and X using the development probability calculation function f. Ψ stands for the function which is used to determine which places are to be developed based on the development probability. Y stands for the urban form, and y_{ij} stands for the land occupation status at the location ij. $y_{ij} = 1$ means the location ij is developed as urban built-up from non urban built-up, while $y_{ij} = 0$ means undeveloped. n stands for the total number of policies. T_s stands for the starting time of the scenario analysis, and T_e the ending time. Y corresponds to the urban form at the ending time T_e. q and p, respectively, stand for the variation times of A and X from T_s to T_e. T^q and T^p, respectively, stand for the value changed time of A and X. The urban form of the future time t is the accumulative influenced result of A and X from the base time to t.

We draw three basic premises for the urban growth SA process, considering the limitations of the dataset available and the calculation time needed. (1) The function f is based on the multi-criteria evaluation (MCE), and the function Ψ is comparing the development probability with the development threshold. (2) X and A remain static from T_s to T_e (in most conditions equals to the value of the-time T_s or T_e), so both p and q are equal to 1. (3) X is homogenous in location.

Based on these three premises, Eq. 2.1 is then transformed into Eq. 2.2:

$$
\begin{aligned}
&X = \{x_k \mid k = 1,2,3,\ldots,n\} \\
&A = \{A_k \mid k = 1,2,3,\ldots,n\} \\
&\quad \text{where } A_k = \{a_{k,ij} \mid k = 1,2,3,\ldots,n; ij \in \Omega\} \\
&p_{ij} = \sum_{k=1}^{n} x_k * a_{k,ij} \\
&y_{ij} = 1, \text{if } p_{ij} \geq p_{\text{threshold}} \\
&Y = \{y_{ij} \mid ij \in \Omega\} \subseteq \Omega
\end{aligned}
\tag{2.2}
$$

where ij is the geographical location, A_k is the spatial distribution of the policy k, $a_{k,ij}$ is the value of the policy k at ij, x_k is the policy parameter of the policy k, Ω is the entire study area, p_{ij} is the development probability at ij, $p_{threshold}$ is the development threshold. If p_{ij} is greater than or equal to $p_{threshold}$, then the space ij will be developed.

These three premises are also currently applied in the urban growth SA. For instance, Klosterman (1999) developed the planning support system "What if?" in which A is the policy of soil condition, flooding control and transportation, and X stands for the corresponding implementation intensity. Landis (1994, 1995) and Landis & Zhang (1998a, b), respectively, developed CUF (California Urban Future Model) and CUF-2, in which A stands for the policy of location, environmental condition, land-use control, zoning, existing development density, accessibility for each development land unit (DLU), and X stands for the policy's implementation intensity. In the routine urban growth SA, X and A, the scenario analysis conditions as input, are applied to generate the corresponding urban form Y. We can get the corresponding urban form based on any development policy and policy parameter set. In most existing traditional urban growth scenarios, A is adjusted to estimate the dynamic change of Y, while X remains constant, to simplify the urban growth SA process.

The FSA approach, proposed in this chapter, can be regarded as the reverse process to standard urban growth SA. Three premises are also considered in FSA, in which the urban form Y as the scenario condition is used to identify development policies required and their parameters (A and X). From the view of solving the equation, for any independent variable Y, dependent variables, including X and A, can be classified as two conditions. The first is No solution, namely there are no policies to yield Y and the second is Multi solutions, namely at least one policy set can be used to realize Y. Constrained CA are adopted to investigate the FSA issue, which regards the desired urban form Y and the already-known policies A as model inputs to solve the equation in order to validate the existing policy parameters X. To simplify the FSA process, we assume the development policies are already known variables, and will focus on the identification of the policy parameters.

2.2.2 Identification of Urban Policy Parameters

Recent literature relevant to the constrained CA has not explored the urban form as the scenario condition. Whether the constrained CA can be adopted to solve FSA and whether the urban form can be regarded as the scenario condition also remains unexplored. First, we will investigate the status transition rule acquisition method of the constrained CA, which is the key procedure in using the traditional constrained CA to simulate urban growth (Wu 1998; Li and Yeh 2000, 2002, 2004). The observed forms (Y) and the known constrained conditions (A), namely policies, in some historical stages are required to identify parameters (X) for the constrained conditions (A). FSA is aimed at acquiring X based on Y and A. In the standard model calibration process, the urban form is based on historical observation, while in FSA, the urban form is based on the predefined urban form. Therefore, the two processes are, in essence, identical.

The key issue for FSA using the constrained CA is the identification of policy parameters (X), which can enable the model calibration method mentioned above. In FSA, T_s can be regarded as the start of the historical stage, and T_e the end. Therefore, the time phase of SA, namely, from $T_s - T_e$, corresponds to the historical stage, and FSA can be transformed into the model calibration issue of standard constrained CA, in which approaches such as logistic regression, artificial neural network, genetic algorithm, and nested loops are widely adopted.

The evaluation indicator for the consistency of the form scenario (Y) and current policies (A) should be established in FSA. As the input of the constrained CA, X is applied to get the simulated urban form (Y'). The Kappa index, the evaluation indicator, is applied to compare the simulated form Y' with the form scenario Y cell by cell and evaluate the goodness-of-fit. Kappa less than 80 % stands for none solution condition (no policy parameters to realize the predefined form), and greater than or equal to 80 % stands for the multi solutions condition, denoting that the calibrated parameters can be used to simulate the designated scenario form. Generally speaking, solutions for FSA can be expressed as $\{X \mid Y' = f(X,A), Kappa(Y,Y') \geq 80\%\}$, which stands for the solution location of X.

2.2.3 Constrained Cellular Automata (CA)

The simulation logic of urban growth in China is influenced by the socialist market economy. First, the government determines at the macro-level the amount of land development in different time phases according to socio-economic conditions. Second, the developers acquire suitable development land from the government during the allocation process, taking account of the comprehensive constraints at micro-level. At the micro-level, a constrained CA model is used to allocate urban developments to locations.

Constraint selection is a core procedure for constrained CA, and existing constrained CA models have distinctive constraints. For instance, White et al. (1997) initiated the concept of constrained CA by taking spatial constraints into account. Engelen et al. (1997) took macro socio-economic constraints into account in a planning support system incorporating CA, GIS and other toolkits. An exclusive layer is set to constrain urban growth in SLEUTH (Clark and Gaydos 1998). In Simland, developed by Wu (1998), constraints for land developments were also considered. Ward and Murray (1999) employed physical constraints, geographic constraints and institutional controls, in addition to the macro socio-economic constraint. Li and Yeh (2000) classified all constraints into three types, local, regional and global. Engelen et al. (2003) and White et al. (2004) employed not only macro constraints, but also physical characteristics, accessibility and zoning as constraints. In Guan et al. (2005)'s CA urban model, macro socio-economic and institutional constraints were included. Social factors, transportation infrastructure, proximity to city centers, facilities infrastructure, neighboring land uses, geographic factors, and spontaneous growth were constraints in the CA-based model LEAM (Sun et al. 2009). Long and Shen (2011) proposed a CA urban growth model for Beijing taking into account spatial and institutional constraints. For our constrained CA model, we selected three types of factors which influence urban growth based on studies in urban economics (especially those using the Hedonic approach (Rosen 1974)) and other research we reviewed. These factors (namely spatial constraints) include the location constraints (standing for market incentives), the neighbor constraint, namely the development ratio within the neighborhood, as well as the institutional constraints.

Based on simulation logic and the selected factors, the conceptual model of the constrained CA for FSA is shown as follows:

$$V_{ij}^{t+1} = f(V_{ij}^{t}, A_{mac}, A_{loc}, A_{ins}, A_{nei}^{t}), \tag{2.3}$$

where V_{ij}^{t+1} and V_{ij}^{t}, respectively, are the cell status at ij of time t+1 and t, and f is the transition rule of the constrained CA.

Constrained conditions in the urban growth process, namely development policies, consist of four types, including the macro socio-economic constraint A_{mac} (non-spatial explicit variable, and thus no corresponding policy parameter X for it), location constraints A_{loc}, institutional constraints A_{ins}, and neighbor constraints A_{nei}^{t}. Location and institutional constraints are assumed to remain fixed during the future urban growth process, and the macro socio-economic constraint reflects the total number of built-up cells to be developed. The neighbor constraint is defined as the development intensity in the neighborhood of each cell, and equals the ratio of developed cells to all cells in the neighborhood (excluding the cell itself). The neighbor constraint keeps changing with iterations of the constrained CA since its value is recalculated based on the simulated urban form in each iteration. In addition, the configuration of the neighborhood of the constrained CA is the Moore type, with eight adjacent cells for each cell, and the discrete time of CA is 1 month in the real world.

The status transition rule of the constrained CA is expressed as follows:

$$\begin{aligned}
&LandDemand = \sum_t stepNum^t \\
&\text{In iteration } t+1: \\
&s^t = x_0 + \sum_{k=1}^{n-1} x_k * a_k + x_n * a_n^{\,t} = s_0 + x_n * a_n^t \\
&p_g^{\,t} = \frac{1}{1+e^{-s^t}} \\
&p^t = \exp\left[\alpha * \left(\frac{p_g^{\,t}}{p_{g\,\max}^{\,t}} - 1\right)\right] \\
&\text{if } p_{ij}^t \ge p_{threshold}\left(p^t, stepNum^{t+1}\right) \text{then } y_{ij}^{t+i} = 1 \\
&\qquad \text{otherwise } y_{ij}^{t+i} = 0
\end{aligned} \tag{2.4}$$

where LandDemand is the total number of cells to be developed, $stepNum^t$ is the number of cells developed in iteration t reflecting the land development demand as the macro constraint, ij is the cell's coordinate, s^t, calculated from the sum of weighted spatial constraints, is the urban development suitability of cell ij, p_g^t is the initial transition potential, $p_{g\,max}^t$ is the max value of p_g^t across the whole lattices, α is the dispersion parameter ranging from 1 to 10, indicating the rigid level of urban development, p^t is the final transition potential, p_{ij}^t is the final transition potential of cell ij, x_0 is the constant item, a_n is the neighborhood development policy, x_n is the weight of a_n, a_k is the spatial constraint (the neighborhood effect excluded), x_k is the weight of a_k, s_0 is the constant part (except the neighborhood effect) of s_{ij}^t among all iterations, y_{ij}^{t+i} is the cell ij's status at iteration t+1, and $p_{threshold}(p^t, stepNum^{t+1})$ is the development threshold to control the development speed and quantity which varies from the value of p^t and $stepNum^{t+1}$ to guarantee $stepNum^{t+1}$ cells will be developed in iteration t+1. In general, this equation is used to allocate future development land into cells using constrained CA, which evaluates the transition potential based on the traditional land use suitability.

We will discuss how to identify model parameters for the constrained CA. The parameters needing to be calibrated include $stepNum^t$, x_k, and x_n.. Various approaches can be adopted. The calibration of $stepNum^t$, reflecting the total amount of land required for economic developments which is assumed to be constant throughout the simulation period, can be calculated as follows:

$$stepNum = \frac{C_{T_e} - C_{T_s}}{(T_e - T_s)/t_0}, \tag{2.5}$$

where C_T and C_t are the total number of developed cells, respectively, in scenario form and current form, T_e and T_s are, respectively, the future and current time, and t_0 is the time in the real world corresponding to one iteration of CA.

Regarding the calibration of x_k and x_n reflecting the intensity of policies, the logistic regression and heuristic approaches can both be applied. In our constrained CA, we integrate the methods of Wu (2002) and Clark and Gaydos (1998), combining their benefits, to identify the parameters of the MCE formatted status transition rule. The weights x_k forocational constraints can be retrieved by logistic regression, in which whether a cell is developed from non-urban to urban is the dependent variable (1 for developed and 0 for non-developed), and location constraints are the independent variables (for details see Wu 2002). The binary logistic regression progress can be represented as $\begin{matrix} P = \frac{1}{1+e^{-z}} \\ z = x_0 + \sum_k x_k * a_k \end{matrix}$, where x_0 is constant, x_k is the regression coefficient, a_k is the influencing factor, and P is the transformation probability (from non-urban to urban, namely developed).

Keeping the identified x_k static, x_n can be calibrated using the MonoLoop method (for details see Long et al. 2009), with x_n continually sampled from 0 to $x_{n,max}$ with an interval of $x_{n,max}$ /*M*. $x_{n,max}$, which can be set based on the user's experience. M is set as 100 in this chapter. The sampled x_n and the already calibrated weights x_k are used as the input variables for the constrained CA model. x_k and x_n calibrated (X^*) represent the policy implementation intensity for the scenario form. Our proposed approach can both identify the overall historical urban growth trend and reduce the time consumed for the model calibration. The Kappa index is calculated by comparing the simulated form (Y') and form scenario (Y). X^* stands for the maximum consistency between Y and Y'. X^*, as one solution, is not the only parameter for the realization standard of the form scenario. However, we consider only X^* in this chapter, and other solutions will be investigated in future research.

Policy implications can be drawn from the calibrated policy parameters. When the policy parameter X is positive, the corresponding policy A should be encouraged, otherwise A should be rejected. The parameters for various policies can also be compared in parallel to show the policy tendencies. Meanwhile, we can compare the parameters with the historical ones to "visualize" the policy implications. In the following sections, the constrained CA will be empirically applied to the 2020 Beijing master plan to examine the consistency of four planning scenarios with already-known spatial policies as well as to identify the policy parameters required.

2.3 Study Area and Data

2.3.1 Study Area

The study area for constrained CA is the Beijing Metropolitan Area (BMA) as shown in Fig. 2.1. With an area of 16,410 km^2, it lies in northern China, to the east of the Shanxi altiplano and south of the Inner Mongolian altiplano. The southeastern part of the BMA is a plain, extending east for 150 km to the coast of the Bohai

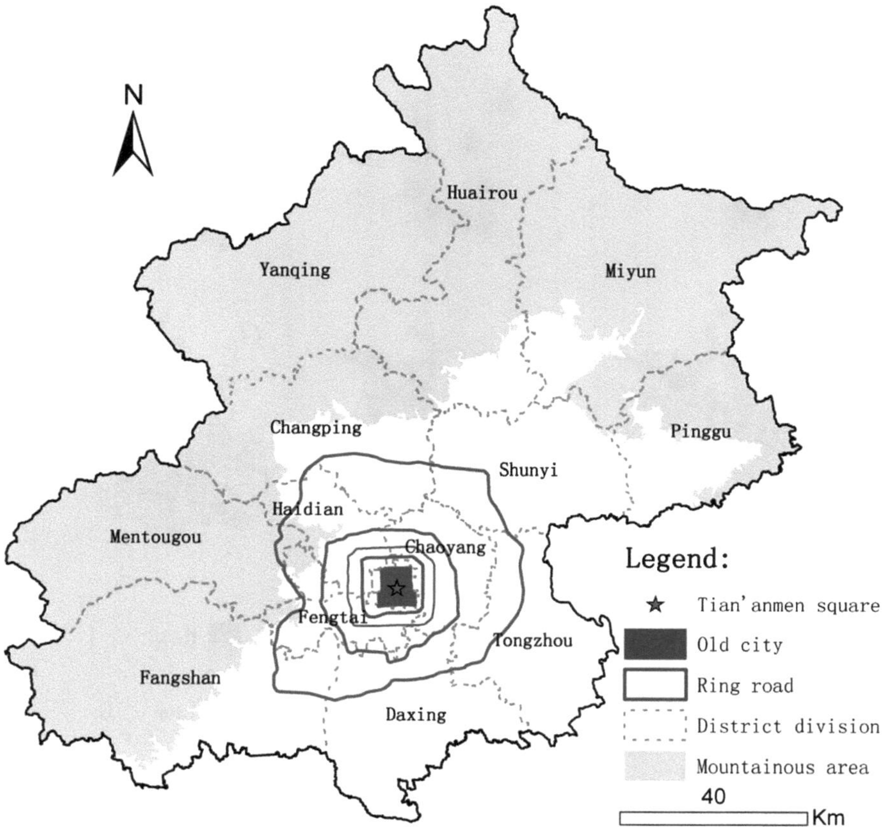

Fig. 2.1 The Beijing metropolitan area map

Sea. Mountains cover an area of 10,072 km^2, 61 % of the whole study area. See Yang et al. (2013) for more background information on Beijing.

The BMA has experienced rapid urbanization in terms of GDP and population growth since the 1978 Reform and Opening by the central government. The GDP in 2006 was ¥787 billion, an 83.7 fold increase from 1976 when it was ¥9.4 billion. The population in 2006 was 15.81 million, 1.9 times greater than in 1976 when it was 8.29 million (Beijing Municipal Statistics Bureau and NBS Survey Office in Beijing 1987 and 2007). Using Landsat imageries, the built-up area in 2006 was 1,324 km^2 (Fig. 2.2), nearly three times larger than in 1976, when it was 495 km^2. Urban growth is predicted by the BMA government to continue for another two decades. Therefore, scholars and decision makers are concerned about the urban growth pattern, especially about how to develop towards the future predefined planned urban form.

To address this concern, we tested the form scenario analysis approach in the BMA to identify policy implications for planning alternatives. The constrained CA

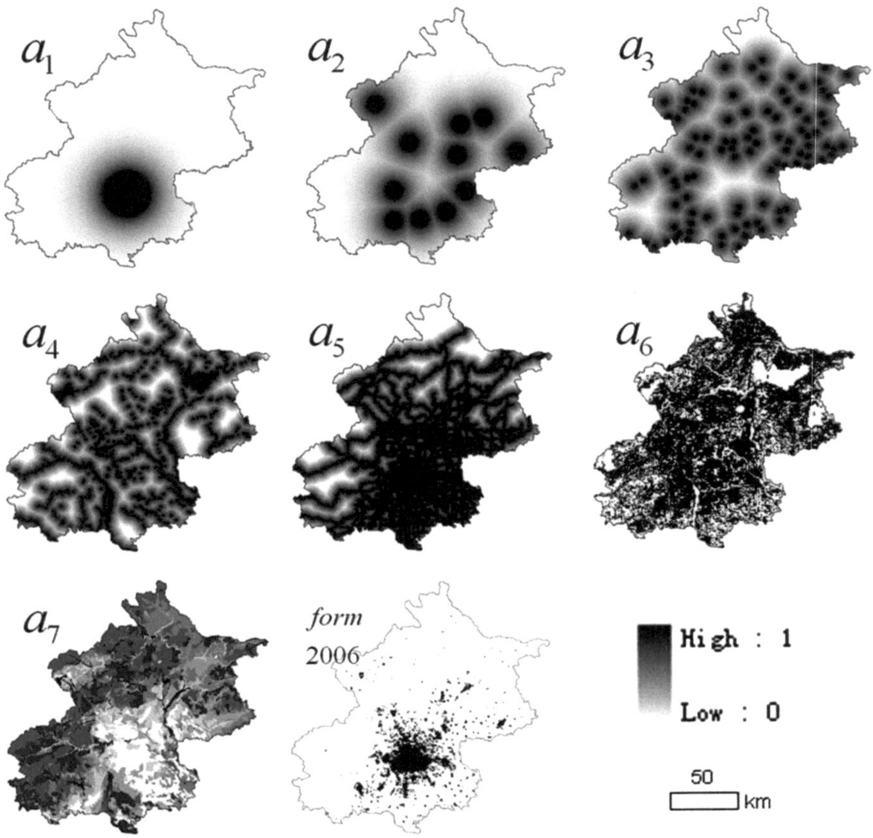

Fig. 2.2 Maps of spatial policies and the urban form of the BMA in 2006

proved suitable for simulating urban growth of the Pearl River Delta, where urban growth is also quite rapid (Li and Yeh 2000, 2002). The driving forces for urban growth of the BMA and the Pearl River Delta are similar with the same domestic socio-economic development background as both these metropolitan areas are the dominating economic growth clusters. Therefore, we also used the constrained CA in the form scenario analysis of the BMA.

2.3.2 *Constraints in Cellular Automata (CA)*

We used a constrained CA based model BUDEM for the FSA using the Python script language based on the ESRI Geoprocessing module (for more detail see Long et al. 2009). BUDEM was established for analyzing historical urban growth and simulating future urban growth in the BMA using cellular automata. In BUDEM,

Table 2.1 Datasets of the constrained CA in the BMA

Type	Variable	Description	Value range	Data source
Macro constraint	*stepNum*	Socio-economic development	>0	Socio-economic Development plan
Location constraints	a_1	Attractiveness of Tiananmen	0–1	Derived from spatial dataset
	a_2	Attractiveness of new cities	0–1	Derived from spatial dataset
	a_3	Attractiveness of towns	0–1	Derived from spatial dataset
	a_4	Attractiveness of rivers	0–1	Derived from spatial dataset
	a_5	Attractiveness of roads in 2006	0–1	Interpreted from TM image of 2006-11-01
Institutional constraints	a_6	Construction prevention policy	0 or 1	Beijing Municipal Planning Committee 2007[a]
	a_7	Suitability for agricultural development	0–1	Beijing Planning Commission 1988
Neighbor constraint	a_n^t	Development intensity in neighborhood	0–1	Calculated by CA

[a]For details of the calculation approach, see Long et al. (2010)

the logistic regression is applied to deriving the transition rule from historical datasets, similar to those in this chapter. As the cell size of BUDEM is 500 m * 500 m, there are 65,628 cells in the BMA. The precision of BUDEM is 87.5 % in terms of Kappa using datasets from 2001 to 2006 (see Table 2.2). This indicates that the model can accurately replicate historical urban growth in Beijing and therefore can be applied for FSA in this paper.

The policies and the corresponding dataset of the constrained CA in the BMA are listed in Table 2.1. The spatial distribution of various policies is shown in Fig. 2.2. StepNum, the macro constraint, as a global control parameter, reflects the total amount of future land development. This constraint is related to the key objective of the socio-economic development plan. The location constraints denote the special plans or policies, e.g., the city hierarchy, flood control and transportation development. The institutional constraints denote the ecological protection, disaster prevention, and farm protection. The neighbor constraint represents the connected development control policy. The three types of spatial constraints are weighted for in Eq. 2.4. The weights are the policy parameters to be identified in this paper.

For convenience in comparing parameters of various policies in parallel and vertically (namely within and across the time periods), policies (A) are standardized to range from 0 to 1, with 1 denoting the greatest probability of development, and 0 the least. The location constraint a_k, using the corresponding spatial feature class as the data source, is processed by the "Distance/Straight Line" toolbox of the Spatial Analyst module in the ESRI ArcGIS package to acquire $dist_k$ followed by $a_k = e^{-\beta * dist_k}$ to calculate the attractive potential, where $\beta = 0.0001$ based on the BUDEM model calibration results. Regarding institutional constraints, when the construction prevention policy a_6 and the suitability for agricultural development policy a_7 are equal to 0, there will be the least probability of development, and when they are equal to 1, there will be the greatest probability of development.

The spatial distributions of various constraints in the BMA, shown in Fig. 2.2, influence the possible future urban growth pattern. Looking at each constraint individually, we can directly and easily recognize the corresponding urban form without considering other constraints. For instance, the construction prevention constraint a_6 stands for the dispersed urban form, with the Miyun Reservoir, partial western mountain area, farmland in the plain area, and the Great Wall protection zone undeveloped. The road constraint a_5 stands for the urban form along the traffic corridors mainly in the plain area due to the high density road networks in this area and the low density in the remote mountain areas. Considering combined constraints and the currently existing urban form featuring the central city sprawl, the future urban form in the BMA is likely to be further sprawl around the central city in the plain area, with scattered developed locations in the mountain areas. Therefore, in the following context, we will come to set form scenarios (namely planning alternatives) and identify policy parameters required for scenarios using constrained CA.

2.3.3 Planning Alternatives

Five versions of the urban master plan for the BMA have been created since 1958, issued separately in 1958, 1973, 1982, 1992 and 2004 (Beijing Municipal Planning Committee et al. 2006). In the 1992 plan for 1991–2010, Han et al. (2009) pointed out that up to 51.8 % of urban developments from 1991 to 2005, within the sixth ring road area (Fig. 2.1), were beyond the planned form. The actual urban developments were significantly inconsistent with the planned form in the BMA. Therefore, form scenario analysis is essential for validating the planned form.

The target of the 2004 plan is for the year 2020, and the plan has estimated a population of 18 million, the developed land to be 2,300 km^2, and specified an urban spatial structure of "Two axes, two belts and multi-sub-centers". In the plan, four planning alternatives, shown in Fig. 2.3, were generated. These alternatives were different in terms of the spatial layout and reflected the preferences of planners. Each alternative reflected different development perspectives of various groups of decision makers and urban planners (details available below). During the creation and approval of the plan, the possibility of realizing each alternative was not considered in detail, and finally Alternative A was approved by the State Council of P. R. China. In this chapter, we apply the FSA approach and develop a constrained CA model to identify the consistency between the planned form and existing policies, thus evaluating the possibility of implementing each alternative.

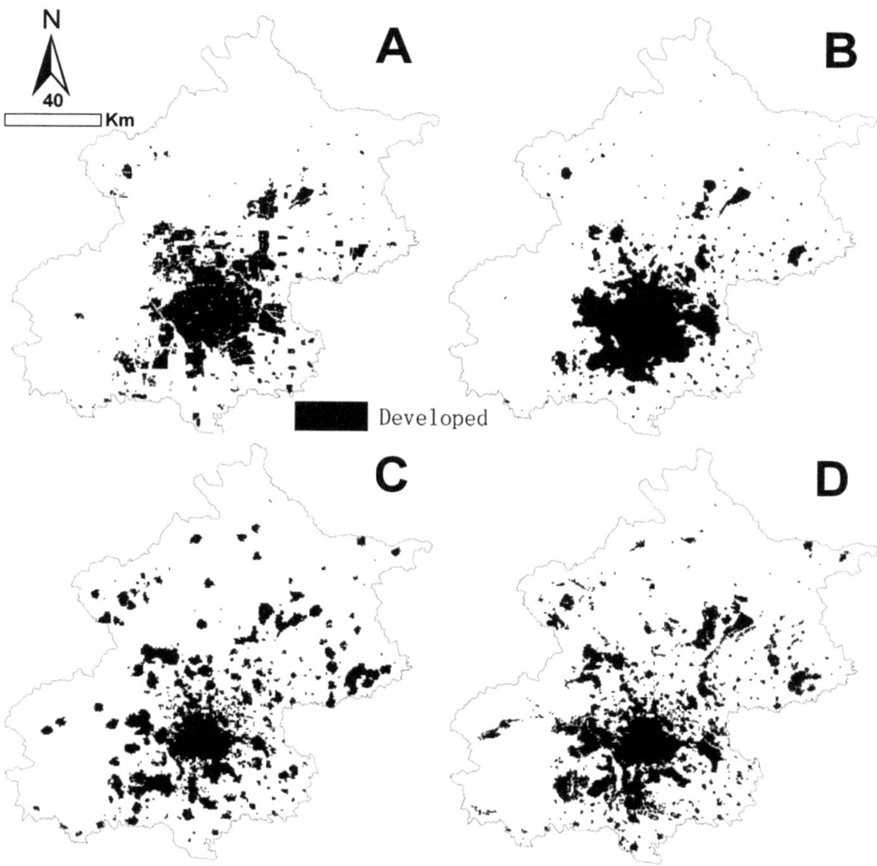

Fig. 2.3 Four planning alternatives in the BMA

Below are the detailed descriptions for the four planning alternatives in the BMA.

- Alternative A (Y_A): The approved alternative by the State Council of P. R. China (see, Beijing Municipal Planning Committee et al. 2006), is characterized by preventing the central city from sprawling further and promoting the development of new cities.
- Alternative B (Y_B): The sprawl alternative, termed Tandabing in Chinese (circle-spread), is characterized by promoting developments in the central city and controlling new developments in new cities.
- Alternative C (Y_C): The grape-cluster alternative, which is characterized by promoting developments both along the transport corridors and around small towns.
- Alternative D (Y_D): The sustainable alternative, which is characterized by preventing construction in specific areas and on high-quality farm land, resulting in a more dispersed form.

2.4 Results

2.4.1 *Identification of Policy Parameters*

To calculate *stepNum*, the total number of iterations is (2020–2006)*12=168. C_{T_s} = 5,011 according to the urban form of 2006 (baseline year), namely Y_{2006}. In Alternative A, C_{T_e} =9,254 is the total number of cells to be developed in the future (2020), and *stepNum* is equal to (9,254–5,011)/168 = 25. The calculation results of *stepNum* for other planning alternatives are listed in Table 2.2, and are slightly different from those of Alternative A. Alternative D has the greatest *stepNum* value, indicating the fastest urban growth among all the alternatives.

The calculation results for the policy parameters, together with the Kappa index, are listed in Table 2.2. All the independent variables are significant at the acceptable 0.001 level. The Kappa index for Alternatives B, C and D are greater than 80 %, indicating that these alternatives have a high probability of implementation. For instance, to implement Alternative B, decision makers proposed a focus on developments around the central city, existing developments and new cities. Other constraints will not be key factors for Alternative B. In contrast, the development route of Alternative C would be different from that of Alternative A, and Alternative C suggested decision makers pay more attention to developments around small towns and along rivers. However, the Kappa index for Alternative A is only 67.5 %, denoting this planning alternative can not be achieved within the current policy context.

Table 2.2 Policy parameters calculation results for four planning alternatives in the BMA

Variable	Alternative A	Alternative B	Alternative C	Alternative D	Historical urban form (2001–2006)
Developed cells number	9254	9270	9895	10,679	5297
stepNum	25	25	29	34	11
x_0 (Intercept)	−8.700	−30.696	−63.599	−55.624	−15.874
x_1 (The city center)	15.268	54.558	15.106	20.849	10.192
x_2 (New cities)	3.575	10.294	10.046	9.701	3.347
x_3 (Towns)	−0.717	5.272	31.639	7.807	−2.839
x_4 (Rivers)	4.105	8.765	24.348	11.622	4.004
x_5 (Roads)	1.368	6.027	7.627	8.113	0.737
x_6 (Construction forbidden policy)	1.193	3.672	4.078	23.000	2.140
x_7 (Agriculture suitability)	−2.396	5.066	6.094	12.003	-3.001
x_n (Neighbor)	15	17	9	7	17
Kappa (%)	69.4	91.8	85.0	85.8	87.5
Valid	False	True	True	True	True

2.4.2 Validation of Planning Alternatives

The MonoLoop's byproduct is the validation process using the Kappa index. When Kappa is greater than or equal to 80 %, the planning alternative is defined as "validated" which means that the parameters identified can be used to realize the planning alternative. When Kappa is less than 80 %, the planning alternative is defined as "not validated". The simulated urban form Y' can be generated using the constrained CA with input parameters listed in Table 2.2, including stepNum, x_{0--7}, and x_n. The cell-by-cell comparison results of Y' and Y (the planning alternative) are shown in Fig. 2.4. The confusion matrix is given in Table 2.3. In each planning alternative, the undeveloped cells substantially outnumber the developed ones, so the overall accuracy is generally high due to the unbalanced dataset.

In addition to the cell-by-cell map comparison method Kappa used in this chapter for comparing simulated and planned urban forms, we further applied other map comparison methods for validating our FSA results using The Map Comparison

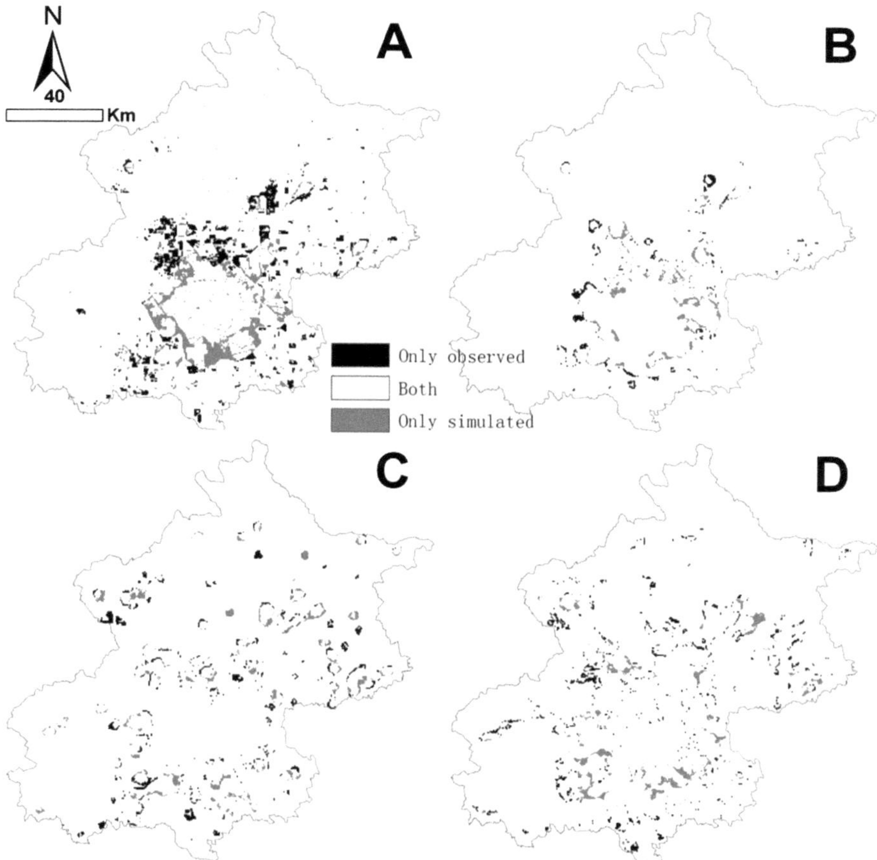

Fig. 2.4 The simulated urban forms compared with the planning alternatives in the BMA

Table 2.3 The confusion matrix of the planning alternatives and simulated urban forms

		Simulated			
Alternative	Planned	0	1	SUM	User accuracy (%)
A	**0**	53,919	2428	56,347	95.7
	1	2440	6826	9266	73.7
	SUM	56,359	9254	65,613	
	Producer accuracy (%)	95.7	73.8	Overall accuracy is 92.6 %	
B	**0**	55,482	620	56,102	98.9
	1	793	8650	9443	91.6
	SUM	56,275	9270	65,545	
	Producer accuracy (%)	98.6	93.31	Overall accuracy is 97.8 %	
C	**0**	54,391	1256	55,647	97.7
	1	1259	8639	9898	87.3
	SUM	55,650	9895	65,545	
	Producer accuracy (%)	97.7	87.3	Overall accuracy is 96.2 %	
D	**0**	53,590	1273	54,863	97.7
	1	1276	9406	10,682	88.1
	SUM	54,866	10,679	65,545	
	Producer accuracy (%)	97.7	88.1	Overall accuracy is 96.1 %	

Table 2.4 Comparing planning alternatives with simulated urban forms using various map comparison methods

Alternative	A	B	C	D
***Kappa* (%)**	**69.4**	**91.8**	**85.0**	**85.8**
Valid	False	True	True	True
Fuzzy Kappa (%)	*67.8*	*88.1*	*81.4*	*81.2*
Fuzzy global matching (%)	*67.0*	*82.3*	*81.2*	*80.9*
Kappa simulation (%)	71.1	91.2	85.0	85.7

Kit 3 (MCK3) developed by Research Institute for Knowledge Systems, The Netherland (Visser and de nijs 2006). These methods are Fuzzy-based and include Fuzzy Kappa, and Fuzzy Inference System. The objective of fuzzy-based map comparison is to propose a method trying to mimics human comparison and gives a detailed assessment of similarity. In Fuzzy Inference System, a global agreement value termed as Fuzzy global matching can be derived by the fuzzy summation of the local matching. The results shown in Table 2.4 indicate that Fuzzy Kappa values for Alternative A to D in percentage are 67.8, 88.1, 81.4 and 81.2, respectively. Fuzzy global matching for Alternative A to D in percentage are 67.0, 82.3, 81.2 and 80.9, respectively. Generally, Fuzzy Kappa and Fuzzy global matching are both slightly lower than Kappa. For each valid alternative in terms of Kappa is also valid in terms of Fuzzy Kappa and Fuzzy global matching in the BMA experiment. This further proves the stability of our approach for FSA.

2.5 Discussion

The Kappa validation and cell-by-cell comparison results show that Alternatives B, C and D match the simulated forms relatively closely. Policy implications for the validated planning alternatives can be presented according to the results of calculation of the policy parameters since the value of each constraint has been standardized to 0 to 1. Three aspects of policy implications are as follows.

1. Parallel comparison of parameters within each planning alternative. For Alternative B, the speed of development of the urban built-up land is 25*12/4=75 km^2 per year from 2006 to 2020 according to *stepNum*. Therefore, the annual population increase is 750,000 based on the assumption of 100 m^2 built-up land per capita. Other parameters can also be compared in parallel, and the greater the parameter is, the more intensely the policy should be implemented to realize the planning alternative. For instance, Alternative B should promote the central city, new city and river side development policies more than other policies.
2. Parallel comparison of parameters across planning alternatives. The different requirements for policy implementation can be easily identified by comparing every parameter for each planning alternative. For instance, x_1 of Alternative B is greater than x_1 of Alternative C. This means that to realize Alternative B, the central city development policy must be implemented much more intensely than in Alternative C. In the same way, the construction protection intensity required to realize Alternative D will be much greater than for Alternative C since the parameter x_6 of Alternative D is obviously greater than that of Alternative C.
3. Parallel comparison of parameters with the historical phase. The policy parameters of 2001–2006, listed in the last column in Table 2.2, can be calculated by the same calibration model approach used with the planning alternatives, using observed forms and historical policies. The speed of urban growth of the four planning alternatives is 2–3 times of that of the historical phase from 2001 to 2006, denoting that, to realize the planning alternatives, urban economic and population developments will need to be promoted much more intensely than during the historical phase. The expansion of neighbor developments should also be controlled to realize the planning alternatives, especially for Alternative C and D.

The calibration result of Alternative A, however, demonstrates that no policy parameter will realize this predefined urban form. For this condition, either Y_A or A should be adjusted to reach the consistency between the predefined planning alternative and the development policies. By adjusting the predefined urban form, a more feasible planned urban from can be established based on constrained CA simulation using different urban growth scenarios. By adjusting the spatial distribution of development policies (e.g. the transportation network, the eco-space distribution), the experiments can also be conducted in constrained CA until the predefined urban form can be realized using the adjusted policies.

2.6 Conclusions and Future Perspectives

In this chapter we attempt to use the urban form as the scenario condition to enable discussions with planners about establishing the possible urban forms within the framework of current development policies from the perspective of development demand, geographical conditions, and institutional controls. Constrained CA is incorporated with the form scenario analysis approach. We use four planning alternatives of the master plan in the Beijing Metropolitan Area as a case study to test our proposed form scenario analysis approach.

FSA using the constrained CA is a breakthrough for CA applications as follows. First, FSA is capable of evaluating the consistency between the planned form and development policies, namely specialty plans. Nowadays in China, planning implementation evaluation is a compulsory requirement in urban planning practices to examine the consistency between the actual urban spatial developments and planned form after several years of planning implementation. The existing reports on planning implementation evaluation are not optimistic as the planned forms have occasionally been exceeded. FSA can detect this in practice. Second, in addition to planning implementation evaluation, FSA can be conducted at the very beginning of the plan creation to assess the possibility of implementing the planned form within the integrated urban development policy environment. FSA can assist planners to design a better layout based on the evaluation results. Third, for a valid plan, the required policies (identified coefficients in this work) can be identified as development pathways to guide decision-makers in implementing the urban plan. Fourth, FSA can be used during the urban planning compilation process to evaluate in terms of spatial constraints the conformity between a spatial plan drafted by planners and corresponding specialized plans proposed by different local government departments. Examples of these specialized plans include hazard-sensitive areas proposed by the geological department and farmland conservation plans proposed by the agriculture department. In sum, FSA can be used as a tool for evaluating the spatial plan compiled by planners and adopted to solve problems faced by planning departments and planners.

Further work is still needed on some aspects of this chapter. First, we drew three premises to simplify the FSA process. To simulate urban growth much more accurately, we suggest further research to focus on the current simplifications. The policy itself (A), beside the policy parameter (X), can be taken into account in FSA to identify not only the required intensity of policy implementation but also the spatial distribution policy required. The spatial heterogeneity of the policy parameter also needs to be considered as various studies indicate that forces driving urban growth in China vary between the sub-regions in the metropolitan area (Liu et al. 2005; Li et al. 2008). Second, agent based modeling can be applied in FSA to represent planners and other decision makers' preferences as the planner agents, to investigate FSA from another perspective (Ligtenberg and Bregt 2001; Saarloos et al. 2005). Third, we set a Kappa value of 80 % as the benchmark for accepting or declining policy parameters for the predefined urban form based on our experience. How an 80 % Kappa insures a good match between two urban forms needs further examination.

References

Beijing Municipal Planning Committee. (2007). *Ecologically limited land-use planning in Beijing (2006–2020) (in Chinese)*. Internal report.

Beijing Municipal Planning Committee, Beijing Institute of City Planning, Beijing Academy of Urban Planning. (2006). *Beijing urban planning atlas (1949–2005) (in Chinese)*. Internal published book.

Beijing Municipal Statistics Bureau, NBS Survey Office in Beijing. (1987). *Beijing statistical yearbook 1987*. Beijing: China Statistics Press.

Beijing Municipal Statistics Bureau, NBS Survey Office in Beijing. (2007). *Beijing statistical yearbook 2007*. Beijing: China Statistics Press.

Beijing Planning Commission. (1988). *Beijing land resources (in Chinese)*. Beijing: Beijing SciTech Press.

Clark, K. C., & Gaydos, L. J. (1998). Loose-coupling a cellular automation model and GIS: Long-term urban growth prediction for San Francisco and Washington/Baltimore. *International Journal of Geographical Information Science, 12*(7), 699–714.

Couclelis, H. (2005). "Where has the future gone?" Rethinking the role of integrated land-use models in spatial planning. *Environment and Planning A, 37*(8), 1353–1371.

Engelen, G., White, R., & Uljee, I. (1997). Integrating constrained cellular automata models, GIS and decision support tools for urban and regional planning and policy making. In H. Timmermans (Ed.), *Decision support systems in urban planning* (pp. 125–155). London: E&FN Spon.

Engelen, G., White, R., & Nijs, T. D. (2003). Spatial support system for the integrated assessment of socio-economic and environmental policies in the Netherlands. *Integrated Assessment, 4*, 97–105.

Ewing, R., Schmid, T., Killingsworth, R., Zlot, A., & Raudenbush, S. (2003). Relationship between urban sprawl and physical activity, obesity, and morbidity. *American Journal of Health Promotion, 18*, 47–57.

Guan, G., Wang, L., & Clark, K. C. (2005). An artificial-neural-network-based, constrained CA model for simulating urban growth. *Cartography and Geographic Information Science, 32*(4), 369–380.

Han, H., Lai, S., Dang, A., Tan, Z., & Wu, C. (2009). Effectiveness of urban construction boundaries in Beijing: An assessment. *Journal of Zhejiang University SCIENCE A, 10*(9), 1285–1295.

Kahn, M. (2000). The environmental impact of suburbanization. *Journal of Policy Analysis and Management, 19*, 569–586.

Kahn, J., & Wiener, A. J. (1967). *The year 2000: A framework for speculation on the next 33 years*. New York: MacMillan Press.

Klosterman, R. E. (1999). The what if? Collaborative planning support system. *Environment and Planning B: Planning and Design, 26*(3), 393–408.

Landis, J. D. (1994). The California urban future model: A new generation of metropolitan simulation models. *Environment and Planning B: Planning and Design, 21*(4), 399–420.

Landis, J. D. (1995). Imaging land use futures: Applying the California urban future model. *Journal of American Planning Association, 61*(4), 438–457.

Landis, J. D., & Zhang, M. (1998a). The second generation of the California urban future model, Part 1: Model logic and theory. *Environment and Planning B: Planning and Design, 25*(5), 657–666.

Landis, J. D., & Zhang, M. (1998b). The second generation of the California urban future model, Part 2: Specification and calibration results of the land-use change submodel. *Environment and Planning B: Planning and Design, 25*(6), 795–824.

Li, X., & Yeh, A. G. O. (2000). Modeling sustainable urban development by the integration of constrained cellular automata and GIS. *International Journal of Geographical Information Science, 14*(2), 131–152.

Li, X., & Yeh, A. G. O. (2002). Neural-network-based cellular automata for simulating multiple land use changes using GIS. *International Journal of Geographical Information Science, 16*(4), 323–343.
Li, X., & Yeh, A. G. O. (2004). Data mining of cellular automata's transition rules. *International Journal of Geographical Information Science, 18*(8), 723–744.
Li, X., Yang, Q., & Liu, X. (2008). Discovering and evaluating urban signatures for simulating compact development using cellular automata. *Landscape and Urban Planning, 86*(2), 177–186.
Ligtenberg, A., & Bregt, A. K. (2001). Multi-actor-based land use modelling: Spatial planning using agents. *Landscape and Urban Planning, 56*(1–2), 21–33.
Liu, X., Wang, J., Liu, M., & Meng, B. (2005). Spatial heterogeneity of the driving forces of cropland change in China. *Science in China Series D-Earth Sciences, 48*(12), 2231–2240.
Long, Y., & Shen, Z. (2011). An urban model using complex constrained cellular automata: Long-term urban form prediction for Beijing. *International Journal of Society Systems Science, 3*(1–2), 159–173.
Long, Y., Mao, Q., & Dang, A. (2009). Beijing urban development model: Urban growth analysis and simulation. *Tsinghua Science and Technology, 14*(6), 787–794.
Long, Y., Shen, Z., & Mao, Q. (2010). An urban containment planning support system for Beijing. *Computers, Environment and Urban Systems, 35*(4), 297–307.
Pearman, A. D. (1988). Scenario construction for transportation planning. *Transportation Planning and Technology, 12*(1), 73–85.
Ratcliffe, J. S. (1999). Scenario building: A suitable method for strategic property planning? *Property Management Bradford, 18*(2), 127–144.
Ringland, G. (1998). *Scenario planning: Managing for the future*. New York: John Wiley.
Rosen, S. (1974). Hedonic prices and implicit markets: Product differentiation in pure competition. *Journal of Political Economy, 82*, 34–55.
Saarloos, D., Arentze, T., Borgers, A., & Timmermans, H. (2005). A multiagent model for alternative plan generation. *Environment and Planning B: Planning and Design, 32*(4), 505–522.
Schoemaker, P. J. H. (1995). Scenario planning: A tool for strategic thinking. *Sloan Management Review, 36*(2), 25–40.
Sun, Z. L., Deal, B., & Pallathucheril, V. G. (2009). The land-use evolution and impact assessment model: A comprehensive urban planning support system. *URISA Journal, 21*(1), 57–68.
Tian, L., & Shen, T. (2011). Evaluation of plan implementation in the transitional China: A case of Guangzhou city master plan. *Cities, 28*(1), 11–27.
Visser, H., & de Nijs, T. (2006). The map comparison kit. *Environmental Modelling & Software, 21*, 346–358.
Ward, D. P., & Murray, A. T. (1999). An optimized cellular automata approach for sustainable urban development in rapidly urbanizing regions. *International Journal of Geographical Information Science, 7*(5), 235–250.
White, R., Engelen, G., & Uijee, I. (1997). The use of constrained cellular automata for high-resolution modelling of urban land-use dynamics. *Environment and Planning B: Planning and Design, 24*, 323–343.
White, R., Straatman, B., & Engelen, G. (2004). Planning scenario visualization and assessment – a cellular automata based integrated spatial decision support system. In M. F. Goodchild, D. G. Janelle, & Z. G. Shrore (Eds.), *Spatially integrated social science* (pp. 420–442). Oxford: Oxford University Press.
Wu, F. (1998). Simland: A prototype to simulate land conversion through the integrated GIS and CA with AHP-derived transition rules. *International Journal of Geographical Information Science, 12*(1), 63–82.
Wu, F. (2002). Calibration of stochastic cellular automata: The application to rural–urban land conversions. *International journal of Geographical Information Science, 16*(8), 795–818.
Yang, Z., Cai, J., Ottens, H. F. L., & Sliuzas, R. (2013). Beijing: City profile. *Cities, 31*, 491–506.

Chapter 3
Urban Expansion Simulation and Analysis in Beijing-Tianjin-Hebei Area Based on BUDEM-BTH

3.1 Introduction

At present, the world is experiencing an unprecedented process of urbanization. The developed countries and developing countries are more or less facing the huge pressure of urban expansion, thus a lot of researches on urban expansion driving force are under way. In China, one of the fastest-developing and most-populous countries in the world, the negative impact of urban expansion has emerged, such as farm land occupation, ecological environment destruction (Lu 2007; Kahn 2000). Beijing-Tianjin-Hebei (BTH) is one of the main areas of China's economic development, which is in the core status of north regional economic development pattern. Therefore, by simulating the urban expansion, identifying the urban expansion driving force, and their impacts on the spatial layout and spatial integration direction, there will be a very important theoretical and practical meaning for promoting the regional economic strength and enhancing the overall competitiveness of our country.

It has been a long history to analyze and simulate urban expansion. The researches mainly concentrated on the process detection of time and space, analysis of driving mechanism, process characterization and simulation, macro ecological effect evaluation, etc.(Liu and Deng 2009). The urban model which used for evaluating the urban expansion mainly by the following types of urban models in different steps of urbanization in BTH area: spatial interaction model, system dynamics and Laurie model, and cellular automata and multi-agent systems (Zhou et al. 1999). CA, as an important research tool of complexity science, is suitable for simulating urban expansion as a space-time dynamic process. In recent years, many researches on CA have been conducted, For example, Batty and Xie (1994, 1997) simulated the suburb expansion process of Buffalo, New York city; White and Engelen (1993; White et al. 1997) simulated the land use change in the city of Cincinnati. Without considering the interaction relationship between economic and social indicators, CA model based on spatial interaction, in the aspect of urban simulation, can reflect the

Y. Long, Z. Shen, *Geospatial Analysis to Support Urban Planning in Beijing*, GeoJournal Library 116, DOI 10.1007/978-3-319-19342-7_3

change of urban spatial pattern. Meanwhile, the urban scale in the simulation model can be designed flexibly, so the change of urban spatial structure can be simulated on the fine scale. In addition, the model can reflect the urban change from generation, development to the demise on a longer time scale. Finally, CA model can well simulate the urban spatial pattern as complex phenomena according to CA theory, such as mutation, self-organization and chaos, as an open dissipative system (Zhou et al. 1999; Chen 1999; He et al. 2002).

Urban simulation based on CA is facing the some challenges discussed as follows (He et al. 2002). As the interactions of the microcosmic individuals represented in CA model is usually simply defined as interaction between neighbor grids that stand for certain land size, which can be designed according the different needs of research interesting regarding urban scale. Aiming at adding geographic constrains as conditions of CA simulation, some scholars began to add constraints into CA model (namely, constraint CA) to control the urban form in simulation (Ward et al. 2000; White et al. 1997), which improved the model precision and achieved good simulation effect. Long et al. (2009b) developed the Beijing Urban Development Model (BUDEM) in which the space constraint, institutional constraint, and neighborhood constraint were considered to simulate the urban form of Beijing at the end of 2020 and 2049 (the 100th anniversary of the founding of the People's Republic of China). Some other scholars integrated CA model with other models or tools to simulate the urban dynamic change, such as the integrating of CA model and GIS platform (Yeh and Li 2001; Wu 1998). He et al. (2005) developed Land Use Scenarios Dynamics Model (LUSD) based on system dynamics model and cellular automata Model, simulating land use change of 13 provinces in northern China over the next 20 years and making a research on the land use change at a regional scales.

BTH has attracted a large number of experts and scholars' concerning because of its important strategic position. Wu Liangyong (2002a, b) from Tsinghua university, Fan Jie (2008) from Institute of Geographic Sciences and Natural Resources Research, China Academy of Urban Planning & Design (2007), etc., have done many research projects related to spatial strategy on BTH region. Xie et al. (2011) studied the earthquake disaster and risk assessment in the Beijing-Tianjin-Tangshan region. Through integrating artificial neural network (ANN) model and cellular automata (CA) model, Kuang Wenhui et al. (2011) have developed an urban growth dynamic model, in which simulation based on different scenarios at regional scale have been conducted for urban dynamic growth in the Beijing-Tianjin-Tangshan region. He Chunyang' teams put forward the optimal urban layout and suggest studying urban expansion process under the double pressure of drought and earthquakes.

This chapter, on the basis of summarizing predecessors' work, learning from the experience of BUDEM (Long et al. 2009a), makes use of BTH spatial data, and develops the BTH Urban Development Model (BUDEM – BTH) using constrained CA model and logistic regression analysis method for retrieving parameters of constrains factors of CA model. The model scientifically analyzes the change of BTH urban spatial development pattern under different development scenarios, forecasts BTH urban form in urban growth process under different scenarios. It can provide a decision-making basis for urban plan, construction and management.

This chapter is organized as follows. In Sect. 3.2, the study area and data are described in detail. In Sect. 3.3, BUDEM – BTH model's conversion rule is mainly introduced. In Sect. 3.4, the method to identify the model parameter is introduced, and the model accuracy is verified by the identified parameters as the basis of scenario analysis. In Sect. 3.5, construction land layout of BTH at the end of the plan 2020 is simulated, and then compared with the planed land layout. Seven scenarios are set up to analyze BTH urban construction land layout in 2049. In Sect. 3.6, the research is summarized and the future research consideration is put forward.

3.2 Study Area and Data

The study area for BUDEM-BTH model is BTH area, as shown in Fig. 3.1. It locates between 113.15° and 119.70° east longitude and between 36.04° and 42.56° north latitude, with an area of 21.71*104 km^2 (16,410 km^2 for Beijing, 11,919 km^2 for Tianjin, 18.88*104 for Hebei province).

Fig. 3.1 Location of BTH area

The data required for BUDEM-BTH model is shown in Table 3.1, including land use data in the year of 2000, 2005 and 2010; topographic map 1:50,000 containing information for administrative centers at all levels; topographic map 1:250,000 containing information for railway, highway, state road, provincial road; BTH master plan data for 2020; the construction forbidden areas data; urban statistical yearbook of 1995–2008, etc.

The data of administrative centers at all levels can be extracted from topographic map 1:50,000, and all kinds of road and boarder line information can be extracted from topographic map 1:250,000. They are as the location variable, and the construction forbidden areas data is as the institutional variable. All the data is normalized to 0–1. The nearer to 1, the more probable to be developed; and the nearer to 0, the less probable to be developed. The value of neighbor variable can be acquired by logistic regression method.

The land use data includes mainly urban construction land, rural construction land, forest land, grassland, farmland, water area and unused land. The simulation at this time is just the change of urban construction land, therefore the research needs only two kinds of data: urban construction land and non-urban construction land. The rural construction land, forest land, grassland, farmland, water area and unused land will be amalgamated into non-urban construction land. The result is as Fig. 3.3.

3.3 The Build of BUDEM-BTH Model

Because of the complexity and dynamic nature of spatial system, the development of urban has a "bottom-up" self-organization characteristic (Long et al. 2008a, b), which is similar to CA's "bottom-up" self-organization evolution characteristic, therefore it is available to simulate the urban expansion by BUDEM-BTH model which is based on CA model.

3.3.1 The BUDEM-BTH Model

A standard CA is composed of cell, cell space, neighborhood state and conversion rules, which can be expressed as a formula: A=(Ld, S, N, f) (Amorosos and Patty 1972), in which Ld stands for the cell space, S stands for a set of finite and discrete cell states, N stands for all the cell muster within neighborhood, and f stands for a conversion rules. BUDEM-BTH is also composed of these elements, because it is converted from standard CA model.

The aim to build BUDEM-BTH model is to simulate the conversion from the non-urban construction land to urban construction land, so two states need to be set up: urban construction land and non-urban construction land. A collection of two states is defined as S:={ urban construction land, non-urban construction land}, in a

Table 3.1 Dataset of the BUDEM-BTH model

Type	Variable	Min	Max	Mean valve	Standard	Description	Data	Data resource
LOCATION constraint	*f_ctr_bj*	0.0000	1.0000	0.0029	0.0269	Attractiveness of Beijing	Fig. 3.2a	1:50,000 Topographic map
	f_ctr _tjsjz	0.0000	1.0000	0.0059	0.0380	Attractiveness of Beijing and Tianjin	Fig. 3.2b	1:50,000 Topographic map
	f_ctr _other	0.0000	1.0000	0.0254	0.0767	Attractiveness of prefecture-level	Fig. 3.2c	1:50,000 Topographic map
	f_ctr _cty	0.0000	1.0000	0.2082	0.1931	Attractiveness of the town	Fig. 3.2d	1:50,000 Topographic map
	f_rail	0.0000	1.0000	0.3797	0.3141	Attractiveness of railway	Fig. 3.2e	1:250,000 Topographic map
	f_r _hig	0.0000	1.0000	0.2662	0.3043	Attractiveness of highway	Fig. 3.2f	1:250,000 Topographic map
	f_r _nat	0.0000	1.0000	0.3690	0.2925	Attractiveness of state road	Fig. 3.2g	1:250,000 Topographic map
	f_r _pro	0.0000	1.0000	0.5519	0.2711	Attractiveness of provincial road	Fig. 3.2h	1:250,000 Topographic map
INSTITUTIONAL constraint	*constrain*	0	1	0.1300	0.3350	Construction forbidden areas	Fig. 3.2i	Beijing Municipal Institute of City Planning & Design
NEIGHBOR constraint	*neighbor*	0.0000	1.0000	–	–	The neighbor cell number/8	–	Model calculation

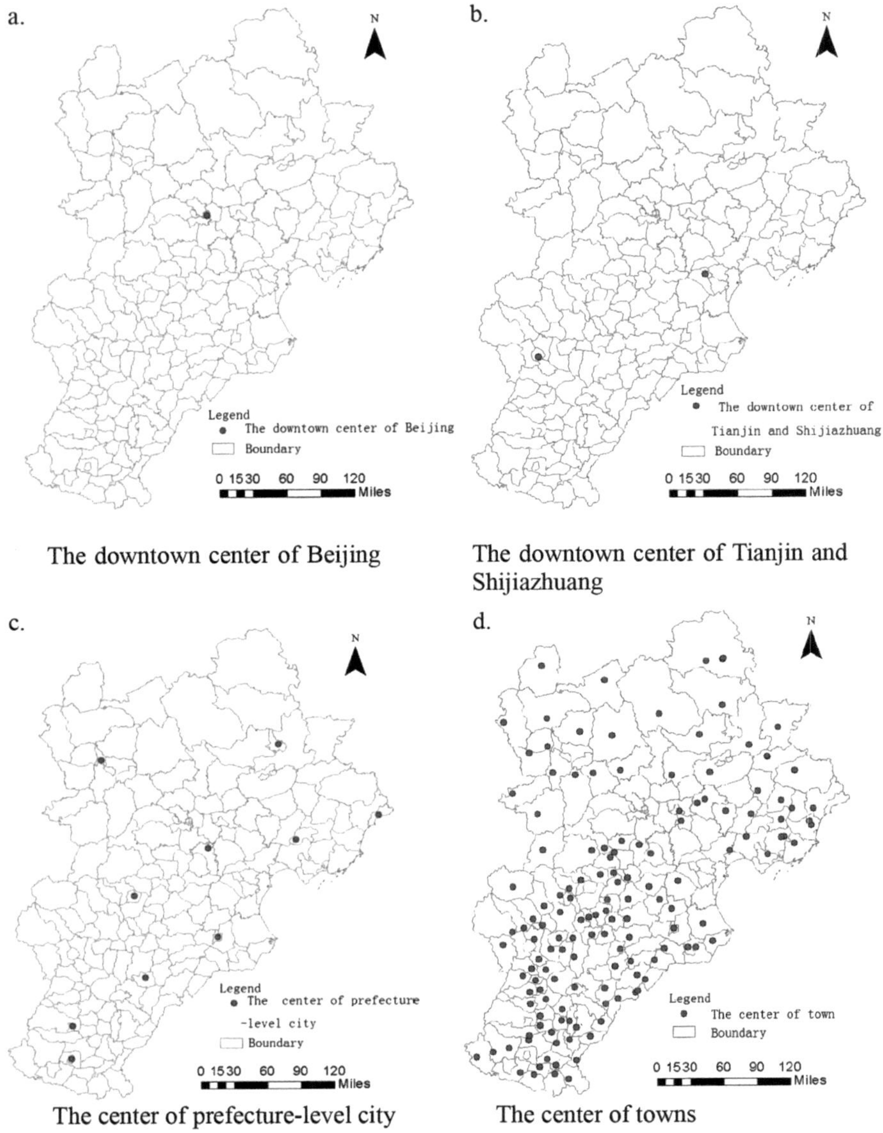

Fig. 3.2 Spatial data for the BUDEM-BTH model

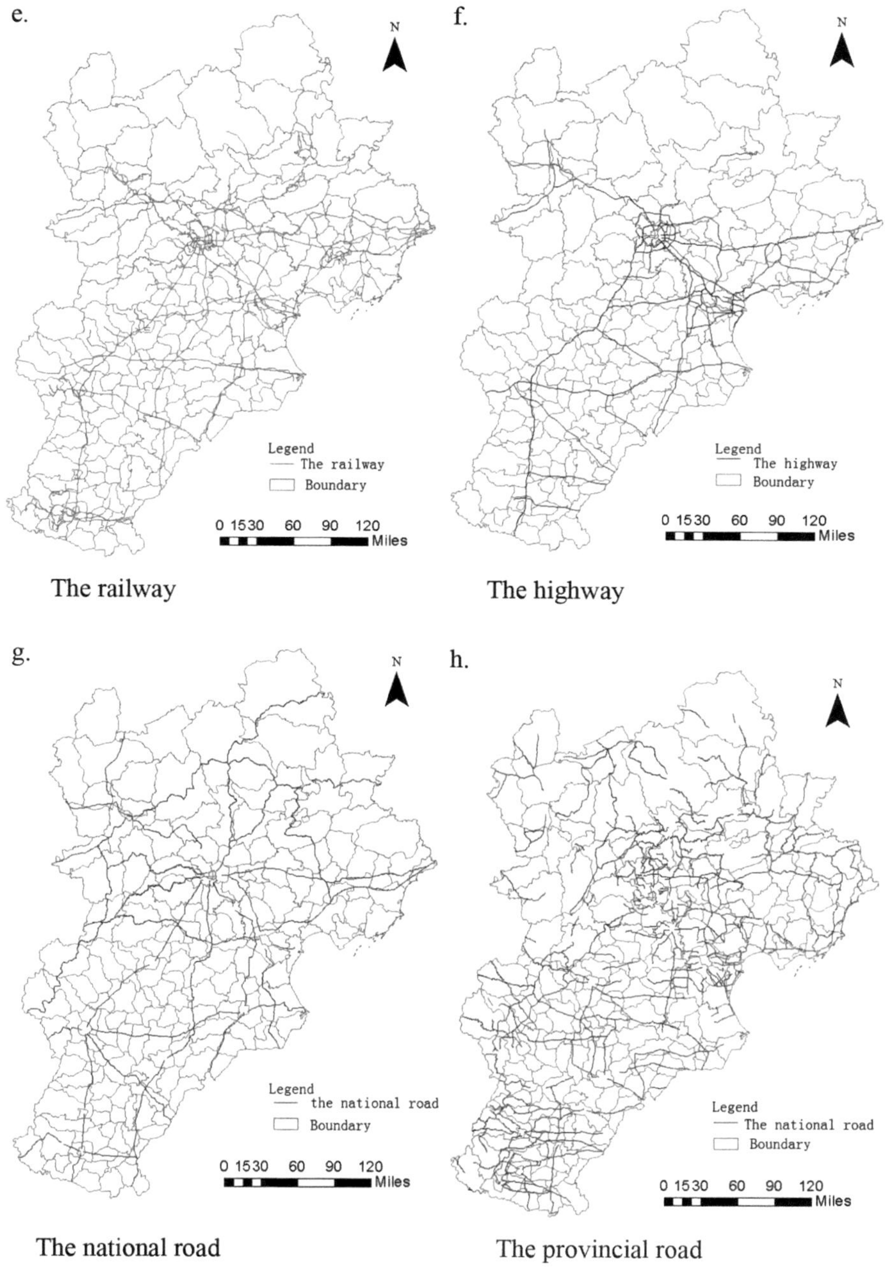

Fig. 3.2 (continued)

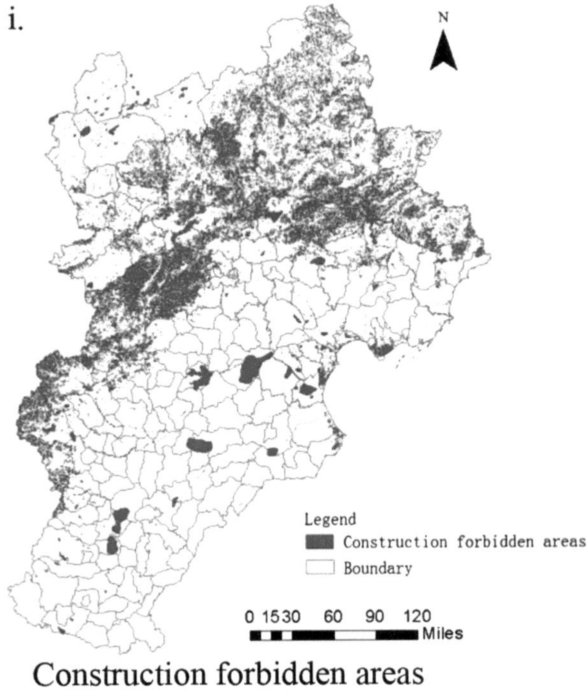

Fig. 3.2 (continued)

mathematical set:S = {1, 0}, in which 1 stands for urban construction land, and 0 stands for non-urban construction land.

BUDEM-BTH model's Neighborhood uses Moore neighborhood (3*3, eight adjacent cells), and the cell size for BUDEM-BTH model is 500 m * 500 m. The BUDEM-BTH model acquires Multi-criteria Evaluation (MCE) as conversion rule.

3.3.2 The Status Transition Rule for BUDEM-BTH

BUDEM-BTH Model is developed from the idea of Hedonic Model. Hedonic model (Lancaster 1966) considers that commodities consist of various different characteristics or qualities (e.g., the area of the real estate, floor, orientation and whether there is a security service, etc.), and the price is the reflection and performance of all these characteristics. Commodities with different characteristics or qualities would have different price. For example, Butler (1982) thinks that housing prices can be decided by three key factors: location factors, the structure of the building and neighborhood environmental factors. Therefore, the housing price can reflect the consumer's preferences for these factors, and the development of urban

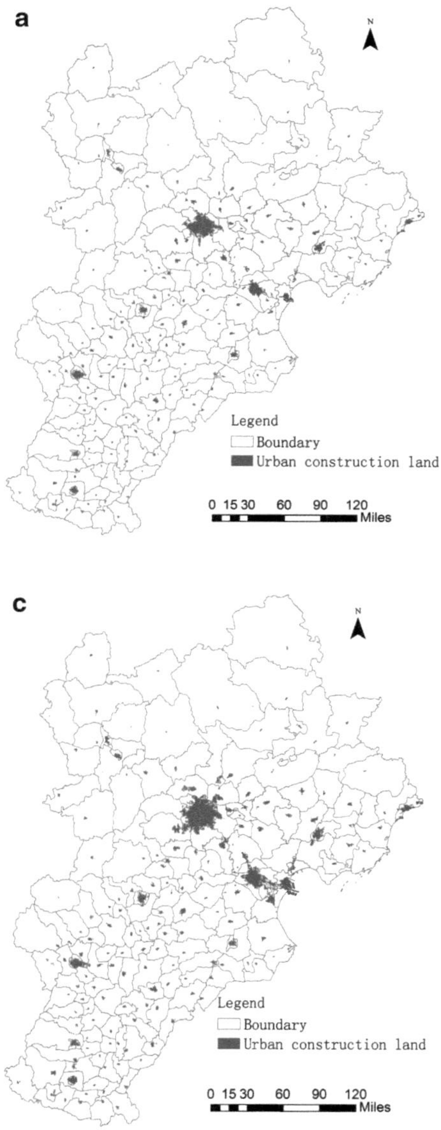

Fig. 3.3 Land use reclassification based on multi-temporal remote sensing data

construction land can also reflect the developers' preferences for land attributes. Referring to the theoretical framework of Hedonic model, considering the availability of data at the same time, the following elements are selected as spatial variable of BUDEM-BTH model:

1. LOCATION constraint:
   The shortest distance to the administrative center at various levels (the center of Beijing, the center of Tianjin and Shijiazhuang, the center of prefecture-level cities, the center of towns) ,the distance to railway (f_rail), the distance to highway (f_r_hig), the distance to national road (f_r_nat), the distance to provincial road (f_r_pro).
2. NEIGHBOR constraint:
   The neighborhood development intensity (neighbor).
3. INSTITUTIONAL constraint:
   Due to the restriction of data resource, only abrupt slop, wet land and blood-proof land should be considered as constrain area.

Urban growth not only be controlled by government on macro level, but also be spontaneous on microcosmic stratums. Hence, the simulation is divided into two steps: First, the government determines at the macro-level the amount of land development in different phases according to socio-economic conditions. Then at the micro-level, constrain CA can be used to simulate the urban growth, and allocate the cell number according to the simulation result.

The simulation process of BUDEM-BTH is as Fig. 3.4.

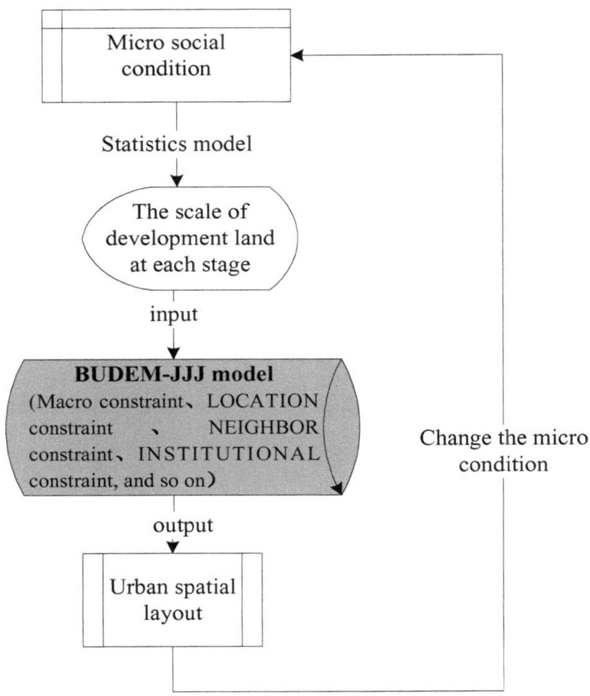

Fig. 3.4 BUDEM-BTH model simulation process

In this research, BUDEM-BTH model only simulates the translation from non-urban construction land to urban construction land, without regarding to the opposite process, urban redevelopment, and urban degeneration process.

Based on the above constraint factors, Multi-criteria Evaluation is considered as the transition rules of CA.

$$V_{i,j}^{t+1} = \{V_{i,j}^{t}, LOCATION\ constraint, INSTITUTION\ constraint, NEIGHBOR\ constraint\}$$

$$= f\left\{\begin{array}{l} V_{i,j}^{t}, \\ f\_ctr\_bj_{i,j}, f\_ctr\_tjsjz_{i,j}, f\_ctr\_cty_{i,j}, f\_ctr\_other_{i,j}, \\ f\_rail_{i,j}, f\_r\_high_{i,j}, f\_r\_nat_{i,j}, f\_r\_pro_{i,j}, \\ constrain_{i,j}, \\ neighbor^{t}_{i,j} \end{array}\right\} \tag{3.1}$$

In the formula, $V_{i,j}^{t}$ is the cell status at i,j at time t, $V_{i,j}^{t+1}$ is the cell status at i,j at time $t+1$, and f is the transition rule of the cell status.

The status transition function based on status transition rule is expressed as follows:

$$\begin{array}{l}
1.\ LandAmount = \sum_{t} stepNum^{t} \\
2.\ s^{t}_{i,j} = w_0 + w_1 * f_ctr_bj_{i,j} + w_2 * f_ctr_tjsjz_{i,j} + w_3 * f_ctr_cty_{i,j} \\
+w_4 * f_ctr_other_{i,j} + w_5 * f_rail_{i,j} + w_6 * f_r_high_{i,j} + w_7 * f_r_nat_{i,j} \\
+w_8 * f_r_pro_{i,j} + w_9 * constrain_{i,j} \\
+wN * neighbor^{t}_{i,j} \\
3.\ p^{t}_{g} = \dfrac{1}{1+e^{-s^{t}_{i,j}}} \\
4.\ p^{t} = \exp[\delta(\dfrac{p^{t}}{p^{t}_{g\max}} - 1)] \\
5.\ for\ instepID = 1\ to\ stepNum \\
if\ p^{t}_{i,j} = p^{t}_{\max}\ then\ V^{t+1}_{i,j} = 1 \\
p^{t}_{i,j} = p^{t}_{i,j} - p^{t}_{\max} \\
p^{t}_{\max}\ update
\end{array} \tag{3.2}$$

In the formula: *LandAmount* stands for the total increased cells number; $stepNum^{t}$ stands for the increased cells number in each cycle; $s^{t}_{i,j}$ stands for the land use suitability; w stands for the weight coefficient of spatial variable; p^{t}_{gx} stands for the global probability after conversion; p^{t}_{gmax} stands for the global probability maximum at each cycle; δ stands for the diffusion coefficient (1–10); p^{t} stands for the ultimate probability; p^t_{max} stands for the maximum probability for each sub-cycle at different loop, constantly updating the value in the cycle.

As for the total land development parameter *stepNum* at each phase, the traditional demographic method has been used to predict the urban construction land. According to the formula $stepNum = P \times A/10000$, P stands for the urban population (person) and A stands for the per capita urban construction land indicators (m^2/person). Urban population at each phase obtains by time trend extrapolation method. According to the model $P = a + bY$, Y stands for the time (year), a and b stand for constants obtained by regression analysis.

The weight coefficient $w_{1\text{–}8}$ of 8 spatial constrain variables can be obtained by logistic regression analysis of historical data. Logistic regression analysis is a probabilistic nonlinear regression model, which is a multivariate analysis method to study the relationship between classification observations result *(Y)* and certain influence factors *(X)*. In medical science, researchers often investigate whether the certain result will happen under certain factors, and what is their relationship (Literature?). For example, it can be used to judge whether the coronary heart disease (CHD) has something to do with the presence of hypertension history, history of hyperlipidemia and smoking history. If the value Y is defined as the result, $Y = 1$ means yes, $Y = 0$ means no, and number between 0 and 1 means the probability of happening.

Its concrete form is shown in the formula (3), which is a semi-logarithm equation, and the regression coefficient b reflects the sensitivity of the variables, namely the impact of the variable changing 1 unit on the overall probability. The larger its absolute value, the more sensitive the corresponding variables.

$$p_{logistic} = \frac{1}{1 + e^{-z_{ij}}}$$
$$z_{ij} = a + \sum_k b_k x_k \tag{3.3}$$

In the above formula: a is the constant in the logistic regression model; b_k stands for the model coefficient; x_k stands for spatial variables; $p_{Logistic}$ stands for transition probability based on logistic regression. Moreover, *w9 is* binary coefficient: $w9 = \{0, 1\}$, if $w9 = 0$, this cell cannot be developed; on the contrary, if $w9 = 1$, this cell can be developed.

When $w9 = 1$, the neighbor weight coefficient *wN* can be obtained by MonoLoop, and the largest coefficient of goodness – of – fit(GOF) will be chosen as the weight coefficient for *wN* value (Butler 1982; Long et al. 2010). The MonoLoop method is as follows: Firstly, obtain the spatial variable weight parameters except *neighbor* variable by the method of logistic regression and keep these parameters unchanged; Secondly, add a loop process in the CA model and constantly adjust the weight coefficient for neighbor variable *wN*; Thirdly, compare the different simulation values and observation values for *wN*; Then choose the best match value *wN* and logistic regression value w_{0-8} into the state transition rules of MCE; Finally, the function of urban spatial form simulation will be realized. As for the choice range of *wN* value, in order to reduce the running time of MonoLoop process, firstly try constantly various numbers in a larger scale, and then adjust the number in a more detailed scale according to the calculation results. In the process of simulation of different *wN* values, the rest parameters and total scale of the target construction land remain unchanged.

3.4 BUDEM-BTH Model Parameter Identification

3.4.1 *The Urban Expansion Analysis for BTH*

The urban expansion in BTH is very apparent in recent years. From Figs. 3.5 and 3.6, we can find that the construction land in each region has an outward expansion trend with different degree from the year 2000 to 2005, and from the year 2005 to 2010.

The statistical results are as follows:

① The growth area in BTH is 2767.25 km^2 from the year 2000 to 2010, and annual growth area is 276.725 km^2.
② The growth area in BTH is 1445.04 km^2 from the year 2000 to 2005, and annual growth area is 289.008 km^2.
③ The growth area in BTH is 1322.21 km^2 from the year 2005 to 2010, and annual growth area is 264.442 km^2.

From the above statistics, we can find the urban expansion speed of the two periods is a relatively stable, but the later period is relatively lower than the former period.

3.4.2 *History Parameters Identification*

Identifying the history parameter is the key to the model prediction and the basis to forecast the future urban construction land layout. The longitudinal comparison of history parameters between the same stage can find out the main driving force for affecting urban expansion at this stage, and the horizontal comparison between the independent variable in different historical stages can find out the independent variable' influence to the urban expansion (Long et al. 2008a, b).

In this model, the parameters are identified through logistic regression method by the statistical analysis software SPSS. Due to the limitation of data source, only the parameters of the year 2000–2005 and 2005–2010 can be identified.

From the statistical analysis, we can find the date is imbalanced, for the sample number of urban construction land is far less than non-urban construction land. The regression precision will be influenced using the data directly. As shown in Table 3.2,

Fig. 3.5 Construction land expansion from 2000 to 2005

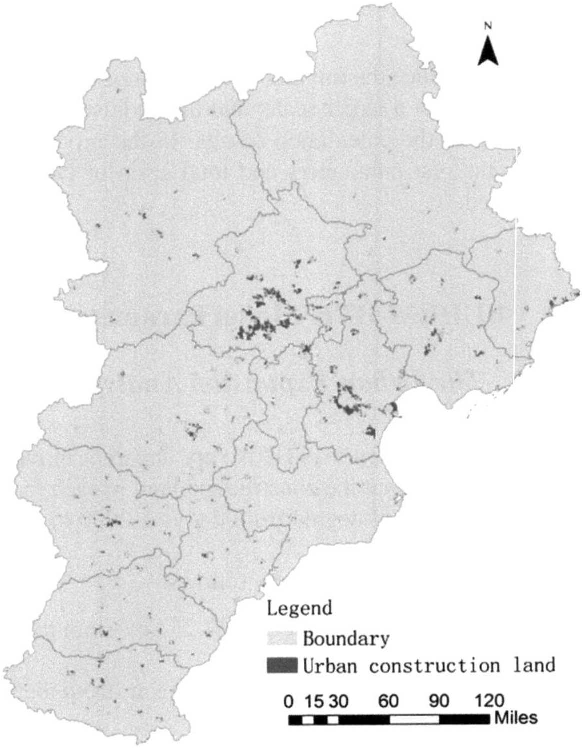

Fig. 3.6 Construction land expansion from 2005 to 2010

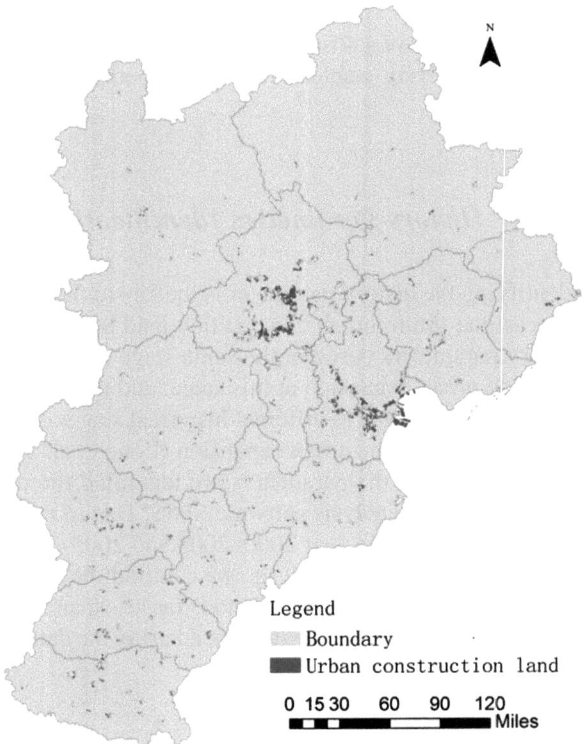

Table 3.2 Logistic classification table in the year of 2000–2005 and 2005–2010

Period of time		Observation (0)	Observation (1)	Total	Production precision (%)
2000–2005	Observation (0)	856,697	0	856,697	100
	Observation (1)	4942	0	4942	0
	total	861,639	0	861,639	
	User precision (%)	99.4	100		Total precision 99.4 %
2005–2010	Observation (0)	857,661	0	857,661	100
	Observation (1)	3970	0	3970	0
	total	861,631	0	861,631	
	User precision (%)	99.5	100		Total precision 99.5 %

no processing the imbalanced data, all the construction land was predicted as non-construction land. It doesn't accord with the reality. Therefore the imbalanced data need to be processed before the logistic regression.

At present, the research for imbalanced data's classification method mainly focuses on three aspects: the data level, the algorithm level and criterion level (Yang et al. 2008). This research adopts price sensitive approach at the algorithm level to process data, which balances the number of data by giving the small sample data greater weight, and giving the large sample data less weight. This research sets sample weight with the weight cases tool in SPSS software. Mainly seven kinds of weight ratio are set. The weight of large sample 0 is constantly set to 1, and the weight of small sample 1 is set to 50, 100, 200, 300, 500, 1000, 10,000 respectively. The matching degree and Kappa value are calculated by regression method, and the results are as follows in Table 3.3.

According to Table 3.3, when the weight is 200, the number gap between 0 and 1 is the smallest, the data gets balanced, and the GOF and Kappa value is also optimal, thus the weight 200 is chosen for regression. And all effective variables of correlation coefficient matrix are close to zero, whereby we can judge the effective coefficients are uncorrelated, and the significance of all variables is at the 0.000 level. It can further prove the validity of the regression results as shown in Table 3.4.

3.4.3 Model Validation

By logistic regression analysis, we can acquire the model variable coefficient except neighbor constraint variables (neighbor). Neighbor constraint coefficient can be acquired by MonoLoop method. This study uses the urban construction land data from the year 2005–2010 to obtain the neighbor coefficient and to verify the model. By using dichotomy to take an integer value from 0 to 100 and repeating experiments, the result proves when $wN = 8$ point-to-point matching degree is highest, at 99.3 % and Kappa value was 0.87, which can prove the model is of high precision and can be used to predict the urban construction land expansion in the later stages for BTH (Fig. 3.7).

Table 3.3 Weight variation table in the year of 2000–2010

	Weight	The number of 0	The number of 1	ratio	GOF (%)	Kappa
2000–2005	1	856,697	4942	17,335/100	99.40	0.000
	50	856,697	247,100	347/100	86.90	0.613
	100	856,697	494,200	173/100	85.30	0.686
	200	856,697	988,400	87/100	87.00	0.737
	300	856,697	1,482,600	58/100	88.30	0.741
	500	856,697	2,471,000	35/100	90.40	0.733
	1000	856,697	4,942,000	17/100	93.50	0.714
	10,000	856,697	49,420,000	2/100	98.30	0.577
2005–2010	1	857,661	3978	21,560/100	99.50	0.000
	50	857,661	198,900	431/100	87.60	0.580
	100	857,661	397,880	216/100	85.60	0.671
	200	857,661	795,600	108/100	86.70	0.734
	300	857,661	1,193,400	72/100	87.10	0.730
	500	857,661	1,989,000	43/100	89.60	0.741
	1000	857,661	3,978,000	22/100	93.20	0.741
	10,000	857,661	39,780,000	2/100	98.70	0.612

Table 3.4 Logistic regression coefficient in different period

	2000–2005	2005–2010
Variable	B (regression coefficient)	B (regression coefficient)
f_ctr_bj	.911	−2.188
f_ctr_tjsjz	−1.989	−2.903
f_ctr_other	−.328	2.016
f_ctr_cty	1.435	1.021
f_rail	2.940	2.276
f_r_hw	3.451	4.051
f_r_nat	1.520	1.524
f_r_pro	3.981	4.025
Constrain	−1.197	−.813
Constant	−6.953	−7.485

3.5 Model Application

3.5.1 *BTH2020: The Urban Expansion Study for the Year 2020*

The latest round of master plan is at the end of 2020 in BTH. Simulating the urban layout in 2020 based on the present situation and comparing the simulation result with the actual plan result can help judge whether the master plan can be

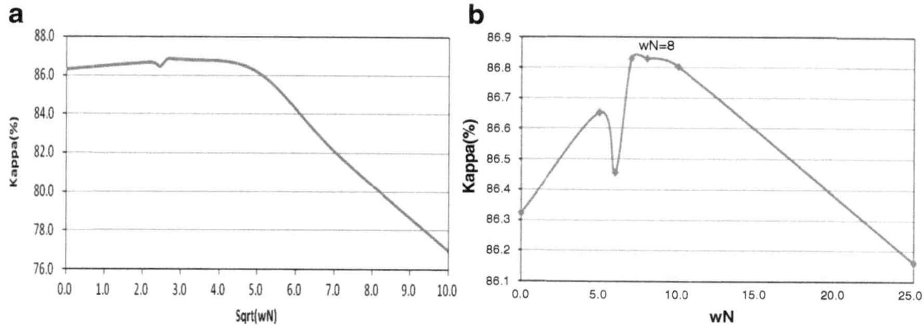

Fig. 3.7 The process curve of MonoLoop

accomplished. Suppose the total scale of urban construction land at the end of 2020 can achieve the planned scale, and the speed of urban growth from 2010 to 2020 is the same as the speed from 2005 to 2010 (namely, all weight coefficient of spatial variables, including the neighbor variable, remain unchanged as a baseline scenario). By simulating the urban layout by BUDEM-BTH model in 2020, and comparing the simulated result with plan data, we can make a comparison and evaluation for current policy and master plan. Due to the limitation of data source, we only simulate BTH metropolitan circle (refer to Beijing, Tianjin and several cities in Hebei province, such as Shijiazhuang, Baoding, Qinhuangdao, Langfang, Cangzhou, Chengde, Zhangjiakou, Tangshan eight cities, excluding Handan, Xingtai, Hengshui) at this stage.

The simulated urban form for BTH in 2020 is shown in Fig. 3.8a, and the compared result with the plan is shown in Fig. 3.8b. The kappa value is 0.54. The calculation result and contrast show there is a large difference between the simulation results and the master plan. If the trend of 2005–2010 continues, BTH metropolitan circle will not meet the planned urban form.

In order to achieve the planned urban spatial layout, the coefficient of the model needs to be adjusted. The dependent variable of logistic regression can be acquired using the subtracted planned urban construction land in 2020 by the urban construction land in 2010. Because the independent variables keep unchanged, the weight coefficient of independent variables can be acquired through regression analysis again. With regression results shown in Table 3.5. The correlation coefficients of effective variables are close to zero and the significant level of all variables is at the 0.000 level, whereby the regression results are acceptable. Comparing the regression coefficient of 2010–2020 with that of 2005–2010 horizontally, it can be seen that the driving force of affecting urban growth has significant differences: firstly, the appeal of Beijing increases apparently; secondly, Tianjin, Shijiazhuang' appeal changes from the original negative correlation to positive correlation, which is opposite to towns' appeal; finally, the rail's appeal changes inconspicuously. The other variables' influence is reduced.

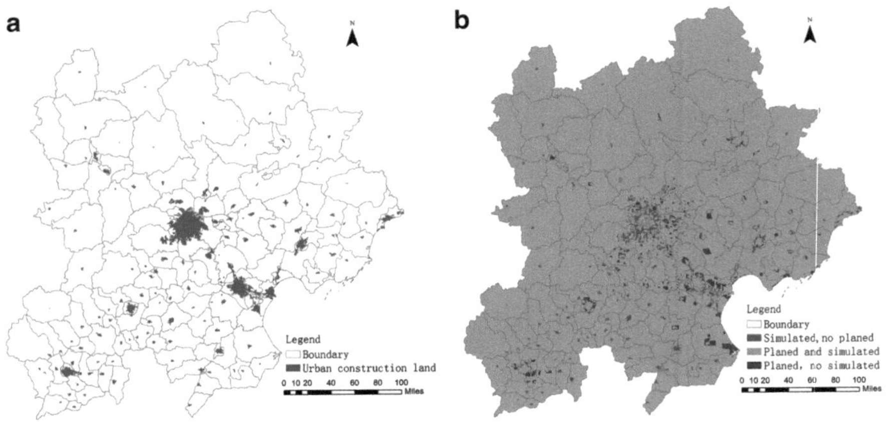

Fig. 3.8 Simulated urban form of 2020 (**a**) and its comparative image with planned form (**b**)

Table 3.5 Logistic regression coefficient from 2010 to 2020 and comparison with the coefficient from 2005 to 2010

	2005–2010	2010–2020			
Variable	B (regression coefficient)	B (regression coefficient)	S.E.	Wald	Df
f_ctr_bj	−2.188	−13.311	.315	1787.257	1
f_ctr_tjsjz	−2.903	2.407	.045	2858.183	1
f_ctr_other	2.016	1.699	.030	3291.351	1
f_ctr_cty	1.021	−4.079	.466	76.636	1
f_rail	2.276	2.449	.009	67087.738	1
f_r_hw	4.051	3.195	.008	157835.516	1
f_r_nat	1.524	.849	.009	9054.795	1
f_r_pro	4.025	1.436	.009	25879.339	1
Constrain	−.813	−.593	.010	3833.048	1
Constant	−7.485	−4.301	.009	214297.382	1

From the regression result, we find the matching rate between simulation result and urban form at the end of 2020 has improved, but there is still large differences (Kappa value was 0.55) as shown in Fig. 3.9. The result shows that the plan cannot be finished under the current policy.

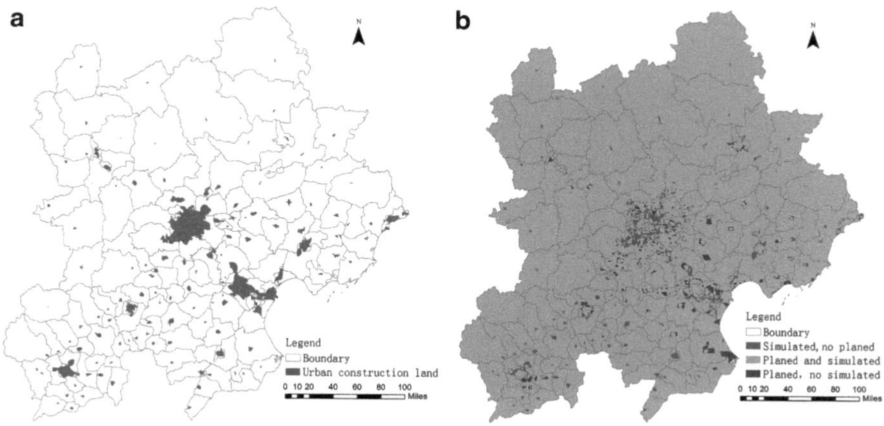

Fig. 3.9 Optimized simulated urban form of 2020 (**a**) and its comparative image with planned form (**b**)

3.5.2 BTH2049: The Scenario Analysis for 2049

3.5.2.1 Scenario Analysis

In order to prepare for the next round of master plan, it is necessary to simulate the urban spatial form from 2020 to 2049.

Simulating BTH urban spatial form in the future, the total amount of urban construction land should be predicted firstly. In the first place, the total urban population in BTH need to be extracted by city statistical yearbook from 1995 to 2008, then the total urban population in 1949 can be calculated by the population forecast measurement formula $P = -228812 + 116y$. The formula is acquired by regression analysis method. According to the regression result, R2 is equal to 0.929 and Sig is 0.000, which demonstrates the formula is available. According to the formula and per capita urban land-use area, the urban construction land area for 2049 can be calculated.

This study sets up seven kinds of scenarios to analyze the urban layout in 2049. Each scenario is based on the urban construction land layout in 2010 (including all variable coefficient, per capita urban land area). By changing the enforcement strength of various policies, different spatial development scenarios will be formed.

1. The trend scenario (namely, the stable development scenario). This scenario will happen if continue the development trend from 2005 to 2010. Under this scenario, there will be not too much of development policy, and the supply of land as well as economy and population are all stable. In this situation, the urban development area in BTH can reach 16,423 km^2 (65,692 cells) in 2049. The simulation result is shown in Fig. 3.10.
2. High-speed growth scenario. This scenario reflects that the rapid economic development leads to a large demand for urban construction land. Under this scenario, the urban development area in BTH can reach 20,000 km^2 (80,000 cells) in 2049. The simulation result is shown in Fig. 3.11.

Fig. 3.10 The trend scenario

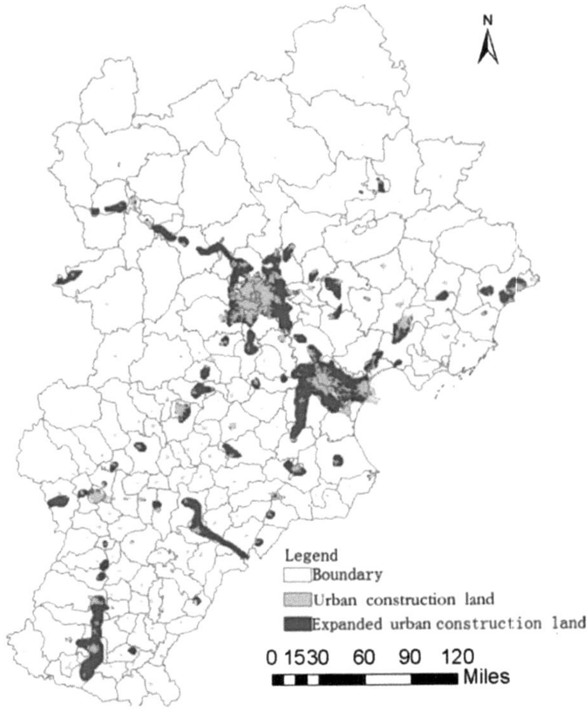

3. Low-speed growth scenario. Under this scenario, the economic development leads to the large increase of urban construction land, even beyond the resources and environment carrying capacity, so the supply of land will be decreased necessarily. In this situation, the urban development area in BTH will reach 12,500 km^2 (50,000 cells) in 2049. The simulation result is shown in Fig. 3.12.

Changing different variable coefficient of the trend scenario, more scenarios can be obtained, for example:

4. Highway finger growth scenario. This scenario will improve the development along the highway. The simulation result is shown in Fig. 3.13.
5. Town promoting growth scenario. When metropolis develops to a certain extent, it will be restrained by resources and environment, and then the regional industrial structure will be adjusted to promote the development of small towns around. The simulation result is shown in Fig. 3.14.
6. Developing forbidden area growth scenario. Due to the improvement of human productivity and the capability to convert nature, the ability to develop the forbidden area is strengthened. The simulation is shown in Fig. 3.15;
7. Traffic leading growth scenario. The scenario will enhance the development along the railway, the highway, the national road, and the provincial road. The simulation result is shown in Fig. 3.16.

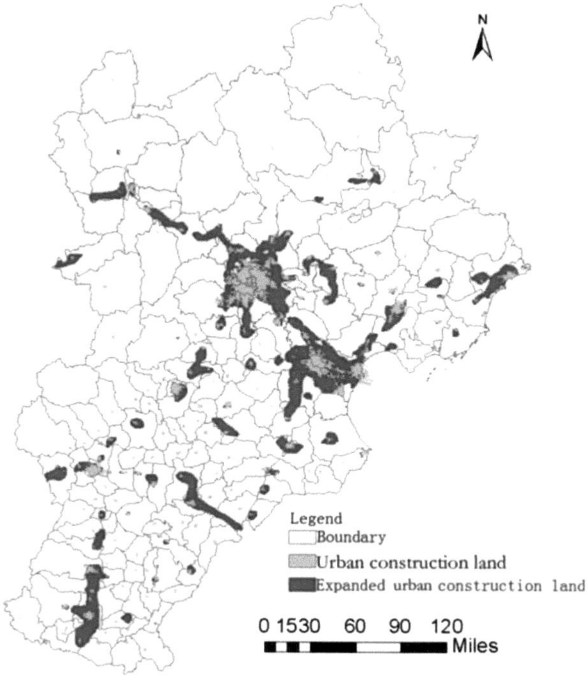

Fig. 3.11 High-speed growth scenario

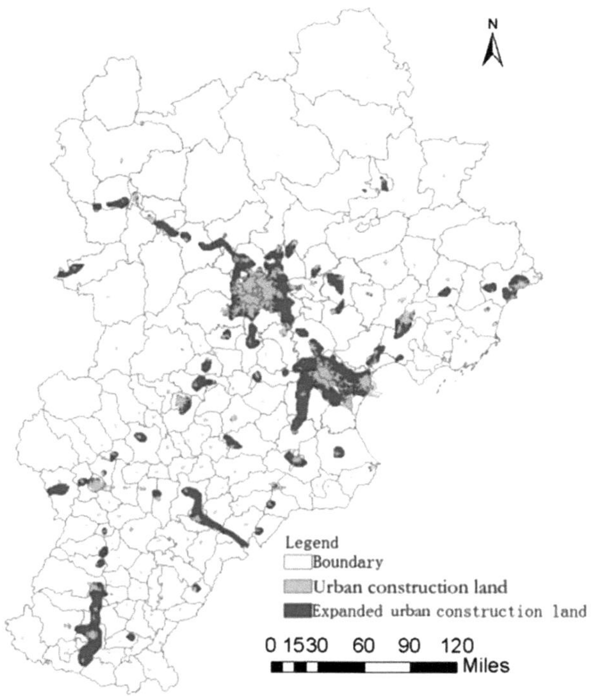

Fig. 3.12 Low-speed growth scenario

Fig. 3.13 Highway finger growth scenario

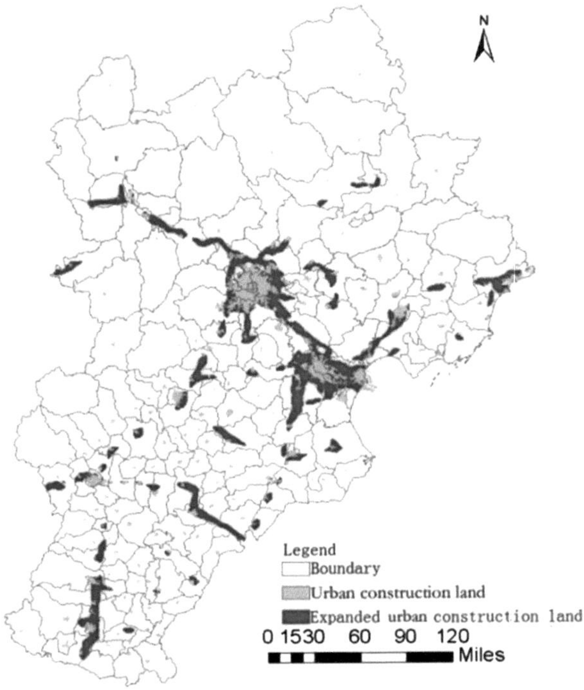

Fig. 3.14 Town promoting growth scenario

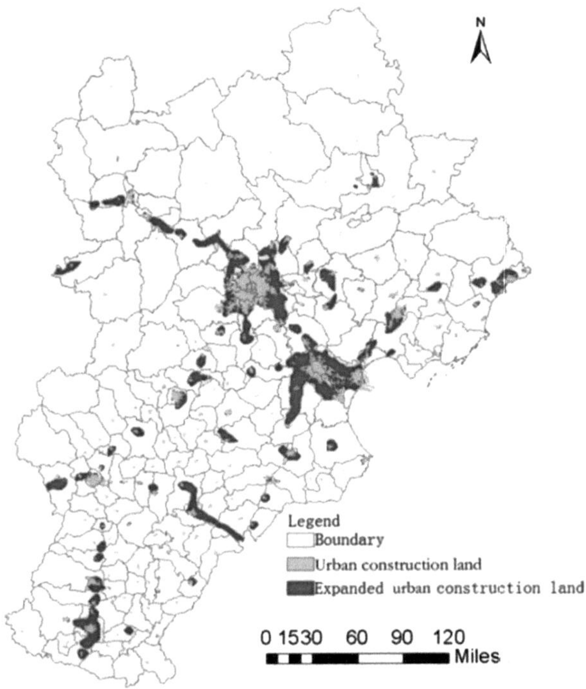

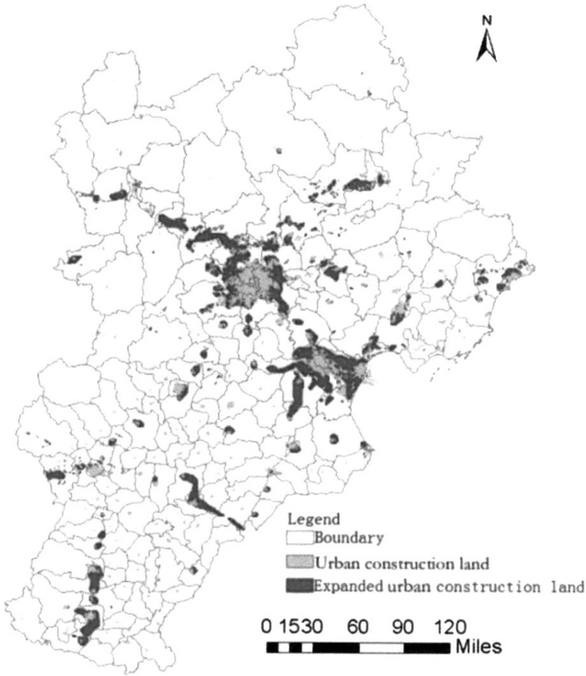

Fig. 3.15 Developing forbidden area growth scenario

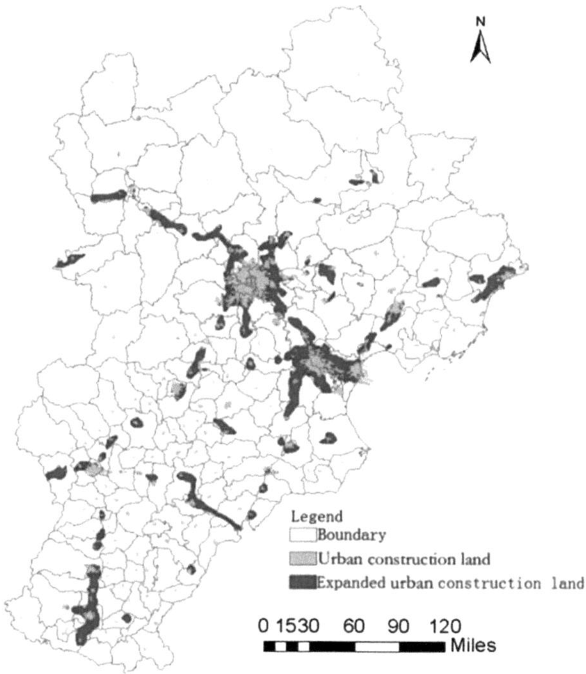

Fig. 3.16 Traffic leading growth scenario

The conversion coefficient of each scenario is as follows (Table 3.6).

3.5.2.2 Scenarios Comparison

In order to compare all the scenarios for 2049, the research selects the following two ways to analyze:

1. The cellular number and proportion of urban construction land for each municipal administrative area in BTH: It is used to describe spatial distribution of urban construction land for each scenario, as shown in Table 3.7 and Fig. 3.17.
   From Table 3.7 and Fig. 3.17, it can be seen that the quality and proportion of urban construction land in different district and different scenario are significantly different. When the total urban construction land increases, the land at each municipal administrative area will increase. The growth rate of urban construction land is directly proportional to the existing urban area. Highway finger growth scenario has a more obvious promoting effect to those area along the highway. Traffic leading growth scenario has a more obvious promoting effect to the area whose transportation is more developed. Developing forbidden area growth scenario has a more obvious promoting effect to the area in the northwest. Town promoting growth scenario has a more obvious promoting effect to the area containing more towns.
2. Firstly, from the perspective of preventing the spatial risk, the seven kinds of simulated scenarios are superimposed respectively with current situation layers: farmland, grassland, forest land, water area and unused land, then calculate the potential spatial risk during the process of urban growth by the spatial analysis function of GIS tools. Secondly, from the perspective of spatial form analysis, BTH urban spatial aggregation forms are analyzed by Patch number, mean area of patch (m^2), Aggregation index, Splitting index, Contagion index. Those indexes are chosen by landscape analysis method. The comparison of urban expansion is shown as Table 3.8.

   It can be seen from the above comparison in Table 3.8 as follows:

High-speed growth scenario: In this scenario, the occupation for farmland is significantly more than the other scenarios. The average patch area and Splitting index are the largest. The Contagion index is the lowest. It shows that High-speed growth scenario will occupy farmland too much, which is unfavorable for the protection of farmland. The excessive dispersive urban construction land goes against the gathered economic theory, which is to the disadvantage of the sustainable development of the society.

Low-speed growth scenario: The occupation for non-construction land is significantly less than the other scenarios. Aggregation index and Contagion index are the highest in all scenarios. Therefore low-speed growth scenario is beneficial to the sustainable development of the society.

Table 3.6 Form scenario analysis results

Type	Name	The trend scenario	High-speed growth scenario	Low-speed growth scenario	Highway finger growth scenario	Town promoting growth scenario	Developing forbidden area growth scenario	Traffic leading growth scenario
Macro constraint	*stepnum*	65,692	80,000	50,000	65,692	65,692	65,692	65,692
LOCATION constraint	*F_ctr_bj*	−2.188	−2.188	−2.188	−2.188	−2.188	−2.188	−2.188
	f_ctr_tjsjz	−2.903	−2.903	−2.903	−2.903	−2.903	−2.903	−2.903
	f_ctr_other	2.016	2.016	2.016	2.016	2.016	2.016	2.016
	f_ctr_cty	1.021	1.021	1.021	1.021	5.021	1.021	1.021
	f_rail	2.276	2.276	2.276	2.276	2.276	2.276	2.276
	f_r_hw	4.051	4.051	4.051	8.051	4.051	4.051	6.051
	f_r_nat	1.524	1.524	1.524	1.524	1.524	1.524	3.524
	f_r_pro	4.025	4.025	4.025	4.025	4.025	4.025	6.025
INSTITUTIONAL constraint	*constrain*	−0.813	−.813	−.813	−.813	−.813	4.813	−.813
NEIGHBOR constraint	*neighbor*	8.000	8.000	8.000	8.000	8.000	8.000	8.000
CONSTANT	*constant*	−7.485	−7.485	−7.485	−7.485	−7.485	−7.485	−7.485

Table 3.7 Form scenarios spatial structure verification

Cities	The trend scenario		High-speed growth scenario		Low-speed growth scenario		Highway finger growth scenario		Town promoting growth scenario		Developing forbidden area growth scenario		Traffic leading growth scenario	
	Cell	%	Cell	%	Cell	%	Cell	Cell	Cell	%	Cell	%	Cell	%
Beijing	16,501	25.2	18,810	23.7	13,781	27.7	15,996	24.0	16,481	25.1	18,513	28.2	15,754	24.3
Tianjin	15,329	23.4	17,877	22.5	12,029	24.2	15,541	23.3	15,656	23.8	15,035	22.9	14,461	22.3
Shijiazhuang	3366	5.2	4157	5.2	2701	5.4	3505	5.3	3516	5.4	3415	5.2	3948	6.1
Handan	4504	6.9	5324	6.7	3052	6.1	3684	5.5	4089	6.2	3211	4.9	4303	6.6
Xingtai	2994	4.6	3632	4.6	2501	5.0	3013	4.5	2713	4.1	2600	4.0	2847	4.4
Hengshui	3580	5.5	4123	5.2	2771	5.6	3218	4.8	3424	5.2	2921	4.4	3012	4.7
Baoding	3689	5.6	4528	5.7	2784	5.6	3990	6.0	3720	5.7	3023	4.6	3717	5.7
Langfang	2210	3.4	2861	3.6	1451	2.9	2058	3.1	2260	3.4	2144	3.3	2218	3.4
Qinhuangdao	2205	3.4	2861	3.6	1451	2.9	2655	4.0	2331	3.6	2239	3.4	2218	3.4
Cangzhou	3499	5.4	4820	6.1	2379	4.8	4051	6.1	3558	5.4	2619	4.0	3791	5.9
Tangshan	3086	4.7	4033	5.1	2382	4.8	3569	5.4	3235	4.9	2949	4.5	3202	4.9
Zhangjiakou	3809	5.8	5126	6.5	2086	4.2	4313	6.5	4027	6.1	4485	6.8	4443	6.9
Chengde	615	0.9	1299	1.6	432	0.9	1002	1.5	659	1.0	2598	4.0	919	1.4

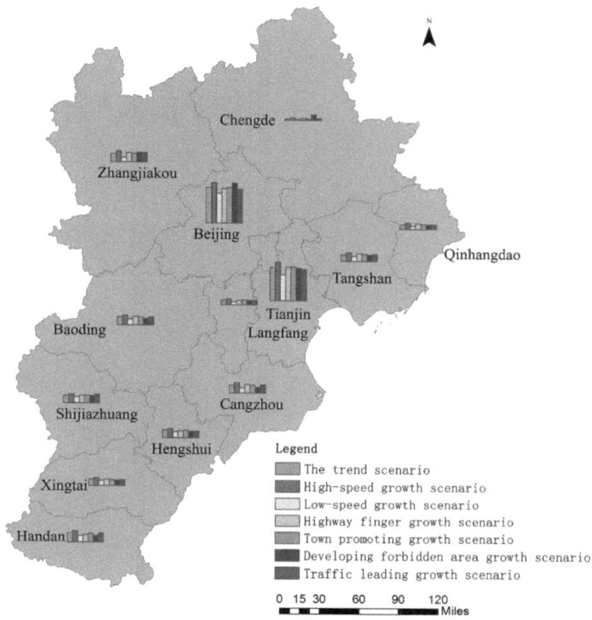

Fig. 3.17 Urban form scenarios in BTH Area

Developing forbidden area growth scenario: The occupation for forest land and water area is more than other scenarios, and the occupation for farmland is less than other scenarios. Patch number is significantly more than other scenarios. Although developing forbidden area growth scenario is beneficial to the protection of farmland, it occupies more of other agricultural land and water area. From the perspective of environmental diversity, it is to the disadvantage of the sustainable development of the society.

In the trend scenarios, town promoting growth scenario, highway finger growth scenario, traffic leading growth scenario occupy more farmland, forest land and grassland than the trend scenario. Town promoting growth scenario and highway finger growth scenario are lower than the trend scenario in average patch area, aggregation index and Contagion index. Traffic leading growth scenario have the least number of patches, the largest average patch area and most least Contagion index. All these results verify respectively the influence of different scenarios to urban layout.

3.6 Conclusion and Discussion

This research establishes BUDEM-BTH model by adjusting BUDEM model from the parameter identification, model structure adjustment and scenario design, etc. BUDEM-BTH model can be used to support the development of BTH regional planning. The results are as followings:

Table 3.8 The comparison of urban expansion

Spatial form	The trend scenario	High-speed growth scenario	Low-speed growth scenario	Highway finger growth scenario	Town promoting growth scenario	Developing forbidden area growth scenario	Traffic leading growth scenario
Farmland occupation (km^2)	7229.54	9733.45	4654.94	7515.81	7251.04	5593.21	7327.31
Forest land occupation (km^2)	530.64	770.09	313.77	530.63	543.06	1796.59	563.18
grassland occupation (km^2)	182.99	290.72	81.84	221.73	192.12	435.64	221.66
water area occupation (km^2)	690.28	841.00	72.88	680.40	696.70	1017.73	643.07
unused land occupation (km^2)	27.02	28.32	17.49	16.62	26.82	25.05	24.03
Patch number	237.00	217.00	255.00	255.00	238.00	836.00	234.00
mean area of patch/m^2	90,890.19	99,267.17	84,474.41	84,474.41	90,508.30	25,766.72	92,055.45
Aggregation index	99.05	99.02	99.08	98.91	99.04	98.42	98.98
Splitting index	1.17	1.22	1.18	1.18	1.17	1.17	1.17
Contagion index	77.59	74.52	81.11	76.90	77.47	75.75	77.30

1. BUDEM-BTH model, based on the constraint CA, integrates spatial constraints, institutional constraints, neighborhood constraints in the transition rules. It has more advantages than pure CA in simulating urban expansion change. BUDEM-BTH model integrates logistic regression and MonoLoop to identify the transition rule to realize the desired urban form, which is more scientific and authentic than manual assignment.
2. The previous researches mainly focus on regional scale or urban scale. But this research considers a regional area as simulation scale. BUDEM-BTH model, based on constraint CA, is suitable for simulating the urban spatial expansion in BTH. According to the simulation results of the year of 2005–2005, the Kappa coefficient suggests a good simulation effect, and the GOF of simulation results and the actual situation is very high.
3. As for the simulation inaccurate problem resulted from imbalance land use data in the same area, this research puts forward a solution to balance the data by increasing the weight of small sample data, and obtains a good simulation effect.
4. From the simulated urban layout at the end of 2020, we can find if the development trend goes on as the year from 2005 to 2010, the expected urban layout in 2020 is unable to achieve. The research provides a reference basis for adjusting the current urban development.
5. According to the actual situation of BTH and the requirements for our research, seven kinds of scenarios are set up to predict urban spatial layout in 2049. It provides a reference and scientific basis for the next round of urban master plan.

Urban system is a complex system, and urban expansion may be influenced by various complex factors. In this study, some of the relevant factors haven't been taken into account. For example, in the institutional constraints aspect, the basic protection farmland constraints isn't considered. In the spatial constraints aspect, this study is confined to Euclidean distance without considering the surface resistance, rivers and ports attraction, and also doesn't simulate respectively the different region according to their own special cases. These questions will be paid attention to in the follow-up study. If the model is used for decision-making support, it is necessary to further improve the constraints condition and constraints variables, and verify in multiple regions so as to obtain a better simulation effect.

References

Amorosos, S., & Patty, Y. N. (1972). Decision procedures for surjectivity and injectivity of parallel maps for tessellation structures. *Journal of Computer and System Sciences, 5*(6), 448–464.

Batty, M., & Xie, Y. (1994). From cells to cities. *Environment and Planning B, 21*, 531–548.

Batty, M., & Xie, Y. (1997). Possible urban automata. *Environment and Planning B, 24*, 175–192.

Butler, R. W. H. (1982). A structural analysis of the Moine Thrust zone between loch Eriboll and Foinaven, NW Scotland. *Journal of Structural Geology, 4*(1), 19–29.

Chen, S. P. (1999). *Urbanization and urban geographic information system*. Beijing: Science Press.

Fan, J. (2008). *Beijing-Tianjin-Hebei metropolitan district comprehensive planning study*. Beijing: Science Press.

He, C. Y., Chen, J., Shi, P. J., & Yu, Z. T. (2002). Study on the spatial dynamic city model based on CA (cellular automata) model. *Advance in Earth Sciences, 17*(2), 188–194.

He, C. Y., Shi, P. J., Chen, J., Pan, Y. Z., Li, X. B., Li, J., Li, Y. C., & Li, J. G. (2005). Developing land use scenario dynamics model by the integration of system dynamics model and cellular automata model. *Science in China (D), 35*(5), 464–473.

Kahn, M. (2000). The environmental impact of suburbanization. *Journal of Policy Analysis and Management, 19*, 569–586.

Kuang, W. H., Liu, J. Y., Shao, Q. Q., He, J. F., Sun, C. Y., Tian, H. Q., & Ban, Y. F. (2011). Dynamic urban growth model at regional scale and its application. *Acta Geographica Sinica, 66*(2), 178–187.

Lancaster. (1966). A new approach to consumer theory. *Journal of Political Economy, 74*(2), 132–157.

Liu, J. Y., & Deng, X. Z. (2009). Progress of the research methodologies on the temporal and spatial process of LUCC. *Chinese Science Bulletin, 54*(27), 3251–3258.

Long, Y., Mao, Q. Z., Shen, Z. J., Du, L. Q., & Gao, Z. P. (2008a). Comprehensive constrained CA urban model: Institutional constrains and urban growth simulation. *Urban Planning Forum, 6*, 83–91.

Long, Y, Shen, Z. J., Du, L. Q., Mao, Q. Z., Gao, Z. P. (2008b) BUDEM: an urban growth simulation model using CA for Beijing metropolitan area. In: Proceedings of the SPIE-geoinformatics. 71431D-1-15.

Long, Y., Han, H. Y., & Mao, Q. Z. (2009a). Establishing urban growth boundaries using constrained CA. *Acta Geographica Sinica, 64*(8), 999–1008.

Long, Y., Mao, Q. Z., & Dang, A. R. (2009b). Beijing urban development model: Urban growth analysis and simulation. *Tsinghua Science and Technology, 14*(6), 782–794.

Long, Y., Shen, Z. J., Mao, Q. Z., & Dang, A. R. (2010). Form scenario analysis using constrained cellular automata. *Acta Geographica Sinica, 65*(6), 643–655.

Lu, D. D. (2007). Urbanization process and spatial sprawl in China. *Urban Planning Forum, 3*, 16–22.

Urban Group Planning for Beijing-Tianjin-Hebei district (2007–2020) (2007) Beijing: China Academy of Urban Planning & Design.

Ward, D. P., Murray, A. T., & Phinn, S. R. (2000). A stochastically constrained cellular model of urban growth. *Computers, Environment and Urban Systems, 24*, 539–558.

White, R., & Engelen, G. (1993). Cellular automata and fractal urban form: A cellular modeling approach to the evolution of urban land –use patterns. *Environment and Planning A, 25*, 1175–1199.

White, R., Engelen, G., & Uijee, I. (1997). The use of constrained cellular automata for high-resolution modelling of urban land-use dynamics. *Environment and Planning B: Planning and Design, 24*, 323–343.

Wu, F. (1998). Simland: A prototype to simulate land conversion through the integrated GIS and CA with AHP-derived transition rules. *International Journal of Geographical Information Science, 12*, 63–82.

Wu, L. Y. (2002a). *The second period report of urban and rural spatial development planning study for Beijing-Tianjin-Hebei*. Beijing: Tsinghua University press.

Wu, L. Y. (2002b). *Urban and rural spatial development planning study for Beijing-Tianjin-Hebei*. Beijing: Tsinghua University press.

Xie, F. R., Wang, Z. M., & Liu, J. W. (2011). Seismic hazard and risk assessments for Beijing-Tianjin-Tangshan, China area. *Pure and Applied Geophysics, 168*, 731–738.

Yang, M., Yin, J. M., & Ji, G. L. (2008). Classification methods on imbalanced data: A survey. *Journal of Nanjing Normal University (Engineering and Technology Edition), 8*(4), 7–12.

Yeh, A. G. O., & Li, X. (2001). A constrained CA model for the simulation and planning of sustainable urban forms by using GIS. *Environment and Planning B: Planning and Design, 28*, 733–753.

Zhou, C. H., Shun, Z. L., & Xie, Y. C. (1999). *Geographical cellular automata research*. Beijing: Science Press.

Chapter 4
Parcel Direction: A New Indicator for Spatiotemporally Measuring Urban Form

4.1 Introduction

The urban form is the integration of geometric elements in the geographical space, including the built-up space, road network, and landscape. The parcel, as the basic component of the urban form, is the container of buildings, green lands, municipal infrastructures, and public services facilities. There is traffic connection to the road network from each parcel in a block, which generally has the same boundary with the road network. The directions of a block can be recognized as those of the roads around itself. Because the parcel has close relationship with roads within the block (see Fig. 4.1), this chapter will propose the parcel direction (PD) as a new spatial metrics to evaluate the urban form at the parcel scale. PD is the geometric analysis for geographical space from the geography perspective. Via comparing the direction variation of the present parcels and planned parcels, the influence of urban planning on the urban form can be identified quantitatively.

This chapter attempts to analyze the urban form through PD resulted from various types of blocks based on road networks. For instance, during the new town development, or downtown redevelopment, the building orientation and its relationship with the road network are accordingly changed due to the re-plan of roads' direction and width, together with the parcels' layout. In the common sense, the road construction project has its benefit in improving the city's total operating efficiency. If the direction of the road network is altered (see Fig. 4.2) instead of being followed as previous road network, a series of undesired issues would occur, such as increased difficulty to protect ecological network for small animals, increased demolition and construction costs, hard to retain well-constructed existing roads, and difficult to keep existing lifelines system. Furthermore, the existing real estate property may be divided into several parts, which increases the difficulty of the urban planning implementation and the cooperation between stakeholders and current property owners. The direction transition of urban road networks results in almost a complete re-placement of the municipal infrastructure, such as the green

Y. Long, Z. Shen, *Geospatial Analysis to Support Urban Planning in Beijing*,
GeoJournal Library 116, DOI 10.1007/978-3-319-19342-7_4

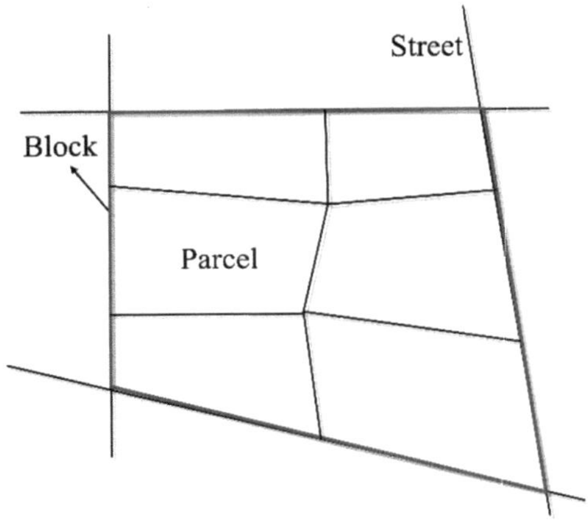

Fig. 4.1 The schematic diagram of the spatial relationship between the road network and parcels

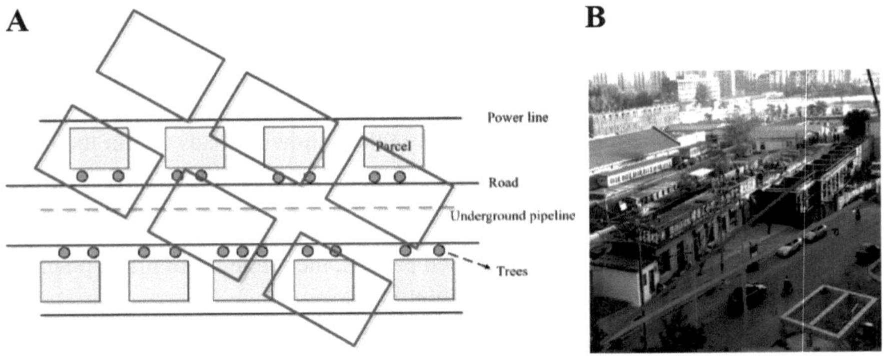

Fig. 4.2 The schematic diagram (**a**) and illustrated photo (**b**) of the changed parcel direction. Note: The oblique parcel in *red* is the planned layout in the left diagram. The *upper line* in the right photo is the original parcel direction, and the *bottom line* is the already implemented parcel direction determined by the urban plan (The photo was taken in Haidian District, Beijing, China)

system, power system, water supply and drainage system. Therefore, evaluating the urban form variation using PD has profound significances in planning practices, especially in rapid-developing countries.

Yang et al. (2010) included the parcel direction as one of indicators to identify the spatial patterns in road networks, which is the only research report regarding the PD. However, they regarded the longest axis's direction of the minimum rectangle enveloping a parcel as the parcel's PD indicator. On one hand, Yang's PD does not equal the direction of any edge for irregular parcels that cannot be directly identified by residents. On the other hand, Yang's PD is only one of indicators for measuring urban form, and is not emphasized in a spatiotemporal manner. In this work, we will attempt to define the PD in another way, and use it to measure urban form in both spatial and temporal dimensions.

Most of urban form related researches focused on the spatial structure analysis without taking PD into account. For instance, the space syntax, as a novel theory for the urban spatial analysis, is the approach based on the graphic theory (Hillier and Hanson 1984; Hillier 1996; Jiang et al. 1999; Jiang et al. 2002), which investigates the relationship between the urban spatial organization and various aspects of human society through the quantitative description of the human settlement spatial structure. Okabe et al. (2001, 2006) developed the Spatial Analysis of Network (SANET) as a toolkit to analyze the relationship between convenience stores as points and the streets as a network. SANET proposed the K function and Voronoi polygon based for the network, but the parcel in it is abstracted into a point instead of a polygon, not to say the parcel direction is considered. As the developer of ArcGIS, Environmental System Research Institute (ESRI) (2001) published the Land Parcels Data Model, which could assist users managing parcel data as a data model without considering the concept of the parcel direction.

Parcel is the fundamental analysis unit of the urban form, and several indicators have already been proposed to evaluate the urban form. Existing indicators such as the area and perimeter mainly focus on its geometric characteristics. Maniruzzaman et al. (1994) proposed the compactness indictor for the parcel in GIS, and analyzed the distribution as well as the relationship between compactness indictor and land use types. Xie and Ye (2007) proposed the Comparative Tempo-spatial Pattern Analysis (CTSPA) approach to investigate the spatio-temporal dynamics of the urban form. CTSPR analyzes the urban form in terms of the amount, shape and size. The parcel direction, however, is not explored in CTSPR. In planning practices, the parcel as the urban built-up land enclosed by urban street is the basic geo-feature, or planning unit in the regulatory detailed planning. Several indicators such as the planned land use type, the maximum building height, the maximum building density, and the floor area ratio (FAR) are proposed to control the urban form. The parcel direction, as one of significant indicators in the spatio-temporal analysis of the urban form in the parcel scale, few researches regarding it has ever been conducted.

The conception, computation approach and application of PD will be elaborated in the following context. In Sect. 4.2, we will introduce the definition of PD and computation approach using GIS. In Sect. 4.3, an empirical study will be conducted based on the dataset of the Detailed Planning of the Beijing Central Metropolitan Area (BCMA), together with the correlation analysis of PD and other parcel indicators. The spatial variation of the indicator and how to measure urban form in three spatial scales in terms of PD are elaborated in Sect. 4.4. The transition between the historical form and the planned form will be evaluated in Sect. 4.5 for exploring the temporal dynamics of PD. Finally, we will come to conclusions, and propose the preliminary agenda of the parcel direction researches in future.

4.2 Approach

4.2.1 Definition

The geometry of the parcel is not unified, and usually is quadrangle. As a special type of polygon, the direction of the longest edge of the parcel is regarded as the parcel direction D (Fig. 4.3), which ranges from −90° to 90° (including −90° and 90°). The degree represents the deviation from east, that 0° denotes east (E), 90° north (N), and −90° south (S). The detailed calculation method of the parcel direction is as follows:

$$
\begin{aligned}
&P = \{A_1, A_2, \ldots, A_n\} \\
&if\ i \neq n, L_i = \overline{A_i A_{i+1}};\ else\ L_i = \overline{A_n A_0} \\
&L = \{L_1, L_2, \ldots, L_n\} \\
&Len = \{length(L_1), length(L_2), \ldots, length(L_n),\} \\
&length(L_k) = \max(Len) \\
&if\ x(A_{k+1}) - x(A_k) = 0\ D(P) = 90\ or\ -90\ \text{(randomly)} \\
&else\ D(P) = arctg\frac{y(A_{k+1}) - y(A_k)}{x(A_{k+1}) - x(A_k)}
\end{aligned}
, \quad (4.1)
$$

where A is the vertex of the parcel P, n is the total number of the vertexes of P, L is the edge of P, length (L_k) is the length calculation function of line segments, y and x, respectively, are the ordinate and abscissa of vertex, and D(P) is the degree of P's parcel direction.

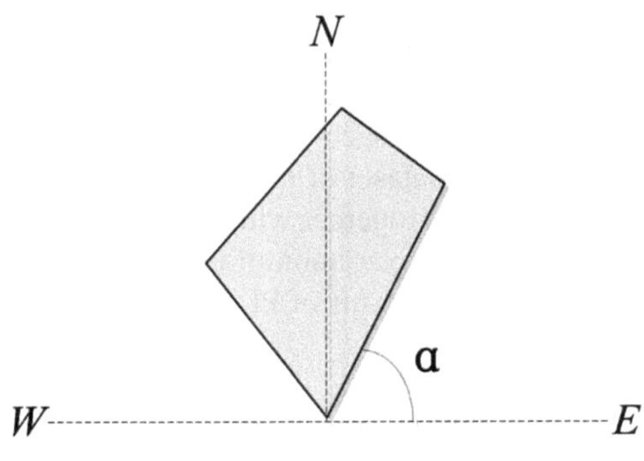

Fig. 4.3 The parcel direction definition diagram

4.2.2 Computational Approach Based on GIS

In order to calculate the parcel direction for large amounts of parcels, we developed a toolkit named PARCTION (PARcel direCTION) using Python to calculate the parcel direction index. PARCTION is based on the GeoProcessing module of ESRI ArcGIS package. The calculation process of PD is as follows:

- (1) Convert the polygon parcel layer "Parcel" with the attribute field "Parcel ID" as the parcel's unique ID into the polyline parcel layer "Parcel Line", divide polylines with their vertexes, and generate the line segments layer "ParcelLineT" (the layer also has the attribute field of ParcelID, indicating which parcel the segment belongs to).
- (2) Analyze "ParcelLineT" based on the field "ParcelID", identify the longest edge of each parcel, mark it, delete all other edges in order to improve the calculation efficiency, and generate the line layer "ParcelLineT2", which only stores the longest edge of each parcel.
- (3) Add X and Y coordination fields respectively for the start point and end point of each edge in "ParcelLineT2", add and calculate the field "D" as the longest edge's direction using Formula 1.
- (4) Add the attribute field "D" to the layer "Parcel" to record the parcel direction, retrieve its value from the field "D" in "ParcelLineT2" matched with the shared field ParcelID.

4.2.3 Measuring Urban Form in Three Spatial Scales

We use the parcel direction to measure the urban form in three spatial scales. **First**, PD can be applied to directly measure the urban form in the parcel scale, and it reflects the orientation of parcels. **Second**, the spatial distribution pattern of PD of parcels within a zone or a parcel's neighborhood can be used to evaluate the urban form in zone/neighborhood scale. Aggregated indicators based on PD will be used for measuring urban form in this scale. **Third**, cluster analysis can be used to evaluate the urban form in the whole region scale based on aggregated indicators in the zone scale. In this scale, various types of urban form can be identified for which the cluster analysis results then can be referred to evaluate the urban form in the whole region scale.

Measuring urban form in the zone scale and regional scale based on classification of parcels and aggregated indicators of parcels respectively. Since measuring urban form using PD in the parcel scale can use the approach detailed in Sect. 4.2.2, we will especially introduce the approach for measuring urban form in the zone scale using aggregated indicators. In the zone scale, we can partition the parcel dataset of the whole study area into multiple zones and calculate aggregated indicators for each zone. The procedure for partitioning the parcel dataset is as follows. First, a polygon dataset is created in terms of the partitioning means, such as dividing the whole study area into several rows and columns, using lattices of the standard map scale, such as 1:25,000, 1:10,000, 1:5000, 1:2,000, and 1:500, or using administrative boundaries as zones. Each polygon in the dataset stands for the spatial extent of a zone (such as the zone 19–20 in Fig. 4.4). Second, the parcels completely within each zone are selected and treated as being contained by the zone, respectively. Thus each zone will contain multi-parcels (the parcels in dark grey within the zone 19–20 in Fig. 4.4). Thus those parcels which intersect with any polygon in the polygon dataset for partitioning do not belong to any zone and are not considered in the zone scale.

Based on the parcels in each zone, aggregated indicators for each zone can be calculated, respectively, using the PD values of all parcels within the zone (termed as PDs), and they can be regarded as the statistic indicator of PDs described as follows:

$$AI_k^m = \underset{P \subset Z_k}{f^m}(D(P)) \tag{4.2}$$

where Z_k are all the parcels within the zone k, f^m is the calculation method for the aggregated indicator m, and AI_k^m is the aggregated indicator m for the zone k. We can investigate how the spatial pattern changing following the aggregated indicators. The neighborhood of one parcel can also be regarded as a zone, and aggregated indicators can also apply for the neighborhood scale urban form evaluation. In this chapter, we select several aggregated for evaluating urban form in the zone scale as

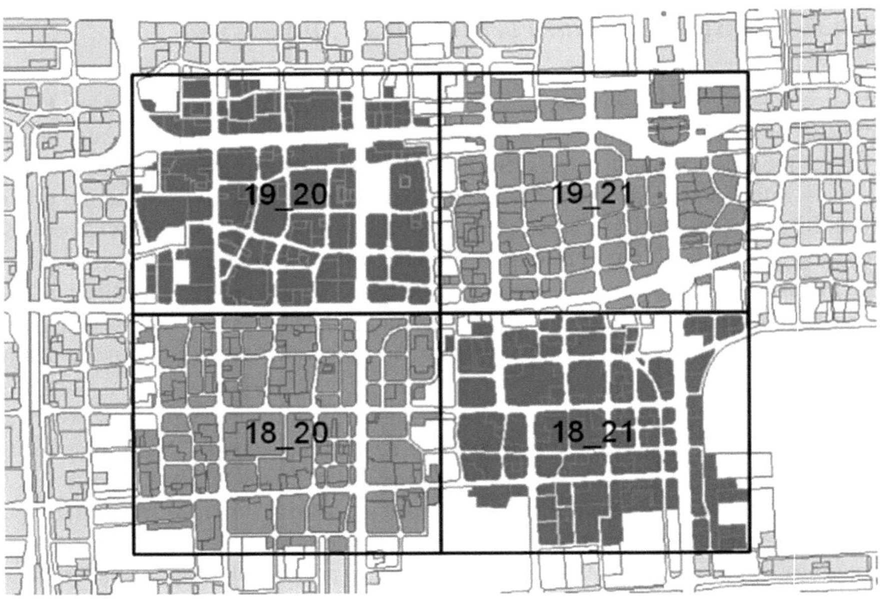

Fig. 4.4 Exemplified zones and their containing parcels

follows (Initially more indicators are considered, and some of them are removed due to the inter-correlation problem.):

1. **AvePD**, the average value of PDs.
2. **StddPD**, the standard deviation of PDs, reflecting the diversity of PDs within the zone.
3. **DetaPD**, equaling the largest PD minuses the least PD in PDs. It reflects the range of PD.
4. **SHDI**, Shannon's diversity index for measuring of relative patch diversity, reflecting the entropy within the zone in terms of PD. The index will equal zero when there is only one patch in the landscape and increases as the number of patch types or proportional distribution of patch types increases.
5. **PROX**, mean proximity index, as the measure of the degree of isolation and fragmentation. It uses the nearest neighbor statistic. This indicator can be achieved by using FRAGSTATS.
6. **ENN**, mean nearest neighbor distance, used to quantify patch isolation. It is defined using simple Euclidean geometry as the shortest straight-line distance between the focal patch and its nearest neighbor of the same class. This indicator can be achieved by using FRAGSTATS.
7. **DIVISION**, landscape division index. This indicator can be achieved by using FRAGSTATS.

4.2.4 Measuring Urban Form in the Temporal Dimension

Since the parcels' spatial distribution often changes with time due to development or re-development, we chose the raster dataset instead of the vector dataset to facilitate the temporal evaluation of the parcel direction. The temporal variation D_{ij}^{dt} of the parcel direction can be calculated by:

$$\begin{aligned} R^s &= \{P_k^s \mid k = 1, M\} \\ R^e &= \{P_k^e \mid k = 1, N\} \\ D_{ij}^{dt} &= \mid D(R_{ij}^e) - D(R_{ij}^s) \mid \\ if\ D_{ij}^{dt} &> 90 \text{ then } D_{ij}^{dt} = 180 - D_{ij}^{dt} \end{aligned} \tag{4.3}$$

where P_k^s is the parcel at the start time point, with the total number of M, forming the corresponding urban form R^s. P_k^e is the parcel at the end time point, with the total number of N, forming the urban form R^e. ij indicates the spatial position. If the variation of the parcel direction is greater than 90°, it can be denoted by its contra-angle. Thus, the final value of D_{ij}^{dt} is between 0° and 90° (including 0° and 90°).

The PD variation across the temporal dimension can be referred to evaluate the urban form. If D_{ij}^{dt} is between 30° and 60°, it can be assumed that the parcel direction is reversed and there is significant urban form variation in terms of the parcel direction. In this condition, the parcels at the start and end time points can be considered inconsistent. The ratio of parcels with reversed PD can also be used as an indicator to evaluate the urban form in the zone scale. In addition, the temporal dynamics of PD can also be evaluated in the same three scales mentioned in Sect. 4.2.3 to measure the urban form. In this condition, the indicator PD will be replaced by D_{ij}^{dt}.

4.3 The Case Study of Beijing

4.3.1 Study Area and Data

Aiming at the Detailed Planning of the BCMA issued in 2006, we computed PD of all the planned parcels based on the previous discussed definition and calculation method. There are totally 24,163 parcels (see Fig. 4.5) within the study area of the BCMA, whose area is 923.2 Km^2; the roads areas between parcels are not included. The statistical description is shown in Table 4.1, in which *compact* is the compactness of the parcel which is calculate using $compact = 4 \times \pi \times \frac{area}{peri^2}$ where *peri* is the perimeter of the parcel, and *area* is the area of the parcel.

4.3.2 Calculation Results

The PD calculation results are shown in Fig. 4.6. The mean value of PD for all the parcels is -11.33°, and the standard deviation is 55.67°. The histogram of the PD calculation results is in the first row of Table4.2. Most of the parcels in the BCMA

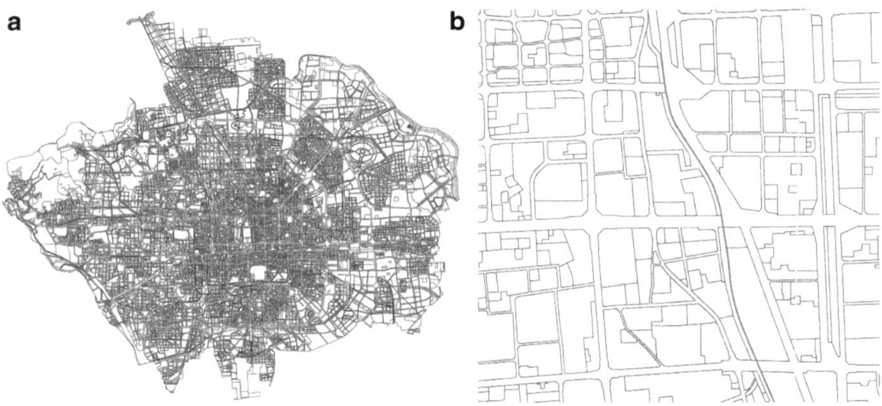

Fig. 4.5 The planned parcels of the BCMA. (**a**) The study area; (**b**) the zoomed-in area of the study area

Table 4.1 Statistical description table of the planned parcels' attributes

	Minimum value	Maximum value	Mean value	Standard deviation
Peri (m)	26	66,534	769.98	1,071
Area (m^2)	10	6,931,627	38,200	142,715
Compact	.000	1.000	.581	.221

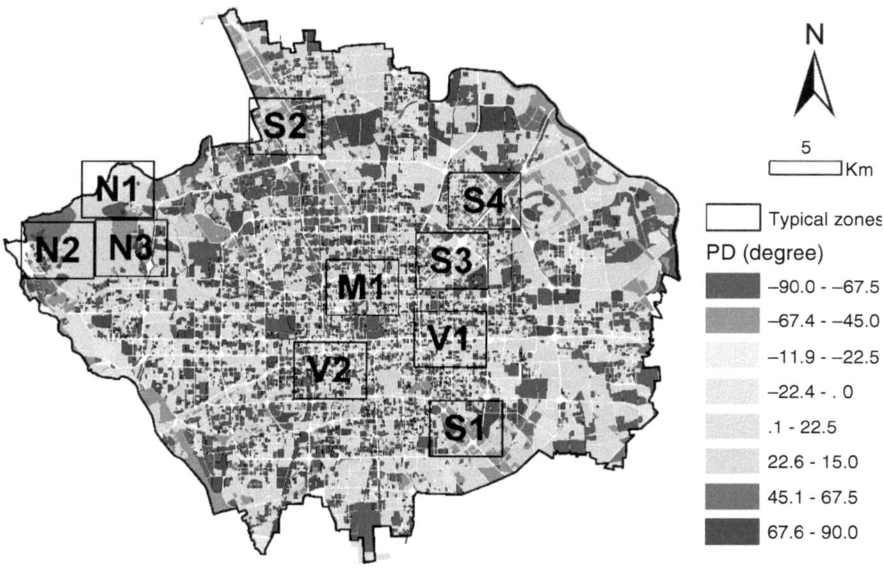

Fig. 4.6 The calculation results of the parcel direction of the planning scheme in the BCMA (the typical zones are used in Sect. 4.4.1)

are with PD value near N–S (90° or −90°) or E–W (0°). The parcels in other directions tend to be in a uniform distribution, and most of them locate in the fringe areas outside the central part of the BCMA.

4.3.3 Correlation Analysis of the Parcel Direction and Other Indicators

We conducted the correlation analysis of four indicators of the planned parcels, including the area, perimeter, compact and PD. The results are listed in Table 4.3. The correlation degree of the parcel direction and other indicators is relatively low. We also found that the PD is not correlated with the land use type of parcels. Besides, we also calculated the correlation efficiency between the PD and the distance to the urban center Tian'anmen Square, and found they were not correlated. Therefore, the parcel direction can be regarded as an independent indicator for the urban form evaluation, thus making it not able to be replaced by other existing indicators.

Table 4.2 Comparisons of urban forms in various typical areas

Type	ID	PD results map	PD histogram
The whole study area			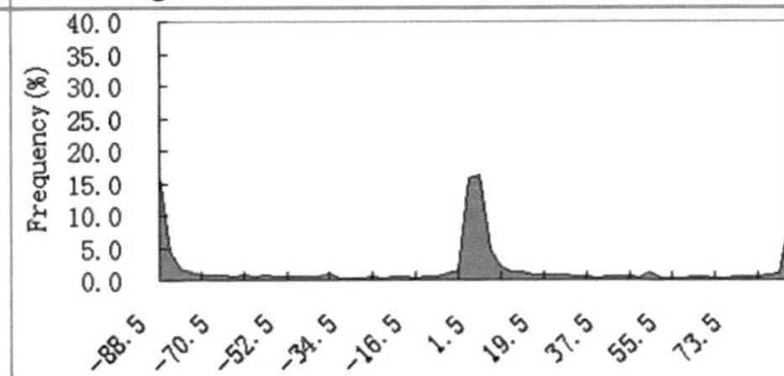
Natural urban form	N1	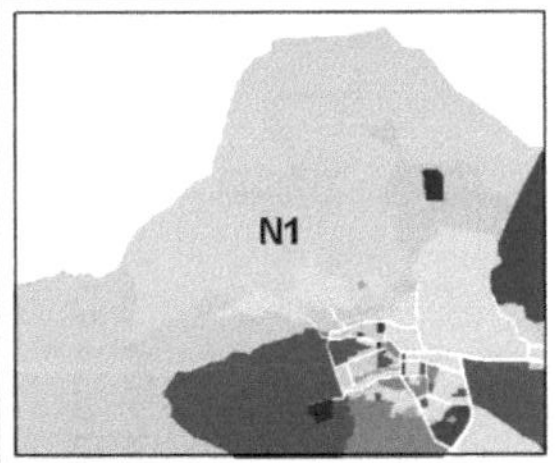	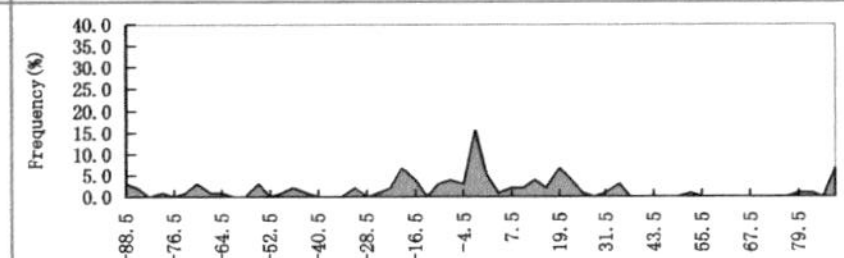
	N2		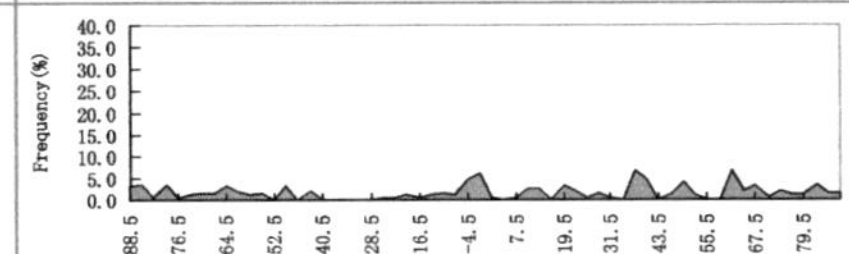

Oblique urban form	S1		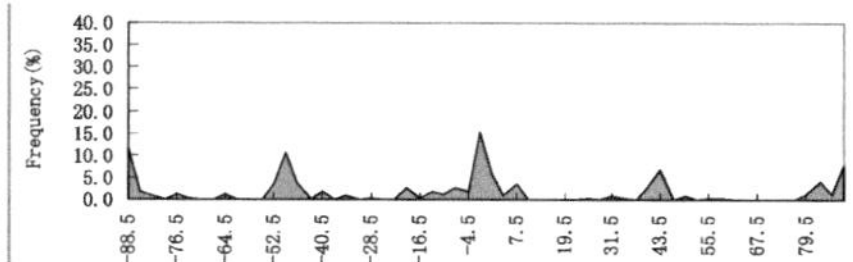
	S2		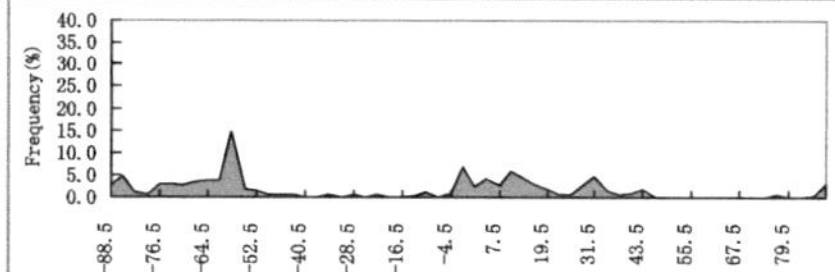
	S3		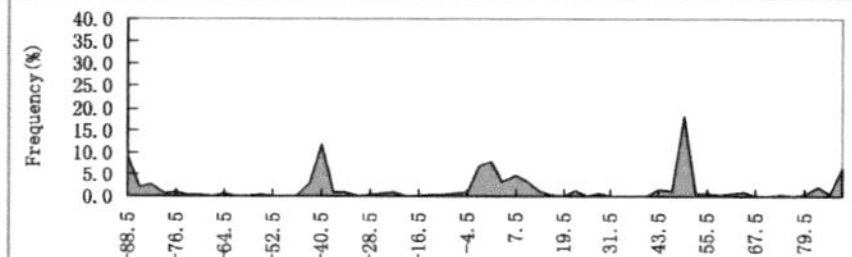
	S4		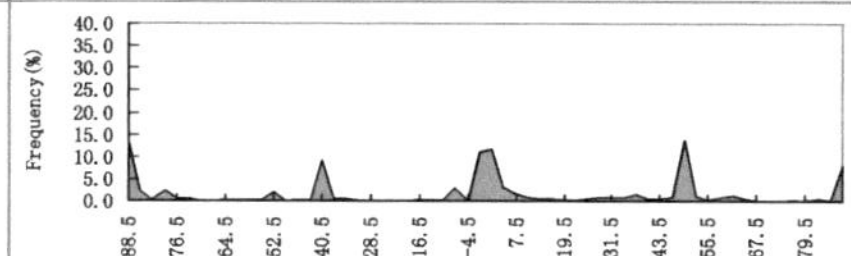

(continued)

Table 4.2 (continued)

Type	ID	PD results map	PD histogram
Regular urban form	V1	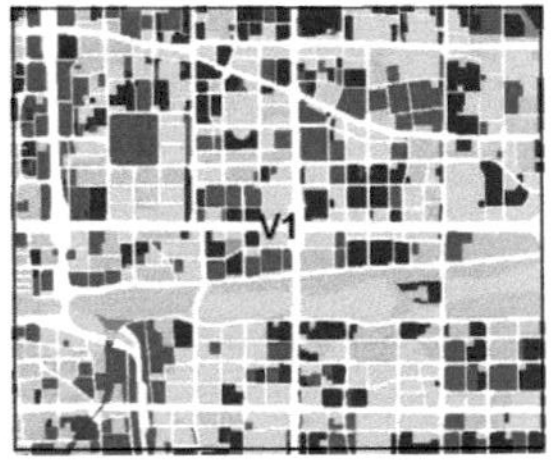	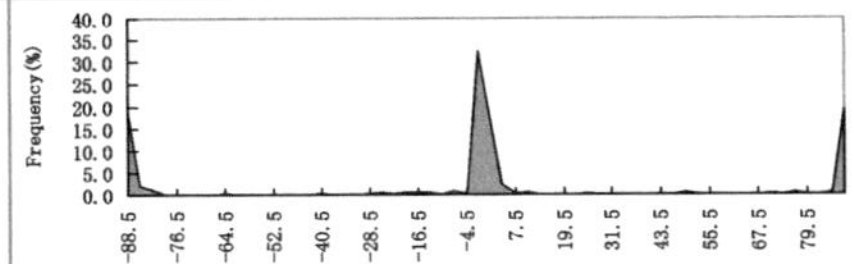
	V2		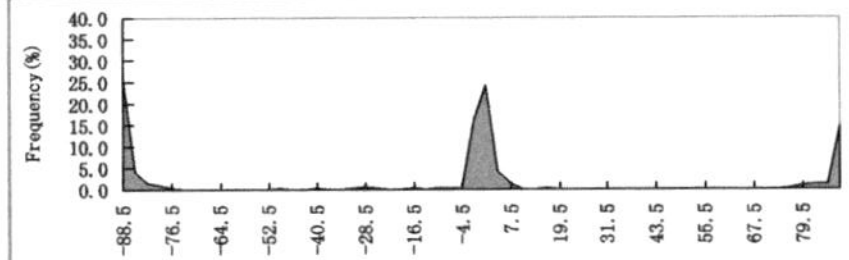
Mixed urban form	M1		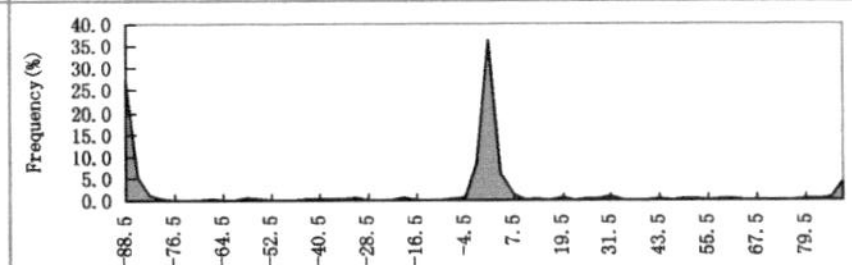

Note: The color of parcels in the maps corresponds to the color classification in Fig. 4.6

Table 4.3 Correlation analysis of the planned parcels' indicators

Name	D	Compact	Length	Area	
D	Pearson correlation	1	−.023**	.000	.004
	Sig. (2-tailed)		.000	.991	.496
Compact	Pearson correlation	−.023**	1	−.309**	−.069**
	Sig. (2-tailed)	.000		.000	.000
Length	Pearson correlation	.000	−.309**	1	.706**
	Sig. (2-tailed)	.991	.000		.000
Area	Pearson correlation	.004	−.069**	.706**	1
	Sig. (2-tailed)	.496	.000	.000	

**Correlation is significant at the 0.01 level (2-tailed)

4.4 Measuring Urban Form in Three Spatial Scales Using the Parcel Direction

4.4.1 The Parcel Scale: Four Types of Urban Forms in Terms of the Parcel Direction

Based on the PD calculation results, there is significantly spatial heterogeneity of PD in the whole study area. According to the morphology of road network as well as the pattern of PD, the whole region can be divided into four patterns of sub-regions, including natural, oblique, regular, and mixed urban forms, in terms of the parcel direction. At least one typical sub-region is selected from each type to analyze the spatial heterogeneity of PD (The sub-region IDs are labeled in Fig. 4.6). Natural urban forms (N1 and N2) mainly locate in the western mountains area. Because the land use types of most parcels within these sub-regions are forest or urban green land, the parcels are relatively irregular in shape with low compact indicator value. Oblique urban forms (S1, S2, S3, and S4) are all cut through by urban expressways, and the direction of the adjacent parcels are significantly correlated to the direction of the expressways, such as the Wangjing area, Fangzhuang area, and Badaling expressway buffer area (the oblique urban form of Qianmen area is formed due to the historical reasons). Such radical urban forms in these sub-regions all result from the contemporary urban planning. The oblique form seldom lies in the old city, but mostly in the new built peripheral areas. Regular urban forms (V1 and V2) generally inherit the original gird texture of Hutong in the old Beijing, mostly with PD in the W–E and N–S directions, and the PD variation among the parcels in the regular urban form is relatively low. Mixed urban form (MI) lie in the regions like Shichahai area and Qianmen Oblique Valley, where the oblique and regular urban forms co-exist. The PD variation among the parcels in the mixed urban form is relatively high, accordingly.

All the analyzed sub-regions with various types of urban forms together with their spatial distribution and the PD histograms are shown in Table 4.2. The PD histograms of each sub-region significantly differ from each other as well as the whole region. Therefore, the parcels in each sub-region can be regarded as the independent sample in terms of the parcel direction. Supposing that there is no significant difference among the four types of independent urban forms (the hypothesis H0), and use Kruskal–Wallis H test to check this hypothesis. The approximate significant level of the hypothesis check is 0.000, which is less than 0.05, and the original hypothesis is then refused. Therefore, there is statistical significance in the PD variation among the four types of urban forms. As a result, the parcel direction can be adopted to evaluate the urban form. Besides, the similarity of the urban forms of sub-regions and the whole region can be evaluated through the comparison between the PD distribution in the sub-region and whole region. The probability density function of PD of parcels in each sub-region can be adopted to evaluate the urban form in terms of the parcel direction.

The spatial autocorrelation is analyzed for the local level to investigate the spatial autocorrelation condition, and the Local Moran's I and their Z score are calculated. This process can identify those clusters of features with very heterogeneous values. The Z indicates whether the apparent similarity (or dissimilarity) in values between the parcel and its neighbors is greater than one would expect simply by chance. The Z score of the transformed PD of each parcel is symbolized in Fig. 4.7. A high positive Z score (in red) for a parcel indicates that the surrounding parcels have similar values (either high or low). A group of adjacent parcels having high Z scores indicates a cluster of similarly high or low values. A low negative Z score (in blue) for a parcel indicates the parcel is surrounded by dissimilar values—that is, if a parcel gets a negative Z score, it has a different value than its neighbors. Most parcels in

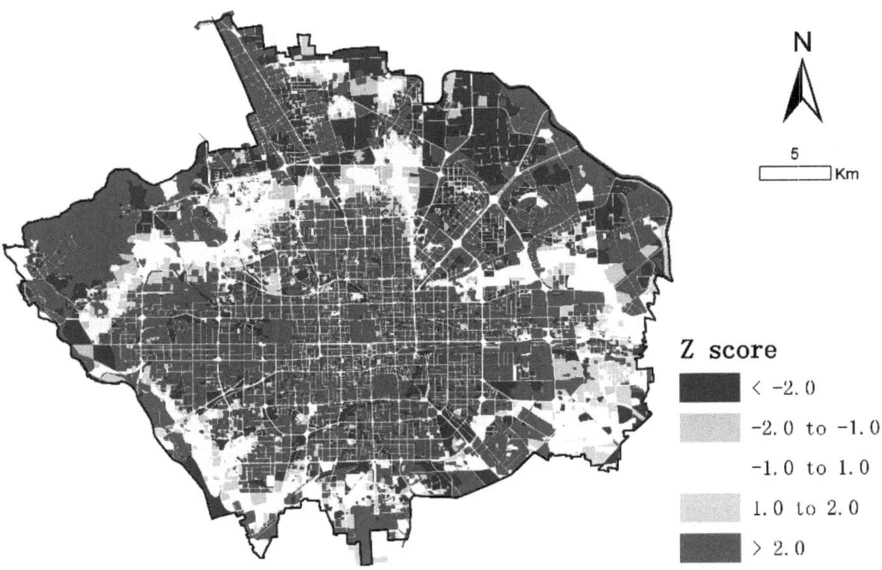

Fig. 4.7 Z score for the transformed PD of each parcel in the BCMA

the central BCMA and highway buffer zones with high Z scores indicate they have similar transformed PD (0 or 45° or so).

4.4.2 The Zone Scale: Aggregated Indicators for Zones

Totally 289 analysis zones (see grids in Fig. 4.8) are generated by merging four adjacent zones in the 1:2000 partitioning map into one zone. In most cases, the shape of the parcel is rectangle, or one edge of the parcel is vertical to the adjacent edge of the parcel. Therefore, in these conditions, we can use transformed parcel direction $D'(P) = 45 - abs(45 - abs(D(P)))$ to reflect the degree to which one parcel is deviating from E-W direction. The first three aggregated indicators mentioned in Sect. 4.2.3, AvePD, StddPD and DetaPD, are calculated based on $D'(P)$ instead of $D(P)$. AvePD based on $D'(P)$ reflects the dominating direction of parcels within the zone, StddPD based on $D'(P)$ reflects the diversity of directions within the zone, and DetaPD based on $D'(P)$ reflects the range of $D'(P)$ within the zone. For the calculation FRAGSTATS related indicators, PD (namely $D(P)$) ranging from −90° to 90° is divided into 18 categories with an interval of 10°. For instance, the parcel with PD ranging from −90 to −80 is set 1, and −80 to −70 is set 2. We then calculate indicators SHDI, PROX, ENN and DIVISION for all the 289 zones.

The calculation results of all aggregated indicators are illustrated in Fig. 4.8, from which we can see that the urban form varies from each zone in terms of these aggregated indicators. The urban form in the zone scale can then be measured accordingly.

4.4.3 The Region Scale: Cluster Analysis of All Zones in the Whole Study Area

We intend to divide all zones in the whole study area into four types of urban forms proposed in Sect. 4.1 to measure the urban form in terms of the parcel direction. The urban form type in terms of PD mainly depends on two indicators, AvePD and StddPD of each zone.

Cluster analysis is applied to divide the whole study area into four types of urban form. The K-Means cluster analysis is conducted using the two indicators, AvePD & StddPD, which are calculated in Sect. 4.4.2 in Fig. 4.9. Five clusters are generated in Fig. 4.8. From the cluster centers of all clusters shown in Table 4.4, Cluster 2 and Cluster 4 correspond to the regular urban form in Table 4.2 with low AvePD and StddPD values of $D'(P)$, Cluster 1 and Cluster 5 correspond to the oblique urban form in Table 4.2 with high AvePD, and Cluster 3 corresponds to the natural and mixed urban forms in Table 4.2 with medium AvePD and high StddPD values of $D'(P)$. Cluster 3 can be further divided into two clusters via introducing the parcels number indicator into the cluster analysis procedure because the natural urban

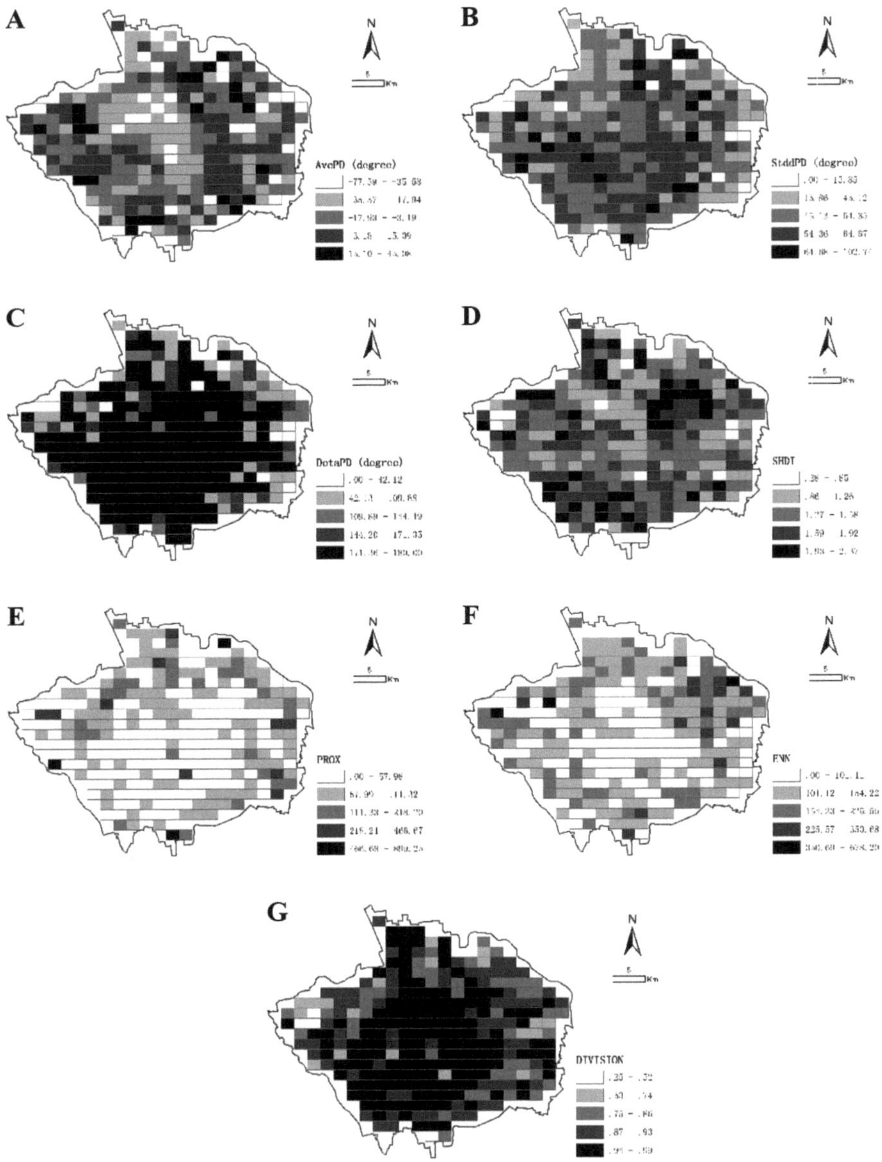

Fig. 4.8 The calculation results of aggregated indicators for all zones in the BCMA. (**a**–**g**) represent the results of AvePD, StddPD, DetaPD, SHDI, PROX, ENN, DIVISION, respectively)

form contains few parcels (also with the low compactness indicator) than that of the mixed urban form.

In addition, the global spatial autocorrelation was conducted for PD of all parcels in the BCMA in the region level. The results show that Moran's I index is 0.06, and

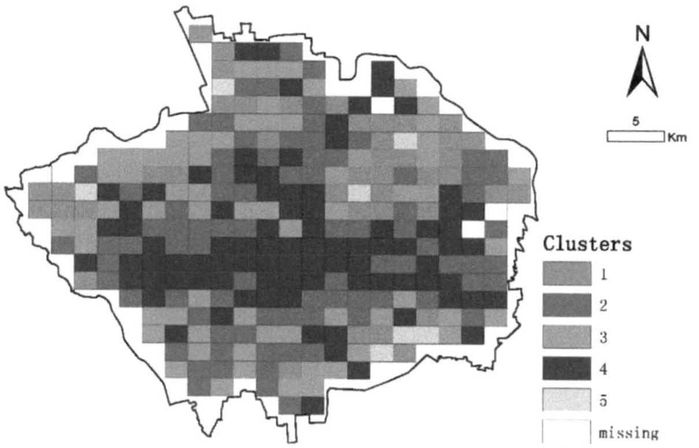

Fig. 4.9 Cluster analysis results map for the BCMA

Table 4.4 Cluster analysis results

Cluster	Count	Cluster center	
		Ave.	Std. D
1	42	20.61	14.58
2	84	5.69	8.32
3	65	11.23	11.98
4	87	2.37	3.81
5	7	32.55	11.03
Valid	285		
Missing	4		

Z Score is 10.41 standard deviations. This indicates that the PD indicator of all parcels in the BCMA is with positive spatial autocorrelation, namely, the PD values of parcels within the neighborhood of a parcel with high PD value also tend to be high.

4.5 Evaluating the Temporal Dynamics of Urban Form Using the Parcel Direction

4.5.1 PD Calculation Results for the Historical Urban Form

In order to analyze temporal dynamics of the planned urban from and historical urban form (or the urban forms in various time points) in terms of the parcel direction, we primarily identified and analyzed the historical urban form of Beijing, and calculated PD using the Beijing historical map of 1947 as illustrated in Fig. 4.10 (the extent of this map does not equal to the planned extent due to limited dataset resources). Most areas in the old map are not urban built-up in 1947, so we treated the polygons divided by road networks as parcels. Based on the historical map, the

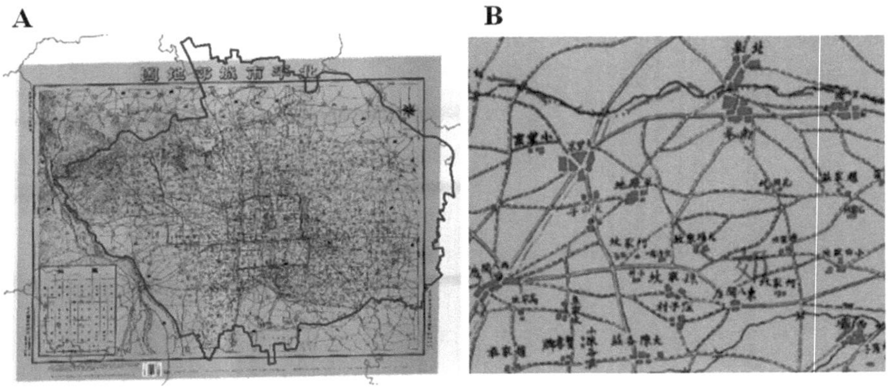

Fig. 4.10 The historical map of Beijing (**a**) and its partial zone (**b**) in 1947 (The *thick outline* in (**a**) is the BCMA, and the *thin line* is the district-scale administrative divisions (Source: Engineering Department of Peiping Municipal Government 2007)

Fig. 4.11 The parcel direction calculation results of the urban form in 1947

urban form in 1947 appeared to be the triangular mesh structure instead of the regular road networks in W–E and N–S directions. The orientation of settlements did not follow the direction of the road network, mainly because settlements were small in scale, and the road between settlements is usually the straight line.

The above historical map is digitalized into the parcel dataset with the vector format. The PD calculation results are shown in Fig. 4.11, from which we can see the historical parcels were more irregular than the planned parcels.

4.5.2 *Comparing the Parcel Direction of the Planned and Historical Urban Forms*

In order to compare the parcel direction of the historical urban form (1947) and the planned urban form (aimed to be fully realized in 2020), the two parcel layers were converted into GRIDs, the raster format supported by ESRI. Due to the difference of the spatial extent between the historical and planned urban forms, the intersection of the two urban forms was determined to be the spatial extent of comparative analysis (namely the non-white area in Fig. 4.12, with the total area of 976.2 km^2). The comparison results of the parcel direction for the two urban forms are shown in Fig. 4.12. The mean value of changed PD is 38.20°, and the standard deviation is 30.16°. In general, parcels with changed PD value of near 0° or 90° are relatively more.

According to the definition in Sect. 4.2.3, the parcels with reversed PD have the total area of 185.6 km^2 (see Fig. 4.13), 19.0 % of the whole analysis area. Based on the comparison results of the parcel direction, the parcels with reversed PD mainly locate in the ecological land use areas, and the new constructed areas in the express way buffers. The urban planning has changed the historical urban morphology within areas of parcels with reversed PD.

In view of the phenomenon of the PD reversed parcels in Beijing, during the compilation of the urban plan, transport planners generally adhere to the form of intersecting the direction of the oblique highways vertically, because it is in accordance with the existing relevant design specifications of China. However, this process is lack of flexibility. The triangle area forms, which are denounced by transport

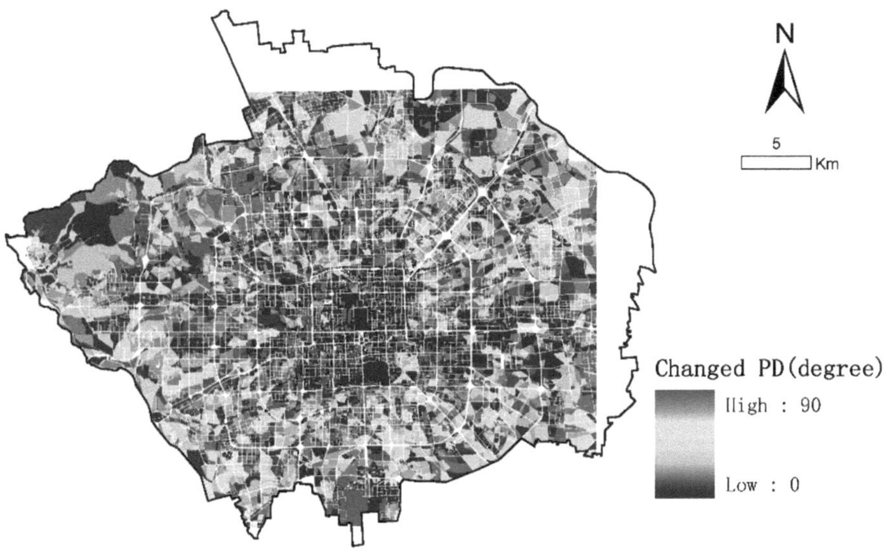

Fig. 4.12 The comparison map of the planned and historical urban forms

Fig. 4.13 The parcels with the reversed parcel direction in the BCMA

planners, can be treated flexibly through modifying the original design scheme, such as delimitating some trapezoidal parcels to solve the problem of oblique intersections, which is quite common in existing old districts, but extremely rare in new districts.

4.6 Conclusions

In this chapter, we proposed the definition of the parcel direction, which was analyzed in both the spatial and temporal dimensions. The conclusions are as follows: (1) In terms of the parcel direction, the urban form of the Beijing Central Metropolitan Area can be divided into the normal, oblique, mixed, and natural types. (2) The parcel direction is not significantly correlated with the perimeter, area, compactness as well as the distance to the urban center indicators of the parcel. (3) The parcel direction is explicitly heterogeneous in space, and the parcel direction in the subregions of the BCMA varies from each other as well as the entire BCMA. The probability density function of the parcel direction can be used to evaluate the urban form. (4) The spatial dimension of the parcel direction can be evaluated in three spatial levels, parcel, zone and region, respectively. In the region level, the parcel direction is with positive spatial autocorrelation. (5) The temporal dimension of the parcel direction can be investigated for evaluating to what extent the planned urban form inherits from the historical urban form, and has the potential to be introduced into the quantitative evaluation of the urban planning scheme.

The parcel direction is considered as a new indicator for the description of the urban form. If urban planners consider not only the indicators like the parcel's area and compactness, but also the parcel direction in planning practices, the economic and ecological influences of the planned road networks can be quantitatively evaluated to promote the sustainability of the planning scheme. Therefore, introducing the parcel direction into urban planning has its important significance.

Although this chapter proposed the definition, the calculation method and the empirical analysis using the novel indicator, the theoretical characteristics of the parcel direction should be analyzed from the perspective of data mining in the near future to promote the indicator's further and wider applications. Besides, the mechanism, driving forces as well as effects of the parcel direction's change should be the future research agenda.

References

Engineering Department of Peiping Municipal Government. (2007). *Beijing suburb map*. Beijing: Chinese Map Press (in Chinese).

Environmental System Research Institute. (2001). Land parcels data model. http://support.esri.com/index.cfm?fa=downloads.dataModels.filteredGateway&dmid=11. Accessed 7 Mar 2010.

Hillier, B. (1996). *Space is the machine*. Cambridge: Cambridge University Press.

Hillier, B., & Hanson, J. (1984). *The social logic of space*. Cambridge: Cambridge University Press.

Jiang, B., & Claramunt, C. (2002). Integration of space syntax into GIS: New perspectives for urban morphology. *Transactions in GIS, 6*(3), 295–309.

Jiang, B., Claramunt, C., & Batty, M. (1999). Geometric accessibility and geographic information: Extending desktop GIS to space syntax. *Computers, Environment and Urban Systems, 23*(2), 127–146.

Maniruzzaman, K. M., Asami, Y., & Okabe, A. (1994). Land use and the geometry of parcels in Setagaya Ward, Tokyo. *Theory and Application of GIS, 2*(1), 83–90.

Okabe, A., & Yamada, I. (2001). The K-function method on a network and its computational implementation. *Geographical Analysis, 33*(3), 271–290.

Okabe, A., Okunuki, K., & Shiode, S. (2006). The SANET toolbox: New methods for network spatial analysis. *Transactions in GIS, 10*(4), 535–550.

Xie, Y., & Ye, X. (2007). Comparative tempo-spatial pattern analysis: CTSPA. *International Journal of Geographical Information Science, 21*(1), 49–69.

Yang, B., Luan, X., & Li, Q. (2010). An adaptive method for identifying the spatial patterns in road networks. *Computers, Environment and Urban Systems, 34*(1), 40–48.

Chapter 5
V-BUDEM: A Vector-Based Beijing Urban Development Model for Simulating Urban Growth

5.1 Introduction

Over last two decades, Cellular Automata (CA) has been widely applied in generating realistic urban growth scenarios for its ability to simulate the dynamic spatial process from a bottom-up perspective (Clarke and Gaydos 1998; Landis 1995; Ward et al. 1999; Guan et al. 2005; Liu 2012; Shafizadeh Moghadam and Helbich 2013). Traditional CA consists of five components, namely a space represented as a regular grid composed of a collection of homogeneous cells, a set of possible cell states, and the transition rules which determine the evolution of the state of each cell based on states of cells within its neighborhood and some external constraints at each time step (Batty et al. 1999; White and Engelen 2000). For most existing CA models, the geographic space is typically represented as a regular raster grid and the neighborhood is defined as an assembly of adjacent cells. However, recent studies have demonstrated that the simulation results of such raster-based CA models are sensitive to the cell size and the neighborhood configuration (Moreno et al. 2009). For example, Jenerette and Wu (2001) used two different cell resolutions in their CA model to study urban expansion and showed that it generated significantly different land use patterns. Chen and Mynett (2003) investigated the impact of cell size and neighborhood configuration in a prey-predator CA model and observed that they affected both the resulting spatial patterns and the system stability. Jantz and Goetz (2005) examined the results of SLEUTH model, in response to different cell sizes and indicated that the cell size at which the land use data were represented could impact the quantification of land use patterns and the ability of the model to replicate spatial patterns. On the other hand, these raster-based CA models will be challenged and have difficulties in generating reliable land-use patterns at fine spatial resolutions (e.g. 5 m resolution) (Wang and Marceau 2013). When the resolution is increased, spatial entities in real world, such as blocks, census tract boundaries, and even individual parcels, can be identified. The use of a grid of regular cells creates areas of assumed homogeneous land use that

Y. Long, Z. Shen, *Geospatial Analysis to Support Urban Planning in Beijing*, GeoJournal Library 116, DOI 10.1007/978-3-319-19342-7_5

may contain variability in reality, thereby cannot precisely represent real entities, which have irregular sizes and shapes (Stevens and Dragicevic 2007).

Some researchers have started to use vector-based, or irregular-based, CA models to avoid the questions mentioned above. The studies of vector-based CA are limited, but have gained much importance recent years. Shi and Pang (2000) presented a Voronoi-based CA in which the CA was extended by using the Voronoi as its spatial framework. Results showed that Voronoi-based CA could simulate local interactions among spatial objects to generate complex global patterns, and further simulate interactions among point, line and polygon objects with irregular shapes and sizes in a dynamic system. Semboloni (2000) used the Delaunay triangle network to represent irregular space. The space is divided into a number of contiguous cells that are generated by randomly distributed nodes connected by Delaunay triangulation, and the cells are represented by polygons whose vertices are located in the centroid of each triangle. O'Sullivan (2001a, b) combined CA and graph theory to present a graph-CA model, where the space is represented as a planar graph composed of vertices and edges. These studies have demonstrated it is workable to develop a CA model using a vector-based representation, but the cells in these models could not represent actual spatial units in real world, composing the landscape a user would perceive as meaningful (Moreno et al. 2009).

In other researches, the irregular cell is able to represent entities in real world. Hu and Li (2004) developed an object-based CA, in which geographical entities, e.g. a land parcel, a block, a road, a school etc. are represented as points, lines, or polygons according to their real shapes and sizes. Geographic automata system (GAS) (Torrens and Benenson 2005) incorporated irregular vector objects as automata to represent spatial entities such as roads, buildings and parks. Stevens and Dragicevic (2007) developed iCity, in which an urban area is partitioned into discrete land use units based on cadastral information and represented as a collection of polygons. Hammam et al. (2007) introduced vector agents (VAs) into GAS (Torrens and Benenson 2005) to realize the change of each agent's geometry while interacting with other agents in its neighbourhood using a set of rules. Shen et al. (2009) developed a geosimulation model using a vector-based CA to visualizing land use patterns in urban partitions. In the entity-based CA model presented by Moreno et al. (2008, 2009), the shape and size of each object can also change and a dynamic neighborhood was semantically implemented. Pinto and Antunes (2010) developed an irregular CA model based on census blocks to determine the land use demand by considering the evolution of population and employment densities over time. Agent iCity, developed by Jjumba and Dragicevic (2012), is an upgraded version of iCity; it can simulate land use change using irregular spatial units and capable of simulating the interactions of different stakeholders involved in the process of land use change. Wang and Marceau (2013) and Marceau et al. (2013) presented a patch-based CA model designed to simulate land use changes at a fine spatial resolution, and tested it in the eastern part of the Elbow River watershed in southern Alberta, Canada.

When simulating urban growth using a vector-based CA method, the parcels are treated as basic spatial units. The parcel, as the basic component of built-up spaces, is the container of various urban infrastructures, such as buildings, green lands, municipal infrastructures, and public services. Since it has traffic connection with the road network, the parcel has close relationship with both of the block and road network. The change of parcel boundaries is common in the urban growth process. For example, when a parcel has a large size initially, locates in the suburban area, but would be developed in the future, it would be highly possible to be subdivided into several small parcels when the urban development happens. When simulating urban growth scenarios using the parcel as the basic spatial unit, it's unavoidable to consider the changing mechanisms or rules of parcel subdivision, which has a significant influence on simulation results.

Land subdivision is a standard practice in land surveying that aims to partition a tract of land into smaller sized lots, which means a small parcel in this chapter (Easa 2008), or a process splitting up of a larger land tract, which means a large parcel here, into streets and smaller subspaces variously called parcels and lots (Dahal and Chow 2014). In general, land subdivision is carried out using expert knowledge and skills, rigorous field work, both spatial and non-spatial data, and through incorporating a range of zoning and development rules (Chen and Jiang 2000; Wakchaure 2001).

The simulation of the land subdivision is very useful (e.g. speeding up the process and saving the cost of field work) in many applied and research areas, including for urban growth prediction. Ko et al. (2006) developed the Fragmented Land Ownership Spatial Simulator (FLOSS) to generate ownership patterns. Alexandridis et al. (2007) used a parcelization algorithm to simulate how a seller agent subdivides its land for sale in an agent-based landscape model. However, these two studies are developed for raster-based landscape models, and are performed at relatively coarse resolutions (Wickramasuriya et al. 2011). Wakchaure (2001) developed an ArcView tool for supporting land subdivision at a single parcel level, but the tool is semi-automated. Vanegas et al. (2009a, b) developed an automated urban layout generation module using the algorithm of recursive binary division, and their solution can create streets, lots arrangements. Wickramasuriya et al. (2011) presented a fully-automated land subdivision tool that uses vector data and is capable of generating layouts with both lot and street arrangements for land parcels of any shape. Vanegas et al. (2012) presented a method for interactive procedural generation of parcels within the urban modeling pipeline using two algorithms. Agent iCity (Jjumba and Dragicevic 2012) has a land subdivision module, which can subdivide land into city blocks, then small cadastral parcels depending on the anticipated growth within the city as determined by the planning agent, but the authors failed to report further details about the corresponding algorithm. Wickramasuriya et al. (2013) improved a method to dynamically subdivide parcels in land use change models. Demetriou et al. (2013) and Demetriou (2014) developed a module called Land Parcelling System (LandParcelS) to automate the land partitioning process by designing and optimizing land parcels in terms of their shape, size and value using a spatial genetic algorithm. Dahal and Chow (2014) developed a GIS toolset, called Parcel-divider, for automated subdivision of land parcels. The highlight of this toolset is that it can provide various subdivision styles based on different geometric attributes of parcels, and can be used either as a module to be integrated into an urban growth model or as a stand-alone application. Except the reports in published literature, CityEngine (ESRI 2013) already provided the parameters for block subdivision to create lots, using recursive OBB (oriented bounding box) algorithm, offset algorithm, and skeleton subdivision algorithm. Although the studies of land subdivision simulation are comparatively mature, most of them have not been integrated into urban growth models. In addition, none of them considers the impacts of urban planning on the land subdivision when urban growth happens.

In the background of the prosperity of macro-economy, Beijing, the capital of China, has achieved unprecedented development of urban construction, especially after adopting the Reform and Open Policy in 1978. To cope with the challenge caused by rapid urban growth, the Beijing Urban Development Model (BUDEM) was developed (Long et al. 2009) to support urban planning and the evaluation of corresponding policies in Beijing Institute of City Planning (BICP), which is a leading and professional institution responsible for establishing various urban plans in Beijing and providing planning-relevant services for Beijing's government. BUDEM is a spatio-temporal dynamic urban model using a combination of the raster-based CA and Agent-based Modelling (ABM) approaches. In the first phase, the model was used to identify urban growth mechanisms in various historical phases since 1986, to retrieve urban policies needed to implement the desired (planned) urban form in 2020, and to predict the urban growth scenario until 2049. The model has been proved to be capable of analyzing historical urban growth mechanisms and predicting future urban growth for metropolitan areas in China, and its vision is to be a comprehensive urban model. Besides the urban growth simulation, some other studies have been conducted, for example, retrieving spatial policy parameters from an alternative plan (Long et al. 2012), calculating transportation energy consumption and environment impact (Long et al. 2013a), and comparing simulation and artwork of urban growth boundaries (Long et al. 2013b).

In this chapter, we aim to improve the initial raster-based BUDEM model into a vector-based one (called V-BUDEM here), using urban parcels as the basic spatial units. At this stage, we are focused on the part of urban growth simulation using this model. The V-BUDEM model is described in detail in Sect. 5.2. Then the model is applied to simulate urban growth scenario of Yanqing Town, a small town in Beijing, rather than testing it in the whole Beijing metropolitan area in Sect. 5.3. Finally, conclusions and discussion are presented in Sect. 5.4.

5.2 The V-BUDEM Model

5.2.1 Constrained Cellular Automata

Considering the complexity of urban growth, many researches have introduced constraints into the CA model, namely constrained CA, thereby rendering the simulation of urban growth closer to real world outcomes. The V-BUDEM is also a constrained CA model, of which the conceptual model is shown as Eq. 5.1.

$$V_i^{t+1} = f\left\{V_i^t, A_{status}, A_{loc}, A_{gov}, A_{nei}\right\} \tag{5.1}$$

In this equation, v_i^t is the status at cell i of iteration t; f is the transition rules of the constrained CA. In the V-BUDEM model, the parcel, with various shape and size, is treated as the cell, and the cell status represents 0 for undeveloped or 1 for developed from non-urban built-up parcel to urban built-up parcel. A parcel's neighborhood is defined as all parcels surrounding it within a certain distance. Constrained conditions in the urban growth process consist of four aspects, including self-status constraints A_{status}, location constraints A_{loc}, government (or institutional) constraints A_{gov},and neighborhood constraint A_{nei}. Locational, and institutional constraints are assumed to remain static during the future urban growth process, and they do not change across simulation iterations. Self-status (e.g. whether the parcel is agricultural land) constraints can also be treated as keeping static, because the parcel will be excluded in the simulation process if its self-status has changed in previous iteration. The neighborhood effect, however, continue to change with simulation iterations of the constrained CA.

The transition rules in constrained CA are illustrated in Eq. 5.2.

$$\begin{aligned}
&1.\ LandDemand = \sum_t stepArea^t \\
&In \text{ iteration } t+1 \\
&2.\ S^t = x_0 + \sum_{k=1}^{n-1} x_k * \mathrm{a}_k + x_n * a_n^t \\
&3.\ p_g^t = \frac{1}{1+e^{-s^t}} \\
&4.\ p^t = \exp\left[\alpha\left(\frac{p_g^t}{p_{g,\max}^t} - 1\right) * RI^t\right] \\
&5.\ RI^t = 1 + \left(\gamma^t - 0.5\right)/k \\
&If\ p_i^t = p_{\max}^t\ and\ Area_i \le \left(stepArea^t - deveArea^t\right) \\
&then\ V_i^{t+1} = 1 \\
&p_i^t = p_i^t - p_{\max}^t \\
&p_{\max}^t update
\end{aligned} \tag{5.2}$$

In this equation, LandDemand is the total area, which will be developed during the whole simulation process; StepArea^t is the total area, which will be developed during the iteration t; s^t is the development suitability; p_g^t is the initial transition probability; ptg ,max is the maximum value of p_g^t in iteration t; p^t is the final transition probability; n is the total number of considered constraint variables; x_0 is the constant item; a_k is the static constraint variable (excluding the neighborhood constraint), x_k is the weight of a_k; a_n^t is the dynamic constraint variable (namely neighborhood constraint); x_n the weight of a_n^t; RI^t is the random item in iteration t; γ^t is the random value varying from 0 to 1; k is the random index used to regulate RI^t; α is the dispersion parameter ranging from 1 to 10, indicating the rigid level of development. The larger α indicates stricter development control and lower development probability with the same suitability. p_i^t is the final transition probability of cell i; p_{max}^t is the maximum value of p_g^t in iteration t, and originally equals ptg ,max; Area$_i$ is the area of cell i; deveArea^t is the developed area in iteration t. In each iteration, the beginning value of deveArea is 0, and the final value is no more than StepArea^t.

The parameters that need to be calibrated in the constrained CA model include LandDemand, StepArea^t, x_k, and x_n. LandDemand is determined by the difference between the total area of existing urban built-up parcels and that of planned urban built-up parcels. StepArea^t is assumed to be constant through the entire simulation period and can be calculated simply as follow:

$$stepArea^t = LandDemand / t \tag{5.3}$$

To calibrate x_k and x_n to reflect the weights of constraint policies, the logistic regression and heuristic approaches can both be applied. The weight of x_k for static constraint variables can be retrieved by the logistic regression in which whether a parcel is developed from non-urban to urban is the dependent variable (0 or 1) and the static constraint variables are the independent variables. Holding x_k static, x_n can be calibrated using the MonoLoop method (for details see Long et al. 2009), where x_n is continually sampled from 0 to $x_{n,max}$ with an interval of $x_{n,max}$/M. $x_{n,max}$ and M can be set based on the user's experience. The sampled x_n and the calibrated x_k are used as the input variables for the constrained CA model. The simulated urban form is compared with the observed urban form (namely the planned urban form) to obtain the Kappa index, which is significantly correlated with overall accuracy for comparing two maps. In this chapter, the Kappa index is calculated to analyze the degree of similarity (goodness-of-fit) between the simulated and the planned urban forms parcel by parcel. The x_n with the maximum Kappa index is specified as the final weight for a_n. Finally, x_k and x_n as the calibrated weights are used in the constrained CA model to simulate future urban growth. Kappa index was initially introduced by Cohen (1960) and adapted for the accuracy assessment in remote sensing applications by Congalton and Mead (1983). The adopted method for calculating the vector-based Kappa is different with the kappa index used in raster-based CA models, and is shown as Eq. 5.4.

$$K = \frac{Area_t \times \left(Area_{dd} + Area_{uu}\right) - Area_{d\_} \times Area_{\_d} - Area_{u\_} \times Area_{\_u}}{Area_t^2 - Area_{d\_} \times Area_{\_d} - Area_{u\_} \times Area_{\_u}} \times 100 \tag{5.4}$$

In Eq. 5.4, K is the Kappa index, $Area_t$ is the total area of the study region; $Area_{dd}$ is the total area of the parcels, which are developed in both simulated and planned urban forms; $Area_{uu}$ is the total area of the parcels, which are undeveloped in both simulated and planned urban forms; $Area_{d\_}$ is the total area of the parcels, which are developed in the simulated urban form; $Area_{\_d}$ is the total area of the parcels, which are developed in the planned urban form; $Area_{u\_}$ is the total area of the parcels, which are undeveloped in the simulated urban form; $Area_{\_u}$ is the total area of the parcels, which are undeveloped in the planned urban form.

5.2.2 *Constraint Variables*

Fourteen constraint variables (means $n = 14$), the same with those in the BUDEM model, are shown in Table 5.1. Some constraint variables may be different with those in other urban growth models, but we think they are suitable in China's or Beijing's situation. For example, variables isagri and isrural stand for the transition from agricultural land and rural built-up land into urban built-up land, which is a big concern for the Beijing Municipal Government. The detailed explanations for the selection of constraint variables can be found in Long et al. (2009).

5.2.3 *The Parcel Subdivision Framework*

Some traditional parcel subdivision tools can automatically subdivide parcels using various algorithms. They highly improve the speed of parcel subdivision process and the corresponding results can provide a good reference to urban planners when

Table 5.1 Constraint variables in V-BUDEM

Type	Name	Value	Description
A_{status}	*isagri*	0, 1	Whether the cell is agricultural land
	isrural	0, 1	Whether the cell is rural built-up land
A_{gov}	*con_f*	0, 1	Whether in the forbidden development zone
	landresource	1–8	Land suitability classified for agriculture
	planning	0, 1	Whether planned as urban built-up land
A_{loc}	*d_bdtown*	≥0	Minimum distance to town boundaries
	d_city	≥0	Minimum distance to new cities
	d_river	≥0	Minimum distance to rivers
	d_road	≥0	Minimum distance to roads
	d_tam	≥0	Minimum distance to Tian'anmen Square
	d_town	≥0	Minimum distance to towns
	d_vcity	≥0	Minimum distance to important new cities
	d_vtown	≥0	Minimum distance to important towns
A_{nei}	*neighbour*	0–1.0	Neighborhood development intensity

establishing urban plans. Unlike these tools, our parcel subdivision framework is used to simulate real parcel subdivision process, which actually would be greatly affected by established urban plans, when urban growth happens. At this stage, we assume the parcel subdivision would be finished before starting the urban growth simulation in our V-BUDEM model, namely not consider the dynamic changing of parcel boundaries. Therefore, the framework of our parcel subdivision methods is used to determine parcel boundaries in the base year, based on which we simulate urban growth.

Specifically, there are four steps for the achievement of our parcel subdivision framework.

(1) Urban plan is one of key instruments for Chinese government to regulate urban development. A typical master urban plan in China can be regarded as a planned urban form, mainly consisting of three parts: road network, urban built-up parcels, and non-urban built-up parcels. Transport infrastructure is crucial to support various urban development, thereby has high priority to be constructed. Compared with other two parts of a planned urban form, road network is more likely to be achieved. In addition, road network is the basic element of an urban form, and the boundaries of parcels (including blocks) are usually determined by it. Therefore, in our parcel subdivision framework, it's reasonable to assume that the existing parcels in the base year are firstly subdivided according to the planned road network (Fig. 5.1a). For example, parcel A would be subdivided into two small parcels if a new road is planned to pass through it.

(2) Main social and economic activities of cities happen in urban built-up parcels. To achieve social and economic goals in a planning duration, it's important to realize the part of a planned urban form consisting of urban built-up parcels. Therefore, the existing parcels then are assumed to be subdivided according to the boundaries of the planned urban built-up parcels (Fig. 5.1b).

(3) When the existing parcels are subdivided according to the planned road network, the effect of the distribution of non-urban built-up parcels has already been considered to some extent, because parcel boundaries are usually determined by road network. On the other hand, the corresponding existing parcels, which covered by the planned non-urban built-up parcels, usually are non-urban built-up ones. Although they are planned to keep undeveloped, when it comes to real situation of China, which is facing a rapid urbanization, some of them may be developed illegally. For

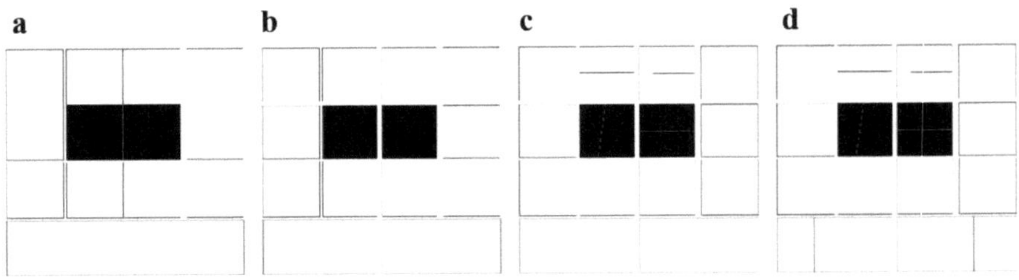

Fig. 5.1 The parcel subdivision framework

example, people may build their own houses on right-owned parcels, as a result, the non-urban built-up parcels may be changed into urban residential parcels, and the corresponding boundaries would be changed. Because we have already considered the influence of non-urban built-up parcels to some extent, and to reflect some issues (e.g. illegal development) which may happen in real world, the boundaries of the existing parcels, which belong to the planned non-urban built-up parcels, are adopted (Fig. 5.1c).

(4) Some existing large non-urban built-up parcels will be subdivided into some small ones, for they may be developed in a smaller size (Fig. 5.1d). It should be mentioned that, after the urban growth simulation, if a non-urban built-up parcel is not developed, the original parcel boundary would be adopted again.

Generally speaking, our approach to subdivide an existing large parcel is to subdivide the Minimum Bounding Rectangle (MBR) of the parent parcel using an algorithm, to clip the subdivided MBR using the parent parcel as a mask, and finally to eliminate the unacceptable generated small parcels. The details are as follows.

Evaluating whether a non-urban built-up parcel is large enough and should be subdivided. The criteria adopted here is simple, shown as Eq. 5.5.

$$\mathrm{Area}_i \geq Area_{average} \times n \tag{5.5}$$

Area_i is the area of parcel i; $\mathrm{Area}_{\mathrm{average}}$ is the average area of all existing urban built-up parcels; n is the integrate number, and can be set according to user's experience. If the area of parcel i is no less than $\mathrm{Area}_{\mathrm{average}} \times \mathrm{n}$, meaning parcel i is a large parcel, and should be subdivided.

1. Determining the area of the subdivided parcel $Area_{sub}$. In our tool, we assume the areas of all subdivided parcels are the same, and equal $Area_{average}$. Existing researches usually generate the road first, then subdivide the parent parcel according to the value of $Area_{sub}$. Different from these, we don't consider the road generation, but assume the created subdivided parcel has contain the road area, which can be used to support the connectivity of subdivided parcels. Therefore, the step of road generation is emitted in our method, but the area of subdivided parcel should contain the road area. Showing as Eq. 5.6.

$$Area_{sub}' = Area_{sub} + Area_{road} = Area_{sub} \times (1 + \partial) \tag{5.6}$$

   $Area_{sub}$ is the area containing the road area, $\mathrm{Area}_{\mathrm{road}}$ is the corresponding road area, and ∂ is the ratio of parcel and road areas. The method has an assumption that all subdivided parcels need to be adjacent to the road. This adoption can speed up the parcel subdivision process, considering the creation of road is comparatively complicated. The simulation result can still represent the urban growth trend based on this method.
2. Creating the MBR of the parent parcel (Fig. 5.2b); the parcel's MBR can reflect its direction.

3. Subdividing the MBR according to $Area_{sub}$ (Fig. 5.2c). The row number (Row_{num}) and column number ($Column_{num}$) of the MBR are determined by Eq. 5.7.

$$\begin{aligned} & t = \sqrt{Area_{MBR} \,/\, Area'_{sub}} \\ & \textit{if } t < \mathrm{int}(t) + 0.5: \\ & \quad Row_{num} = Column_{num} = \mathrm{int}(t) \\ & \textit{else}: \\ & \quad Row_{num} = Column_{num} = \mathrm{int}(t) + 1 \end{aligned} \tag{5.7}$$

Where, $Area_{MBR}$ is MBR's area.
4. Clipping the subdivided MBR using the parent parcel as a mask (Fig. 5.2c).
5. Eliminating unacceptable small parcels, resultant of clipping, by merging them into adjacent parcels (Fig. 5.2d). The criteria to determine whether a small parcel should be eliminated is by Eq. 5.8.

$$Area_{created} \leq Area_{sub} \,/\, m \tag{5.8}$$

In Eq. 5.8, $Area_{created}$ is the area of a created small parcel; m is the integrate number, and can be set according to user's experience. The criteria to determine which adjacent parcel would be selected to merge with the small created parcel is that the parcel which has the longest shared border.

5.2.4 The Simulation Procedure

The V-BUDEM's simulation procedure is shown in Fig. 5.3. $Parcel_{i,max}$ is the parcel i, which has the maximum probability to be developed among all remaining non-urban built-up parcels in iteration t, and $Area_{i,max}$ is its area. In each iteration, the final value of $deveArea^t$ is always less than that of $StepArea^t$ (although very close). As a

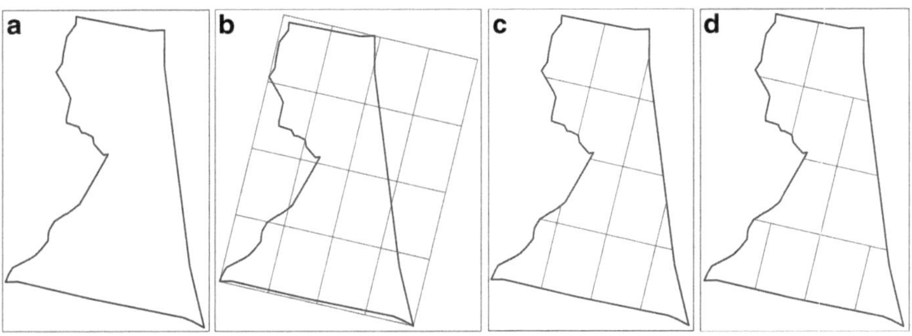

Fig. 5.2 The method to subdivide a large parcel

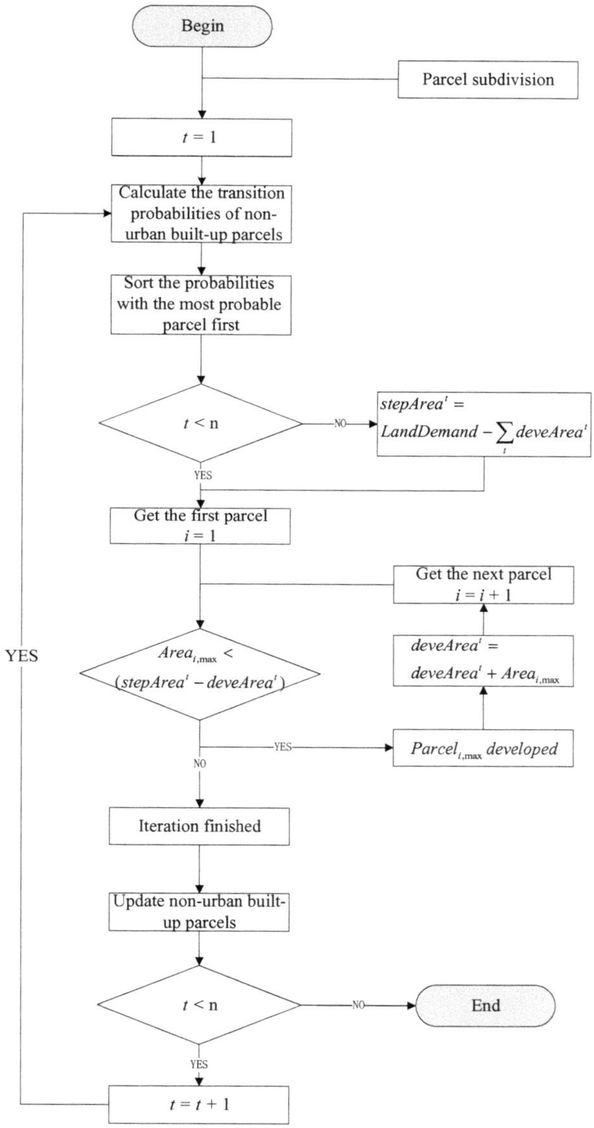

Fig. 5.3 The V-BUDEM simulation process

result, if $t = n$, the value of $\mathrm{StepArea}^t$ is determined by $LandDemand - \sum_t deveArea^t$ in which $\sum_t deveArea^t$ is the total developed area during the finished $n - 1$ iterations. This consideration can promise the final developed area is close to LandDemand.

5.3 Model Application

5.3.1 *Study Area*

The whole study area of V-BUDEM is the Beijing Metropolitan Area (BMA) (Fig. 5.4a), but in this chapter, Yanqing Town (Fig. 5.3) was chosen as a case area to illustrate the model's applicability. The BMA has an area of 16,410 km^2. It has experienced rapid urbanization in terms of GDP and population growth since the Reform and Opening Policy of 1978, established by Chinese central government. There are 16 districts under BMA jurisdiction. In 2013, the total GDP was 1 950.06 billion Yuan, with an average of 93.213 thousand Yuan per capita, and the total residential population was 21.148 million (Beijing Municipal Bureau of Statistics 2014). In 2010, the area of the BMA urban built-up land was 1,758 km^2, about 3.5 times than that in 1976, which was 495 km^2. Yanqing (Fig. 5.4b), a small town located in the northwestern of the BMA, has a distance of about 74 km from Beijing's city center. In 2011, the residential population is 37,000 (Yanqing Government 2012). In 2010. Yanqing has an area of 67.36 km^2 (including 12.22 km^2 urban built-up land), and consists of 1,121 parcels (including 604 urban built-up parcels).

5.3.2 *Data*

The input spatial data is classified into six types, namely LANDUSE, PLANNING, CONSTRAINT, LANDRESOURCE, LOCATION, and BOUNDARY. All the spatial data was converted into Personal Geodatabase Feature Class, using the same coordinate and projection system. Explanations of these data are shown as follows.

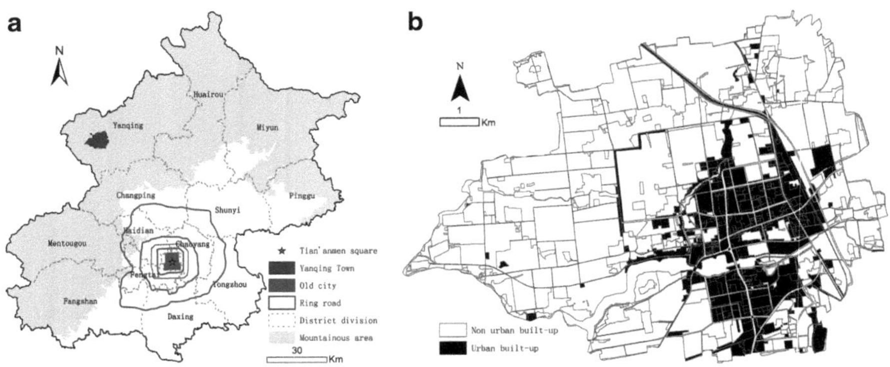

Fig. 5.4 The Beijing Metropolitan Area and Yanqing Town

1. LANDUSE: In this test, we only used Yanqing's land use pattern in 2010 (corresponding to existing urban form, shown in Fig. 5.5a). The data is from the database of Beijing Institute of City Planning (BICP). Originally, there were seven land use types, including urban built-up land, rural built-up land, transport land, agriculture land, forest land, wetland land, and vacant land. Then, they were integrated into three land use types, including urban built-up land, rural built-up, and other land. The variable *isagri* and *isrural* (Fig. 5.6a, b) were derived from the LANDUSE dataset.

Fig. 5.5 Yanqing's land use pattern in 2010 and land use plan in 2020

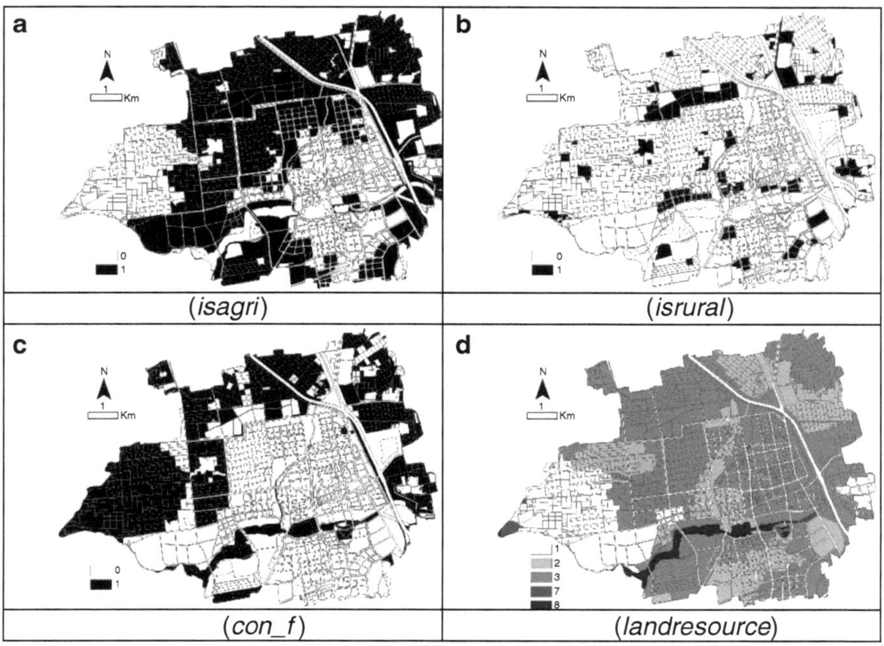

Fig. 5.6 The spatial distributions of static constraints data

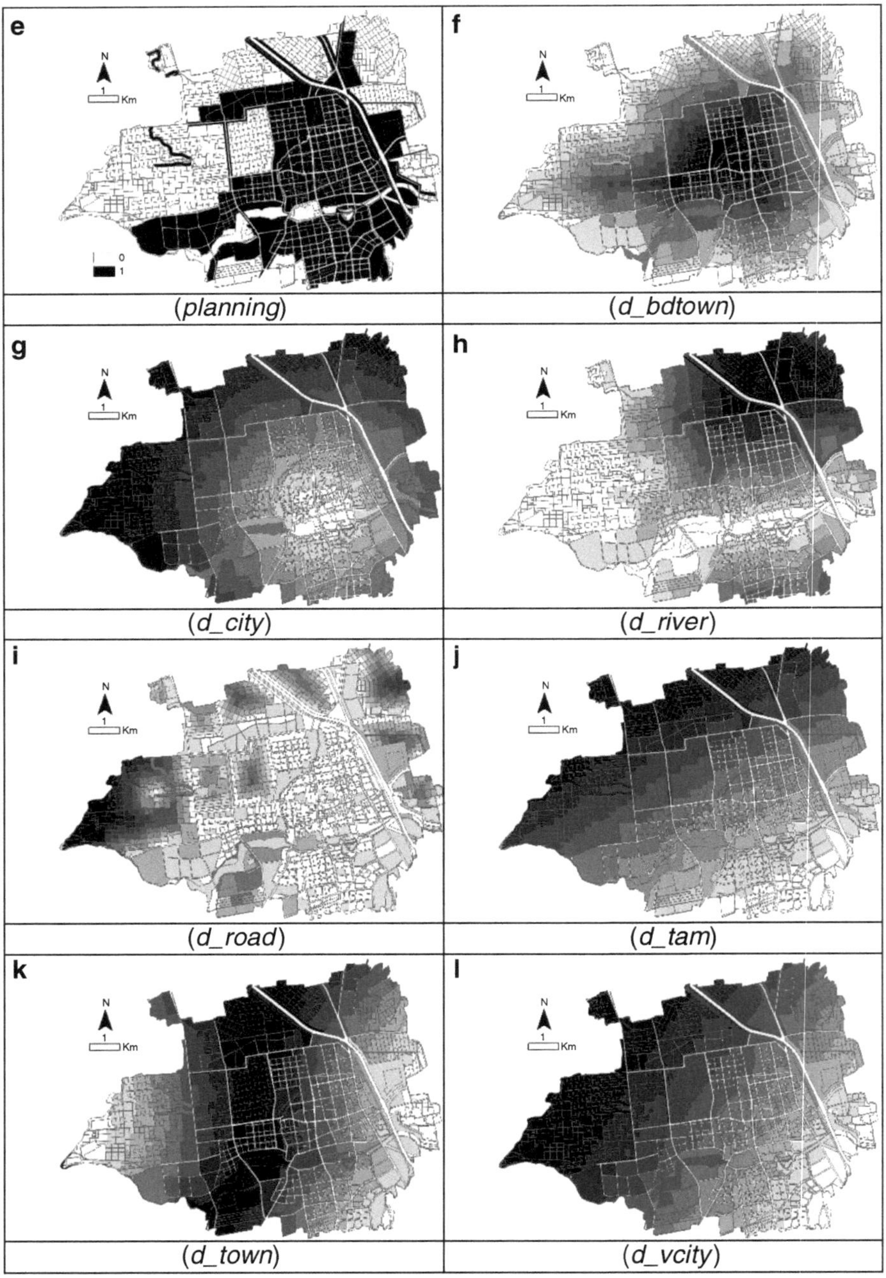

Fig. 5.6 (continued)

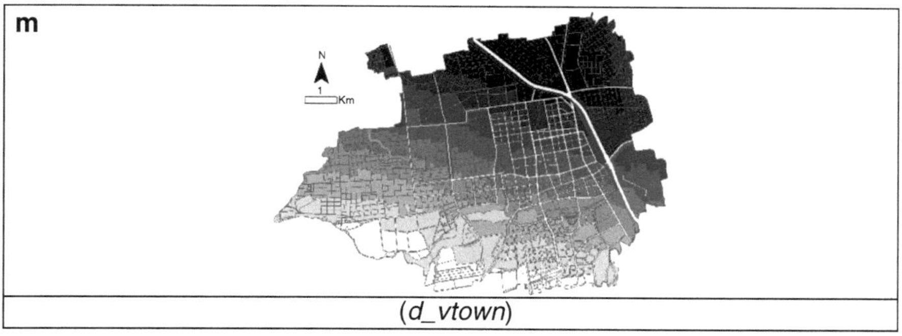

Fig. 5.6 (continued)

2. PLANNING: In this test, Yanqing's planned urban form, shown in Fig. 5.5b, is from the Beijing Urban Master Plan (2004–2020). The variable *planning* (Fig. 5.6e) is derived from the PLANNING dataset with 1 for urban built-up land and 0 for non-urban built-up land. In 2020, there are 30.07 km^2 for urban built-up land, with 975 urban built-up parcels.
3. CONSTRAINT: The data, from the Beijing Limited Construction Zone Plan (Long et al. 2006), reflects urban development constraints derived from 110 spatial layers of natural resource protection and hazard prevention according to current laws, legislations, and standards of China. Beijing was zoned into the forbidden built-up areas, constrained built-up areas, and suitable built-up areas. If the value of variable *con_f* (Fig. 5.6c) is 1, the parcel is located in forbidden built-up areas.
4. LANDRESOURCE: The data, from the BICP's database, indicates the suitability for the agricultural use and was classified into eight types, ranging from 1 to 8. The variable *landresource* (Fig. 5.6d) is derived from LANDRESOURCE with the same value.
5. LOCATION: The data, including the distributions of centers at various levels, road network, and river resource distribution, are from BICP's database. The location variable, e.g. *d_tam*, *d_vcity*, *d_city*, *d_vtown*, and *d_town*, indicates the minimum distance to city-level or town-level centers for various administrative divisions. The location variable *d_road* indicates the minimum distance to the road network and *d_river* as the minimum distance to the rivers. The location spatial variables were derived from LOCATION using the Distance/Straight Line command in ArcGIS 10.2, and shown in Fig. 5.6g–m.
6. BOUNDARY: The data, from BICP's database, includes various boundaries, such as administrative, ring road, eco-zoning, and watershed boundaries. In Yanqing test, the variable *d_bdtown* (Fig. 5.6f) which means the minimum distance to town boundaries, is considered.

5.3.3 Yanqing 2020 Simulation

In this sub-section, we simulate urban growth scenario "Yanqing 2020", from 2010 to 2020. Before the simulation, existing parcels were subdivided according to existing and planned urban forms using our parcel subdivision framework. Figure 5.6 shows the corresponding subdivided existing urban form. After subdivision, there are 9.76 km^2 for urban built-up land, a little smaller than that before subdivision (12.22 km^2). There are 2,201 parcels in total, with 617 urban built-up parcels, which is a little more than that before subdivision (604 parcels).

Secondly, time step was set as five times with a total of 10 years. According to the planned urban form in 2020, there are 30.07 km^2 (namely LandDemand) for urban built-up land. Assuming the area which would be developed in each time step is the same, stepArea is 6.01 km^2. In addition, α was set as 3, and k was set as 20 empirically.

Thirdly, parameters of constraint variables should be estimated according to different historic phases. Considering there are only 1,121 parcels in Yanqing, thereby the parameter set estimation may not significant and convincing. Additionally, the objective of this test is to show the feasibility, especially of the part of parcel subdivision framework in the V-BUDEM model, with no attention to identify suitable and accurate parameter at this stage. Assuming the constraint influence of each x_k is the same between Yanqing and the whole BMA, we adopted the parameter set (excluding the weight of neighbor variable x_n), with the whole accuracy of 96.8 %, directly from Long et al. (2009), see Table 5.2. In the logistic regression for its identification, dependent variables x_k (excluding neighbor variable) were calibrated via algebra operation on the BMA's existing urban form in 2006 and planned urban form in 2020. Each independent variable is significant at the acceptable level. Detailed explanations can be found in Long et al. (2009).

To identify neighbor parameter x_n, the neighbor distance, which may affect simulation result greatly, should be determined at first. We set it as 200 m empirically. In the process of identifying x_n via MonoLoop method, x_n is sampled from 0 to 5, with an interval of 0.5 empirically. According to the simulated results of Kappa index, shown in Fig. 5.7, x_n was set as 2.5, with a Kappa value of 81.16. The weights x_k,

Table 5.2 Policy parameter set for 2006–2020

Name	Coefficient	Name	Coefficient
isrural	6.886 21***	*d_river*	−0.000 52***
isagri	6.971 87***	*d_road*	−0.000 96***
d_tam	−0.000 10***	*d_bdtown*	−0.000 27***
d_vcity	−0.000 03***	*planning*	8.770 71***
d_city	−0.000 10***	*con_f*	−0.200 97*
d_vtown	−0.000 28***	*landresource*	−0.093 55**
d_town	−0.000 11***		

Note: ***p (significance) = 0.001; **p = 0.05; *p = 0.5

obtained by the logistic regression, x_n from MonoLoop, and with other parameters were then input into the established transition rules to simulate urban growth scenario Yanqing2020. The simulated urban form was shown in Fig. 5.8.

The Kappa value is 81.16, showing the simulated urban form was fitted well with the planned urban form. In the simulated urban form in 2020, there are 1,517 parcels

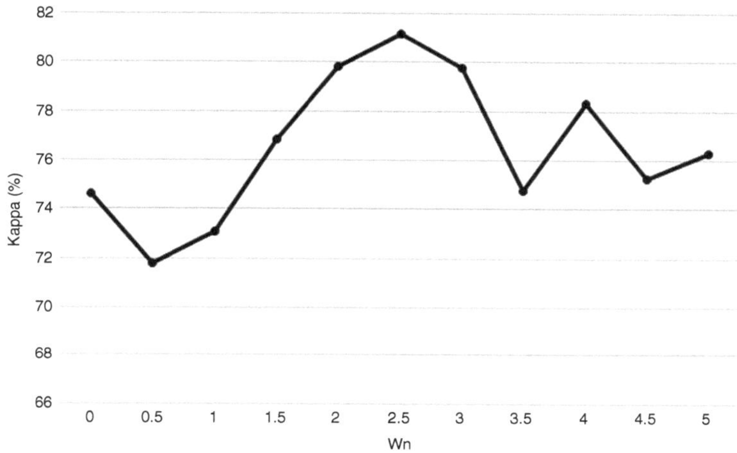

Fig. 5.7 Kappa values of the MonoLoop procedure

Fig. 5.8 Simulated urban form in 2020

in total. Among them, there are 1,066 urban built-up parcels, with a total area of 30.05 km^2, slightly smaller than predicted LandDemand 30.07 km^2. Specifically, 903 urban built-up parcels are located in the planned urban areas, while 163 urban built-up parcels located in the planned non-urban areas. When it comes to the parcels adopting the boundaries of the planned urban built-up parcels, most of them are developed into urban built-up land, except some mainly located in middle-northern part of Yanqing Town. The fact shows it is reasonable to adopt the boundaries of the planned urban built-up parcels when subdividing existing parcels. Additionally, there are some parcels, adopting existing parcel boundaries in 2010 or subdivided from large ones, have been developed into urban built-up land, which would be unlikely to happen if we adopt planned non-urban built-up parcel boundaries, which usually have large size thereby not suitable to be regarded as the basic spatial units when simulating. The simulated result shows the vector-based CA method is feasible and our parcel subdivision framework is reasonable for simulating urban growth.

5.4 Conclusion and Discussion

BUDEM (Beijing Urban Development Model) is a raster-based CA model for supporting city planning and corresponding policies evaluation in Beijing (Long et al. 2009). The model has been proved to be capable of analyzing historical urban growth mechanisms and predicting future urban growth in Beijing. In this chapter, we improved the initial raster-based BUDEM model into a vector-based one, V-BUDEM. In the V-BUDEM model, urban space consists of irregular land parcels, and a parcel's neighborhood is defined as all parcels surrounding it within a certain distance. The model adopted a constrained CA method, and considered four constraints: self-status, government, locational, and neighborhood constraints. Fourteen constraint variables were selected for simulating future scenarios. Additionally, a framework of parcel subdivision was adopted in this model to subdivide existing parcels. At this stage, we assume parcel subdivision would be achieved before starting the urban growth simulation, rather than consider a dynamic subdivision process during the simulation. Specifically, the parcel subdivision framework consists of four steps: (1) subdividing existing parcels according to road distribution in the urban plan; (2) subdividing existing parcels according to the boundaries of planned urban built-up parcels; (3) adopting the boundaries of the existing parcels, which belong to planned non-urban built-up parcels; and (4) subdividing existing large non-urban built-up parcels using an automated parcel subdivision tool developed by us. If some small parcels subdivided from an existing large non-urban built-up one in step (4) are not developed after simulation, they will be merged, and the original boundary will be adopted again. In this chapter, after the description of the V-BUDEM, We tested the model to simulate urban growth scenario of Yanqing Town, located in northwestern Beijing, from 2010 to 2020. The Kappa value is 81.16, showing the simulated urban form was fitted well with the planned urban form. Most of the parcels adopting planned urban built-up parcel boundaries were developed into urban built-up land. And some parcels, adopting existing parcel boundaries in 2010 or subdivided from large ones, have been developed into urban built-up land, which would be unlikely to happen if we adopt planned non-urban built-up parcel boundaries directly. The simulated result shows the vector-based CA method is feasible and our parcel subdivision framework is reasonable for simulating urban growth.

The main contributions of this study are as follows: (1) the model adopts a vector-based CA method using land parcels to represent urban space, composing the landscape a user would perceive as meaningful, and can simulate urban growth in a way more close to real world situation; (2) the model integrates a process of parcel subdivision, and the proposed parcel subdivision framework comprehensively considered the impacts of existing parcel boundaries in the existing urban form and planned parcel boundaries in the planned urban form, and especially developed a straight-forward and automatic parcel subdivision tool, included in the framework as the fourth step, to partition existing large parcels; (3) compared with other urban models, V-BUDEM is developed specifically for urban planning applications, e.g., it can be used to identify policies required for implementing the planned urban form desired by planners and decision makers.

In the future, several works can be done to improve V-BUDEM study. At first, the model application would be expanded from Yanqing Town to the whole Beijing metropolitan area, including identifying urban growth mechanisms in various historical phases, retrieving urban policies needed to implement the planned urban form in 2020, and predicting urban growth scenarios in the future (e.g. 2030) using parameter set identified from historical urban forms. Secondly, the fourth step of our parcel subdivision framework would be improved, so as to be an independent automatic subdivision tool, which can be used to speed up parcel subdivision during the process of planning establishment, just like the tools proposed by other literature (e.g. Wickramasuriya et al. 2011). At present, we are developing this tool. Thirdly, the detailed urban form, showing the distributions of commercial, industrial, and residential parcels, can be established using Planner Agents (Zhang and Long 2013; Long and Zhang 2015). And finally, we would calculate transport energy consumption in the established detailed urban form based on Agenter (Long and Shen 2013) and FEE-MAS model (Long et al. 2013). Comparing the energy consumption among different established forms, a low carbon form, creating the least energy consumption, can be identified, which is valuable to support the construction of low carbon cities.

References

Alexandridis, K., Pijanowski, B. C., & Lei, Z. (2007). Assessing multiagent parcelization performance in the MABEL simulation model using Monte Carlo replication experiments. *Environment and Planning B: Planning and Design, 34*, 223–244.

Batty, M., Xie, Y., & Sun, Z. (1999). Modeling urban dynamics through GIS-based cellular automata. *Computers, Environment and Urban Systems, 23*, 205–233.

Beijing Municipal Bureau of Statistics. (2014). Beijing 2013 national economic and social development statistics bulletin. http://www.bjstats.gov.cn/xwgb/tjgb/ndgb/201402/t20140213_267744.htm

Chen, J., & Jiang, J. (2000). An event-based approach to spatio-temporal data modeling in land subdivision systems. *GeoInformatica, 4*, 387–402.

Chen, Q., & Mynett, A. E. (2003). Effects of cell size and configuration in cellular automata based prey-predator modelling. *Simulation Modelling Practice and Theory, 11*, 609–625.

Clarke, K. C., & Gaydos, L. J. (1998). Loose-coupling a cellular automaton model and GIS: Long-term urban growth prediction for San Francisco and Washington/Baltimore. *International Journal of Geographical Information Science, 12*, 699–714.

Cohen, J. (1960). A coefficient of agreement for nominal scales. *Educational and Psychological Measurement, 20*, 37.

Congalton, R., & Mead, R. A. (1983). A quantitative method to test for consistency and correctness in photointerpretation. *Photogrammetric Engineering & Remote Sensing, 49*, 69–74.

Dahal, K. R., & Chow, T. E. (2014). A GIS toolset for automated partitioning of urban lands. *Environmental Modelling & Software, 55*, 222–234.

Demetriou, D. (2014). *The development of an integrated planning and decision support system (IPDSS) for land consolidation*. Berlin: Springer.

Demetriou, D., See, L., & Stillwell, J. (2013). A spatial genetic algorithm for automating land partitioning. *International Journal of Geographical Information Science, 27*, 2391–2409.

Easa, S. M. (2008). Unified direct method for land subdivision: Circular sides permitted. *Journal of Surveying Engineering, 134*, 55–60.

ESRI. (2013). Creating shapes from graph networks. http://cehelp.esri.com/help/index.jsp?topic=/com.procedural.cityengine.help/html/manual/is/create/block_params.html

Guan, Q., Wang, L., & Clarke, K. C. (2005). An artificial-neural-network-based constrained CA model for simulating urban growth. *Cartography and Geographic Information Science, 32*, 369–380.

Hammam, Y., Moore, A., & Whigham, P. (2007). The dynamic geometry of geographical vector agents. *Computers, Environment and Urban Systems, 31*, 502–519.

Hu, S. Y., Li, D. R. (2004). Vector cellular automata based geographical entity. In: *Proceedings of the 12th International Conference on Geoinformatics – Geospatial Information Research: Bridging the Pacific and Atlantic*, University of Gävle, Sweden, 7–9 Jun 2004, *Geoinformatics*, pp. 249–256

Jantz, C. A., & Goetz, S. J. (2005). Analysis of scale dependencies in an urban land-use-change model. *International Journal of Geographical Information Science, 19*, 217–241.

Jenerette, G. D., & Wu, J. (2001). Analysis and simulation of land-use change in the central Arizona–Phoenix region, USA. *Landscape Ecology, 16*, 611–626.

Jjumba, A., & Dragićević, S. (2012). High resolution urban land-use change modeling: Agent iCity approach. *Applied Spatial Analysis and Policy, 5*, 291–315.

Ko, D. W., He, H. S., & Larsen, D. R. (2006). Simulating private land ownership fragmentation in the Missouri Ozarks, USA. *Landscape Ecology, 21*, 671–686.

Landis, J. D. (1995). Imagining land use futures: Applying the California urban futures model. *Journal of the American Planning Association, 61*, 438–457.

Liu, Y. (2012). Modelling sustainable urban growth in a rapidly urbanising region using a fuzzy-constrained cellular automata approach. *International Journal of Geographical Information Science, 26*, 151–167.

Long, Y., & Shen, Z. (2013). Disaggregating heterogeneous agent attributes and location. *Computers, Environment and Urban Systems, 42*, 14–25.

Long, Y., & Zhang, Y. (2015). Land-use pattern scenario analysis using planner agents. *Environment and Planning B, 42*(4), 615–637.

Long, Y., Mao, Q., & Dang, A. (2009). Beijing urban development model: Urban growth analysis and simulation. *Tsinghua Science & Technology, 14*, 782–794.

Long, Y., Shen, Z., & Mao, Q. (2012). Retrieving spatial policy parameters from an alternative plan using constrained cellular automata and regionalized sensitivity analysis. *Environment and Planning B: Planning and Design, 39*, 586–604.

Long, Y., He, Y., Liu, X., & Du, L. Q. (2006). Planning of the controlled-construction area in Beijing: Establishing urban expansion boundary. *City Planning Review, 30*(12), 20–26 (In Chinese).

Long, Y., Han, H., Lai, S.-K., & Mao, Q. (2013a). Urban growth boundaries of the Beijing Metropolitan Area: Comparison of simulation and artwork. *Cities, 31*, 337–348.

Long, Y., Mao, Q., & Shen, Z. (2013b). Urban form, transportation energy consumption, and environment impact integrated simulation: A multi-agent model. In Z. Shen (Ed.), *Spatial planning and sustainable development* (pp. 227–247). Berlin: Springer.

Marceau, D. J., Wang, F., & Wijesekara, N. (2013). Investigating land-use dynamics at the periphery of a fast-growing city with cellular automata at two spatial scales. In *Modeling of land-use and ecological dynamics* (pp. 51–79). Berlin: Springer.

Moreno, N., Ménard, A., & Marceau, D. J. (2008). VecGCA: A vector-based geographic cellular automata model allowing geometric transformations of objects. *Environment and Planning B: Planning and Design, 35*(4), 647–665.

Moreno, N., Wang, F., & Marceau, D. J. (2009). Implementation of a dynamic neighborhood in a land-use vector-based cellular automata model. *Computers, Environment and Urban Systems, 33*, 44–54.

O'Sullivan, D. (2001a). Graph-cellular automata: A generalised discrete urban and regional model. *Environment and Planning B: Planning and Design, 28*, 687–706.

O'Sullivan, D. (2001b). Exploring spatial process dynamics using irregular cellular automaton models. *Geographical Analysis, 33*, 1–18.

Pinto, N. N., & Antunes, A. P. (2010). A cellular automata model based on irregular cells: Application to small urban areas. *Environment and Planning B: Planning and Design, 37*, 1095–1114.

Semboloni, F. (2000). The growth of an urban cluster into a dynamic self-modifying spatial pattern. *Environment and Planning B: Planning and Design, 27*, 549–564.

Shafizadeh Moghadam, H., & Helbich, M. (2013). Spatiotemporal urbanization processes in the megacity of Mumbai, India: A Markov chains-cellular automata urban growth model. *Applied Geography, 40*, 140–149.

Shen, Z., Kawakami, M., & Kawamura, I. (2009). Geosimulation model using geographic automata for simulating land-use patterns in urban partitions. *Environment and Planning B: Planning and Design, 36*, 802–823.

Shi, W., & Pang, M. Y. C. (2000). Development of Voronoi-based cellular automata-an integrated dynamic model for Geographical Information Systems. *International Journal of Geographical Information Science, 14*, 455–474.

Stevens, D., & Dragicevic, S. (2007). A GIS-based irregular cellular automata model of land-use change. *Environment and Planning B: Planning and Design, 34*, 708–724.

Torrens, P. M., & Benenson, I. (2005). Geographic automata systems. *International Journal of Geographical Information Science, 19*, 385–412.

Vanegas, C. A., Aliaga, D. G., Benes, B. et al. (2009a). Visualization of simulated urban spaces: Inferring parameterized generation of streets, parcels, and aerial imagery. *IEEE Transactions on Visualization and Computer Graphics, 15*, 424–435.

Vanegas, C. A., Aliaga, D. G., Beneš, B., et al. (2009b). Interactive design of urban spaces using geometrical and behavioral modeling. In *ACM Transactions on Graphics (TOG)* (vol. 28, p. 111). New York: ACM.

Vanegas, C. A., Kelly, T., Weber, B., Halatsch, J., Aliaga, D. G., & Müller, P. (2012). Procedural generation of parcels in urban modeling. *Computer Graphics Forum, 31*(2pt3), 681–690. Blackwell Publishing Ltd.

Wakchaure, A. S. (2001). *An ArcView tool for simulating land subdivision for build out analysis*. Blacksburg: Virginia Polytechnic Institute and State University.
Wang, F., & Marceau, D. J. (2013). A patch-based cellular automaton for simulating land-use changes at fine spatial resolution. *Transactions in GIS, 17*, 828–846.
Ward, D. P., Murray, A. T., & Phinn, S. R. (1999). An optimized cellular automata approach for sustainable urban development in rapidly urbanizing regions. *International Journal of Geographical Information Science, 7*, 235–250.
White, R., & Engelen, G. (2000). High-resolution integrated modelling of the spatial dynamics of urban and regional systems. *Computers, Environment and Urban Systems, 24*, 383–400.
Wickramasuriya, R., Chisholm, L. A., Puotinen, M., et al. (2011). An automated land subdivision tool for urban and regional planning: Concepts, implementation and testing. *Environmental Modelling & Software, 26*, 1675–1684.
Wickramasuriya, R., Chisholm, L. A., Puotinen, M., et al. (2013). A method to dynamically subdivide parcels in land use change models. *International Journal of Geographical Information Science, 27*, 1497–1513.
Yanqing Government. (2012). Population statistics. http://www.bjyq.gov.cn/sy/yqgk/rkzk/
Zhang, Y. P., & Long, Y. (2013). *Land use pattern scenario analysis using Planner Agents: A preliminary study*. In 13th international conference on computers in urban planning and urban management, Utrecht, 2–5 July.

Part II
Urban Form: Human Behaviour and Their Spatial Patterns

Chapter 6
Population Spatialization and Synthesis with Open Data

6.1 Introduction

Spatial distribution of population and their socioeconomic attributes are essential to feed micro-level applied urban models (such as spatial micro-simulation and agent-based modeling) for policy evaluation. Population spatialization is to allocate population on a finer scale using public available materials, and population synthesis is to synthesize population attributes using known information. A number of studies have been conducted to generate synthetic individual data by reweighting large-scale surveys (Wu et al. 2008). In developing countries like China, population distribution on a fine scale, as the input for population synthesis, is not universally available. Neither are large-scale surveys for population synthesis. China is also facing an institution-induced digital divide: a gap between data availability and official open datasets, as the government exercises tight control of official data. This situation is common in developing countries like Southeast Asian countries, South American countries, and African countries (Tatem and Linard 2011). We aim to mitigate this gap by illustrating how the collection, analysis, and visualization of big (open) data can open up new avenues for synthesizing micro data in developing countries. In this chapter, we prepare an alternative solution for population spatialization and synthesis simultaneously using open data, in the condition of which researchers are difficult to ask for the dataset of high-resolution population distribution and large-scale surveys.

The existing studies on population spatialization and synthesis are generally separated. There have been extensive studies on mapping population distribution (Langford and Unwin 1994; Mennis 2003; Liao et al. 2010). The most common method used is interpolating population density with spatial factors and population census data. Most of the spatialized products, from statistic aggregated in administrative units (Linard et al. 2012) or nighttime satellite imagery (Sutton 1997; Lo 2001), are associated with a coarse spatial resolution scale, which is not detailed

Y. Long, Z. Shen, *Geospatial Analysis to Support Urban Planning in Beijing*,
GeoJournal Library 116, DOI 10.1007/978-3-319-19342-7_6

enough for micro-level urban models (for instance, 1-km population grids for China by Chinese Academy of Sciences (CAS), 5 km-grid UNEP/GRID for several continents, 2.5′ grids GPW/GRUMP for the world, 1 km-grid Land Scan for the world). With the emerging techniques of high-resolution remote sensing images and datasets for individual buildings, several studies generate population distribution on a fine scale, e.g. 100 m or building level (Silvan et al. 2010; Azar et al. 2013; Silva et al. 2013). However, these methods could be time-consuming, expensive, and labor-intensive, which makes it difficult for developing countries to conduct such studies. Most of the studies are for population at nighttime, with an exception of Bhaduri et al. (2007) for both daytime and nighttime population distribution. In addition, these studies did not consider the disaggregation of population attributes.

Population synthesizing techniques have been developed for disaggregating population attributes, based on known population distribution. Reweighting and synthetic construction are the two major approaches for population synthesis (Hermes and Poulsen 2012). Müller and Axhausen (2010) reviewed a list of population synthesizers, including PopSynWin, ILUTE, FSUMTS, CEMDAP, ALBATROSS, and PopGen. Iterative proportional fitting (IPF) adopted by PopGen, a typical reweighting method, can adjust data cells of a table for both the columns and rows (in two-dimensional cases). The unadjusted data cells are referred as seed cells, and the related totals are referred as marginal totals. Synthetic construction can generate micro dataset with only aggregated information. This approach does not require individual samples. For instance, Barthelemy and Toint (2013) produced a synthetic population for Belgium at the municipality level without a sample. Long and Shen (2013) synthesized individuals with aggregated data, empirical studies and common sense for Beijing. Therefore, the approach of synthetic construction is more appropriate for applications in the research project with the data-sparse issue.

Now, the emerging trend of open and big data provides opportunities for population spatialization and synthesis in developing countries. As one of the most successful volunteered GIS projects, OpenStreetMap (OSM) in developing countries has been encouraging, as the volume of OSM data in China has experienced a ninefold increase during 2007–2013 (Long and Liu 2013). OSM has been proposed as a promising candidate for a quick and robust delineation of parcels (Haklay and Weber 2008), thus providing basic spatial units for allocating population on a fine scale. Points of interest (POIs) available in most of online mapping services are also promising for identifying residence related places (Long and Liu 2013). The coupling of OSM and POIs would be an alternative for mapping population on a fine scale. In addition, as we have reviewed, the sub-district census data can be used to feed synthetic construction method for population synthesis.

In this chapter, we propose an automatic process using open data for population spatialization and synthesis. Specifically, the road network in OSM is used to identify and delineate parcel geometries, while crowd-sourced POIs are gathered to infer urban parcels with a vector cellular automata model referring to our previous study (Long and Shen 2014). Housing-related online check-in records or POIs are then applied for selecting residential parcels from all identified urban parcels. Finally the published population census data at the level of sub-district, in which distribution of attributes and relationships between attributes can be retrieved, is used for synthesizing population attributes to parcels supported by a previously developed tool Agenter (Long and Shen 2013). We focus on urban residents (spatial distribution at night time) in this study and rural ones would be reserved in our future research, since urban residents are the majority of the population in Beijing. In this chapter, the study area and data are described in Sect. 6.2. The applied methodologies are elaborated in Sect. 6.3. We discuss the results and validations in Sects. 6.4 and 6.5, respectively. The concluding remarks are in Sect. 6.6.

6.2 Study Area and Data

6.2.1 Study Area

As the capital of China, the Beijing Metropolitan Area (BMA) with a coverage of 16,410 km^2 has over 20 million residents in 2010 and is becoming one of the most populous cities in the world. The BMA lies in northern China, to the east of the Shanxi altiplano and south of the Inner Mongolian altiplano. The southeastern part of the BMA is a plain, extending eastward for 150 km to the Bohai Sea. Mountains cover an area of 10,072 km^2, 61 % of the whole study area (Fig. 6.1).

According to Beijing Municipal Bureau of Statistics and NBS Survey Office in Beijing (2013), the total urban residents of BMA in 2012 were 17.837 million. According to the land use dataset of Beijing Institute of City Planning, the total urban area of BMA in 2012 was 1,674.9 km^2, with 419.2 km^2 residential areas.

6.2.2 The OSM Road Networks of Beijing

We downloaded OSM road networks for Beijing on October 5, 2013. The OSM dataset contains 43,006 road segments, in a total of 20,904 km (Fig. 6.2). We also collected the 2012 ordnance survey map of Beijing with detailed road networks to verify the results produced by OSM data. OSM data quality is promising, especially in large cities like Beijing as we found in 2013 (Long and Liu 2013).

The city center and main roads are also extracted from the OSM data. We will use them as the spatial features to identify urban parcels for mapping population.

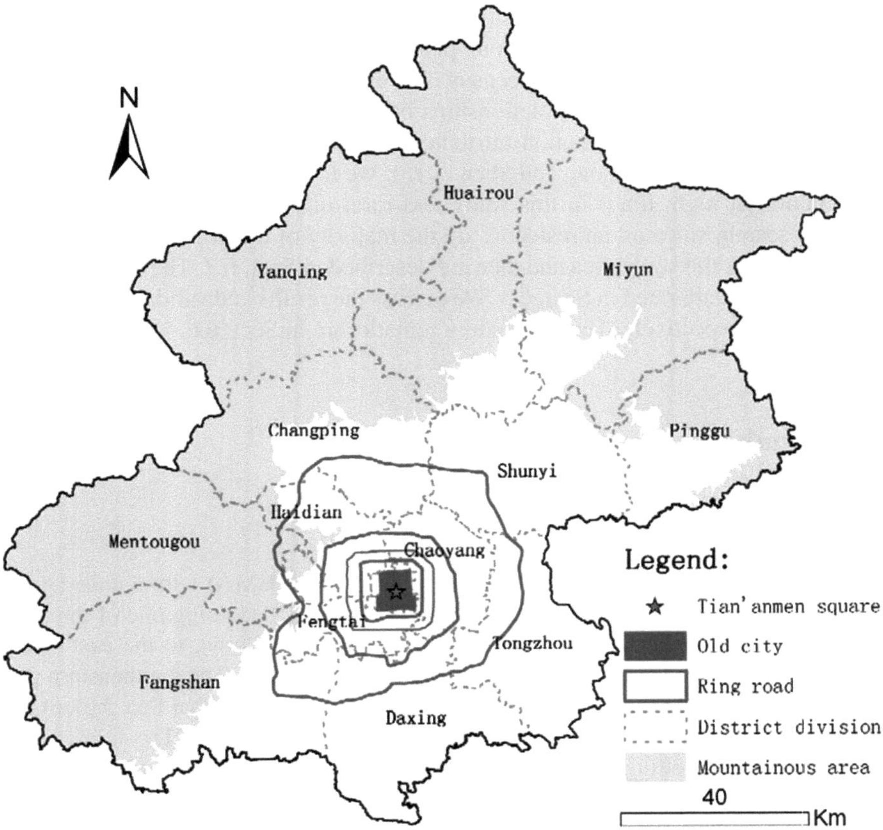

Fig. 6.1 The Beijing Metropolitan Area

6.2.3 *POIs*

A total of 186,437 POIs are gathered and geo-coded by business cataloguing websites, among which 31,189 are residential POIs. The initial 20 POI types are aggregated into eight more general assemblies, including residential communities (RES), commercial sites (COM), business establishments (FIR), transportation facilities (TRA), government buildings (GOV), educational sites (EDU), green space (GRE) and others (OTH). The quality of data is guaranteed through manually checking on the randomly selected POIs. The total amount of POIs associated with each parcel is used for inferring the parcel's density and the residential POIs amount for mapping population (Fig. 6.3).

Fig. 6.2 The OSM road network of Beijing in 2013. (**a**) The whole BMA, (**b**) a part of the central city of Beijing

Fig. 6.3 Spatial distribution of POIs in Beijing by type (in a local area of Beijing)

6.2.4 The 2010 Population Census of Beijing

The main data source for population synthesis in Beijing is the Sixth Population Census Report of the BMA conducted in 2010 (the census, detailed at http://www.bjstats.gov.cn/rkpc_6/pcsj/201105/t20110506_201581.htm). The census was conducted at

Table 6.1 Descriptions and known information for each attribute of residential agents in the BMA

Name	Description	Type	Known information	Data source	Data type	Order
AGE	Age in years	Non-spatial attribute	Frequencies	The census	Ratio	1
SEX	Gender	Non-spatial attribute	Frequencies	The census	Nominal (male, female)	2
MARRIAGE	Marital status	Non-spatial attribute	Frequencies, relationship with AGE	The census	Nominal (married, unmarried, divorced, remarried, widowed)	3
EDUCATION	Level of education	Non-spatial attribute	Frequencies, relationship with AGE	The census	Ordinal (junior middle school, undergraduate, etc.)	4
JOB	Occupation	Non-spatial attribute	Frequencies, relationship with EDUCATION	The census	Nominal	5
INCOME	Monthly income	Non-spatial attribute	Frequencies	The survey	Ratio	6
FAMILYN	Number of family members	Non-spatial attribute	Frequencies	The census	Ordinal (one person, two person, etc.)	7
PARCEL	ID of parcel at which the agent resides	Location	Frequencies	An empirical study	Nominal	8
TAM	Distance to the city center	Spatial attribute	Location of Tiananmen Square	Urban GIS	Ratio	9

Note: The attribute PARCEL is the ID of the parcel used to map the disaggregated residents

the census tract level,[1] which size is between a city block and sub-district. These data were aggregated from the original census tract level to the sub-district level (BMA has 307 sub-districts). The census data provide both the statistical distributions of attributes and relationships among attributes of residents in the form of tables. Many cross-tabulations for various combinations of the attributes are listed in the census report, and were used in this (Table 6.1). These are the attributes and locations of residential agents to be disaggregated within the BMA. The dependent relationships among attributes are illustrated in Fig. 6.4. Note that we are about to spatialize and synthesize urban residents of Beijing in 2012, and we admit that there is a mismatch between the target year 2012 and the census year 2010.

[1]The spatial distribution of census tracts has never been released by the Beijing Municipal Statistical Bureau. Therefore, it is not possible to determine whether census tracts are compatible with TAZs.

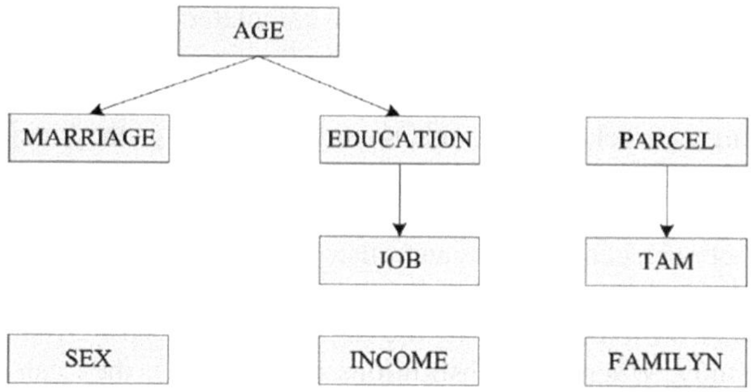

Fig. 6.4 Dependent relationships among attributes of residential agents (*arrows* denote the relationship between two attributes during population synthesis process, e.g. the attribute EDUCATION depends on the attribute AGE)

6.3 Approach

6.3.1 *The Proposed Process*

The process for population spatialization and synthesis is as follow: (1) Generating parcels using road networks (Sect. 6.3.2); (2) Selecting urban parcels using POIs and constraints with vector cellular automata (Sect. 6.3.3); (3) Identifying residential parcels from all urban parcels selected using residential POIs (Sect. 6.3.4); (4) Allocating urban population into identified residential parcels using totals in the sub-district level and residential POI density (Sect. 6.3.5); (5) Synthesizing urban population attributes using the census and empirical research with Agenter (Sect. 6.3.6).

6.3.2 *Generating Parcels*

In this chapter, a parcel is defined as a continuously built-up area bounded by roads. Identifying land parcels and delineating road space are therefore *dual* problems. In other words, our approach begins with the delineation of road space, and individual parcels are formed as polygons bounded by roads.

The delineation of road space and parcels is performed as follows: (1) All OSM road data are merged as line features in a single data layer; (2) individual road segments are trimmed with a threshold of 200 m to remove hanging segments according to our careful detecting the dataset; (3) individual road segments are then extended on both ends for 20 m to connect adjacent but non-connected lines according to our careful detecting the dataset; (4) road space is generated as buffer zones around road networks. A varying threshold ranging between 2 and 30 m is adopted for different road types (e.g. surface condition, as well as different levels of roads); (5) parcels are delineated as the space left when road space is removed; and (6) a final step involving overlaying parcel polygons with administrative boundaries to determine whether individual parcels belong to a certain administrative unit.

Four parameters are further calculated for each parcel. The first two, size and compactness, are determined by the geometric characteristics of each parcel. The third, the accessibility, is taken into consideration as a location variable for describing a parcel. The last one, the urban density, is the functional attribute of a parcel for reflecting its actual use. The numbers of POIs within or close to a parcel are measured as its urban density. Due to the natural unevenness of urban density between central cities and other ones, POIs density is further normalized and placed between 0 and 1 to release the heterogeneity among cities. Because of lacking further attributes of POIs, the popularity of each POI is assumed as the same in this study. When any substitutions are available, they can be expected to approximate the intensity of urban activities explicitly. Hence, in the way of using road network and POIs to describing parcels, the spatial and functional features are incorporated together for further urban parcel selection.

6.3.3 Selecting Urban Parcels

Vector-based constrained cellular automata models are used for picking up urban parcels from the initial ones generated by road network in diverse cities. We suppose this process is similar to that for modelling urban expansion, which sees extensively CA applications. Apart from the conventional raster CA model, vector-based CA model here depends on irregular polygons rather than regular cells. In this research, each parcel is regarded as a cell with a status that is 0 (urban) or 1 (non-urban). This can be illustrated as a formula as the following.

$$S_{ij}^{t+1} = f\left(S_{ij}^{t}, \Omega_{ij}^{t}, Con, N\right) \tag{6.1}$$

Here, a parcel's status at $t + 1$ is considered as a function f of parcel's statues and other factors at t. In this function, S_{ij}^{t} and S_{ij}^{t+1} denote to the statues of parcels at time points of t and $t+1$ respectively; Ω_{ij}^{t} is the neighboring situation; *Con* refers to the constraints and N is the amount of all parcels. This function can be further transferred to a detailed probability formula:

$$P_{ij} = \left(P_{l}\right)_{ij} \times \left(P_{\Omega}\right)_{ij} \times con(\cdot) \times P_{r}. \tag{6.2}$$

In this function, the possibility of transformation of parcel's state at t is illustrated as multiplied product of probabilities of factors. Specifically, $(P_l)_{ij}$ stands for the local potential that a parcel would convert its status from the non-urban to the urban while $(P_\Omega)_{ij}$ denotes the conversion possibility in terms of the neighboring situations; $con(\cdot)$ stands for constraints and P_r is the stochastic term.

The proposed spatial and functional characteristics are reflected in measuring the local potential. This could be explained in the formula below using a logistic regression model:

$$(P_l)_{ij} = \frac{1}{1+\exp\left[-\left(a_0 + \sum_{k=1}^{m} a_k c_k\right)\right]}, \tag{6.3}$$

where a_0 is a constant, a_k is an estimated coefficient responding to the spatial variable c_k and *m* is the total amount of spatial variables. As a result, spatial and functional factors are bonded to reflect the parcel's status in this study. Parcel size is measured in the natural logarithm of area. Compactness is calculated as perimeter square subdivided by area. Accessibility is abstracted as the minimum Euclidian distance to the city center. On the other hand, the functional factor is presented by applying the standardized POIs density, which is calculated as the rate of raw density in the max density among the samples.

The neighboring potential for a parcel is measured by the amount of peripheral urban parcels around it. This can be defined as:

$$(P_\Omega)_{ij} = \frac{\sum con\left(S_{ij}^{\ t} = urban\right)}{n}. \tag{6.4}$$

For parcel *ij*, $\mathrm{con}\left(\mathrm{S}_{\mathrm{ij}}^{\mathrm{t}} = \mathrm{urban}\right)$ stands for the urban parcels within fixed areas while *n* is the sum of all accessible parcels. The adjacent relation is defined as 500 m around the parcel *ij* according to authors' experience.

Two layers – the steep area (a slope over 25°) and various water bodies, are included as the constraints. Urban expansion is forbidden in these areas. The constraints are expressed as $\mathrm{con}\left(\mathrm{cell}_{\mathrm{ij}}^{\mathrm{t}} = \mathrm{suitable}\right)$ with a value of 0 or 1, where 1 indicates that there is no restriction on the parcel's development as urban while 0 indicates that the parcel is forbidden to be urban.

The stochastic disturbance P_r in the model stands for any possible change of local policies and accidental errors. It is calculated using

$$P_r = 1 + (-\ln\gamma)^{\beta} \tag{6.5}$$

where γ is a random number ranging from 0 to 1, and β, ranging from 0 to 10, controls the effect of the stochastic factor.

Furthermore, by comparing the measured probability $(P_l)_{ij}$ with a fixed threshold value P_{thd}, the parcel's status at $t+1$ could be detected. If the measured value is greater than the threshold, the parcel is considered to be urban, if not, the parcel will stay as non-urban. This progress can also be presented as a binary expression:

$$S_{ij}^{t+1} = \left\langle \frac{Urban \text{ for } P_{ij}^{t} \succ P_{thd}}{NonUrban \text{ for } P_{ij}^{t} \leq P_{thd}} \right. . \tag{6.6}$$

Finally, for controlling the total area of all urban parcels, the urban area in 2012 is applied as the upper limits for the total area of selected urban parcels.

6.3.4 Identifying Residential Parcels

We define residential density as the ratio between the counts of residential POIs in a parcel to the parcel area.[2] We further standardized the residential density to range from 0 to 1 using the following equation: standardized residential density = log(*raw*)/log(*max*), where *raw* and *max* correspond to density of individual parcels and the city-wide maximum residential density value.[3] We rank all selected urban parcels in terms of the calculated residential density. Residential parcels are identified via benchmarking the density with the residential area as the total control.

6.3.5 Allocating Urban Population

For each identified residential parcel, we assume its population is proportional to its inferred residential POI density. In each sub-district of Beijing, we then allocate the related number of population into residential parcels.

6.3.6 Synthesizing Population Attributes Using Agenter

Population attributes are divided into two types, non-spatial attributes (such as age, income, and education for a residential agent) and spatial attributes (such as access to subways and amenities, land use, and height of the building that the residential agent occupies). The approach to disaggregate spatial attributes differs from that used for non-spatial attributes.

The probability distribution of an attribute (hereafter referred to as the distribution) and the dependent relationship among attributes (hereafter referred to as the relationship) can be inferred from existing data sources, including aggregate data, small-scale surveys and empirical studies. Aggregate data include the total number, distribution and relationship (such as the cross-tabulation of marriage-age standing for the dependent relationship of marriage and age, and the cross-tabulation of income-education standing for the dependent relationship of income and education) of agents. Small-scale surveys that store samples can also be used to deduce the distribution of an attribute and the relationships among attributes. The probability distribution of an attribute and its relationship with other attributes can also be deduced using empirical studies. To convert aggregate data to individual samples, the probability distribution of the attributes and the relationship between them must

[2] Residential POIs within the buffered road space were accounted by their closest parcels in our experiment.

[3] The unit is the POI count per km^2. For parcels with no residential POIs, we assume a minimum density of 1 POI per km^2.

be estimated. For more information, Long and Shen (2013) discussed this methodology of population synthesis carefully in the previous work.

6.3.7 Model Validation

According to our methods and research questions, validating our proposed framework includes two steps. The first is to validate population spatialization results and the second is to validate population synthesis results.

First, the selected residential parcels are compared with manually prepared data by planners in Beijing Institute of City Planning. The inferred population of each residential parcel is then correlated with each parcel's observed population which we use total floor space of buildings within the parcel using building footprints and the floor number to proxy.

Second, the synthesized population of Beijing is validated by calculating the similarity between disaggregated and observed dataset (using the 2010 Beijing household travel survey). We used the similarity indicator (SI) proposed by Long and Shen (2013) for comparison. The similarity index SI is 100 % if two sets have the same attribute values. To calculate SI, both sets must be sorted by the same rule. The location attribute should be sorted first, followed by the other attributes in increasing order. It is also necessary to disaggregate the same number of population as observed samples.

6.4 Results

6.4.1 Population Spatialization

We generated 52,359 parcels delineated by road networks in the BMA. To select urban parcels from all parcels, logistic regression is conducted for calibrating the weights for constraint factors in the proposed vector CA model. The 2010 parcel dataset in Beijing City, which is manually prepared by urban planners in BICP, is applied to identify rules for inferring residential parcels. It covers an area of 12,183 km^2 at a detailed urban parcel scale (Yanqing and Miyun counties in the Beijing Metropolitan Area are excluded from Beijing City). There are in total 52,330 parcels reported, among which 36,914 are identified as urban parcels.

According to the result of binary logistic regression (Table 6.2), 78.9 % of all parcels can be explained by identified rules. And all factors except compactness have passed *p* test, revealing that they are significantly related to the differences between non-urban and urban ones. By using the logistic regression results, we selected 43,180 urban parcels. In order to test the accuracy of this model, the results of Beijing City with the CA model was compared with the BICP dataset again, and then an overall accuracy of 81.5 % indicates the applicability of our model in delineating urban areas in terms of urban parcels.

Table 6.2 Binary logistic regression results for BICP parcels

Name	Coefficient	S.E.	Sig.
Constant	5.359	0.058	0.000
Natural logarithm of parcel size	−0.306	0.006	0.000
Distance to the city center	−0.099	0.001	0.000
POIs density	3.431	0.085	0.000

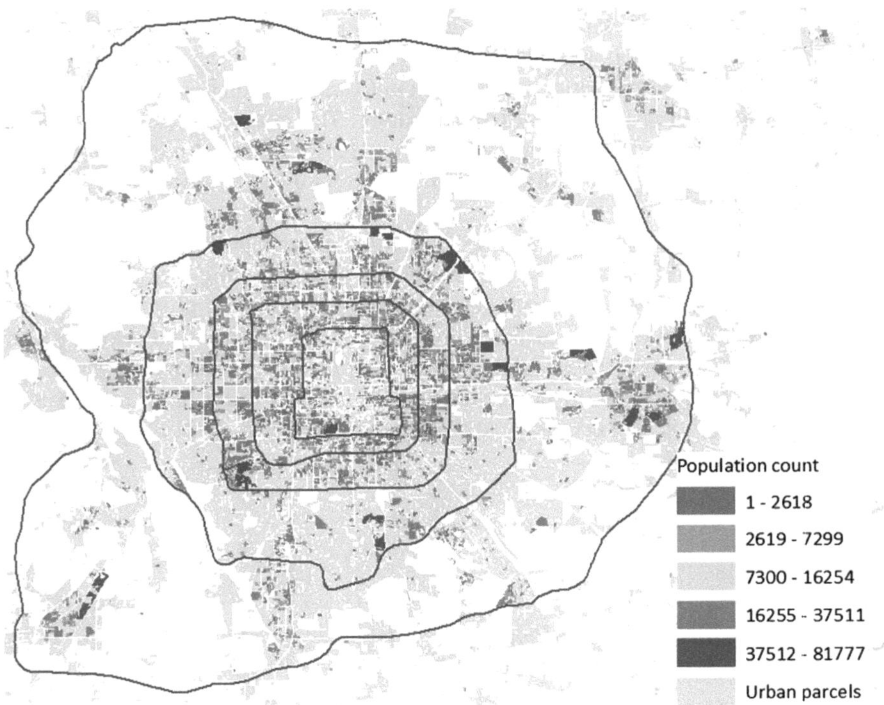

Fig. 6.5 Selected residential parcels and their population count

We then selected 7,161 residential parcels from all selected urban parcels and allocated urban residents on them (see Fig. 6.5).

6.4.2 *Population Synthesis*

The synthesized urban residents are stored as the point Feature Class in the ESRI Personal Geodatabase (Table 6.3). This dataset embeds both the attributes and location information of residential agents, which can be regarded as a primary dataset for urban studies and initializing agent-based models.

Table 6.3 Disaggregated residential agents (partial) in the BMA

AID	AGE	SEX	MARRIAGE	EDUCATION	JOB	INCOME	FAMILYN	PARCEL	TAM
193392	36	Male	Married	Junior high/middle school	Production, transport equipment operator, and related	2,385	Three persons	888	2140
198316	41	Female	Married	High school	Production, transport equipment operator, and related	5,966	Three persons	966	7747
37094	61	Male	Married	High school	Professional technology employee	4,744	Three persons	523	5721
165014	27	Male	Unmarried	High school	Business and service employees	5,559	Five persons	768	4957
344565	41	Female	Married	Elementary school	Production, transport equipment operator and related	5,351	Three persons	18	36739
49808	21	Male	Unmarried	Junior high/middle school	Business and service employees	2,684	Five persons	274	2905
189128	21	Male	Married	Junior high/middle school	Production, transport equipment operator and related	2,578	One person	878	4092
118806	8	Male	Unmarried	Elementary school	Production, transport equipment operator and related	0	Three persons	478	6949
33570	53	Female	Married	Elementary school	Production, transport equipment operator and related	1,304	Five persons	929	23760
179469	50	Male	Married	Elementary school	Farming, forestry, animal husbandry and fishery	4,978	Two persons	804	2286

Table 6.4 Comparison of selected urban parcels in BICP and OSM in Beijing (R = ring road)

Parcels	Parcel count	Average size (ha)	Overlapped with BICP	Spatial distribution (in terms of area, km^2)					
				Within R2	R2–R3	R3–R4	R4–R5	R5–R6	Beyond R6
OSM	7,130	17.2	1,194.2 km^2 (71.2 %)	42.5	74.0	113.4	263.5	666.5	519.9
BICP	57,818	2.9	–	48.6	69.7	99.8	229.5	687.9	544.4
OSM/BICP	0.12	5.93	–	0.87	1.06	1.14	1.15	0.97	0.95

Note: the comparison is only for the BMA, where we have both datasets (Miyun and Yanqing counties not included)

6.5 Validation

6.5.1 *Validating Residential Parcels with Ground Truth from BICP*

Table 6.4 shows a comparison between urban parcels generated by our approach and those contained in the BICP Beijing parcel data. It suggests that OSM-based approach generally produce larger parcels, due to the lack of information about tertiary and detailed roads in the OSM dataset.[4] Nevertheless, the two results are 71.2 % matched with each other, suggesting that they capture similar information on the geographic distribution of urban parcels and land use activities. In addition, we decompose the city of Beijing into sub-regions bounded by major ring roads, and calculate the proportion of parcels falling into individual sub-regions. The proportion of parcels falling into sub-regions between ring roads is consistent across both datasets.

We overlaid residential parcels generated by our approach and BICP, and the overlapping area is 211.5 km^2 (56.3 % out of total 375.6 km^2 OSM-based residential parcels in Beijing City). In other words, the parcel level validation suggests that, despite only using online open data, our OSM-based approach could conduct reasonably good approximations of data produced by and the conventional manual method.

6.5.2 *Validating Population Density with Buildings*

We computed the total floor space for each identified residential parcel using available floor space information in 2010. Here, the floor space is used as a proxy of development density and population density. The Pearson correlation coefficient between floor spaces and inferred population numbers is 0.858 for all residential parcels, suggesting that ubiquitously available POI data could be used as a proxy for population density.

[4] Parcels by ORDNANCE in Beijing were similar with those by planners in BICP in terms of parcel size.

6.5.3 *Validating Population Attributes*

More details on validating population attributes are available in the study by Long and Shen (2013), in which we validated our proposed Agenter model with a household travel survey with 208,291 individuals. The average similarity index of Agenter is 72.6 %, which is significantly greater than that of the null model (43.9 %), indicating that Agenter generates sounder synthesized urban residents.

6.6 Conclusions

Aiming at the paucity of fine-scale population density and their attributes in cities of the developing world, our study proposes a novel and scalable empirical framework for automatic population spatialization and synthesis using ubiquitously available OSM, POIs and censuses. Consecutive steps are embedded in the framework. Our analysis represents an attempt to use open data and combine the two procedures (spatialization and synthesis) that are separated in existing literature. Empirical results also suggest that open data could help produce reasonable population data.

The contribution of this chapter lies in the following aspects: Firstly, we proposed a robust and straightforward approach to delineating parcels, identifying urban parcels, selecting residential parcels, allocating urban population and synthesizing population attributes. Secondly, we employ a novel approach that incorporates a vector-based cellular automata model with the identification of urban parcels, which are associated with a fine spatial scale. Thirdly, considering open data are fast emerging data source, our approach has its potential to be applied to the whole country of China, although here it is developed for Beijing. Our project is also part of the Open Data Initiative, as all our data products will be free and available from the Internet (Beijing City Lab, www.beijingcitylab.com).

The final product of our project is a dataset containing fine scale residents data and their attributes for the BMA. This dataset can be applied to, but not limited to, the following three aspects: first, the parcel-level population density dataset is possible to be referred by urban planners and researchers as an important base map. For instance, issues like quality-of-life, air pollution exposure and population-based urban agglomeration can be estimated by using the dataset. Secondly, as the dataset we generated contain both population distribution and their attributes, it can serve as the direct input, in the form of spatial agents, for emerging agent-based models, which applies coarser dataset as inputs before in the data-sparse environment in China. Lastly, the results have its potential applications in market analysis (e.g. evaluating potential market for retail within the catchment).

With some general limitations of using open data in studying urban dynamics (Liu et al. 2014; Sun et al. 2013), we will conclude with limitations and possible future research avenues that are specific to our population spatialization and synthesis framework. One of the limitations of our approach is that OSM road networks are relatively sparse in many cities (although good enough in our experiment city Beijing) and this will lead to unrealistic large urban parcels (spatial resolution of our study). This should be noted when extending the framework from Beijing to other cities in China or other developing countries. The deficiency of open data is likely to be alleviated by the ever-increasing coverage and quality of OSM data in China. And if possible, more land use/cover datasets should be included in selecting urban parcels out of all delineated parcels. This is expected to increase the precision of selected urban parcels and residential ones.

Acknowledgements We thank Ms Liqun Chen for her proofreading.

References

Azar, D., Ryan, E., Jordan, G., et al. (2013). Generation of fine-scale population layers using multi-resolution satellite imagery and geospatial data. *Remote Sensing of Environment, 130*, 219–232.

Barthelemy, J., & Toint, P. L. (2013). Synthetic population generation without a sample. *Transportation Science, 47*(2), 266–279.

Beijing Municipal Bureau of Statistics, NBS Survey Office in Beijing. (2013). *Beijing statistical yearbook 2013*. Beijing: China Statistic Press.

Bhaduri, B., Bright, E., Coleman, P., & Urban, M. L. (2007). LandScan USA: A high-resolution geospatial and temporal modeling approach for population distribution and dynamics. *GeoJournal, 69*(1–2), 103–117.

Haklay, M., & Weber, P. (2008). OpenStreetMap: User-generated street maps. *IEEE Pervasive Computing, 7*(4), 12–18.

Hermes, K., & Poulsen, M. (2012). A review of current methods to generate synthetic spatial microdata using reweighting and future directions. *Computers, Environment and Urban Systems, 36*(4), 281–290.

Langford, M., & Unwin, D. J. (1994). Generating and mapping population density surfaces within a geographical information system. *The Cartographic Journal, 31*, 21–26.

Liao, Y., Wang, J., Meng, B., & Li, X. (2010). Integration of GP and GA for mapping population distribution. *International Journal of Geographical Information Science, 24*, 47–67.

Linard, C., Gilbert, M., Snow, R. W., et al. (2012). Population distribution, settlement patterns and accessibility across Africa in 2010. *PloS One, 7*(2), e31743.

Liu, Y., Sui, Z., Kang, C., & Gao, Y. (2014). Uncovering patterns of inter-urban trip and spatial interaction from social media check-in data. *PloS One, 9*(1), e86026. doi:10.1371/journal.pone.0086026.

Lo, C. P. (2001). Modeling the population of China using DMSP operational linescan system nighttime data. *Photogrammetric Engineering & Remote Sensing, 67*(9), 1037–1047.

Long, Y., & Liu, X. (2013). Automated identification and characterization of parcels (AICP) with OpenStreetMap and points of interest. arXiv preprint.arXiv:1311.6165.

Long, Y., & Shen, Z. (2013). Disaggregating heterogeneous agent attributes and location from aggregated data, small-scale surveys and empirical researches. *Computers, Environment and Urban Systems, 42*, 14–25.

Long, Y., & Shen, Y. (2014). Mapping parcel-level urban areas for a large geographical area. arXiv preprint.arXiv:1403.5864.

Mennis, J. (2003). Generating surface models of population using dasymetric mapping. *The Professional Geographer, 55*, 31–42.

Müller, K., & Axhausen, K. W. (2010). *Population synthesis for microsimulation: State of the art.* The 10th Swiss transport research conference. Zurich: ETH.

Silva, F. B., Gallego, J., & Lavalle, C. (2013). A high-resolution population grid map for Europe. *Journal of Maps, 9*(1), 1–13.

Silvan, J., Wang, L., Rogerson, P., et al. (2010). Assessing fine spatial resolution remote sensing for small area population estimation. *International Journal of Remote Sensing, 31*(21), 5605–5634.

Sun, L., Axhausen, K. W., Lee, D. H., & Huang, X. (2013). Understanding metropolitan patterns of daily encounters. *PNAS, 110*(34), 13774–13779.

Sutton, P. (1997). Modeling population density with night-time satellite imagery and GIS. *Computers, Environment and Urban Systems, 21*(3), 227–244.

Tatem, A. J., & Linard, C. (2011). Population mapping of poor countries. *Nature, 474*, 36.

Wu, B. M., Birkin, M. H., & Rees, P. H. (2008). A spatial microsimulation model with student agents. *Computers, Environment and Urban Systems, 32*, 440–453.

Chapter 7
Spatially Heterogeneous Impact of Urban Form on Human Mobility: Evidence from Analysis of TAZ and Individual Scales in Beijing

7.1 Introduction and Background

This chapter aims to quantitatively measure the impact of urban form on human mobility in Beijing Metropolitan Area, and to answer the question whether the impact is spatially heterogeneous so as to shed light on policy implications for decision makers in Beijing as well as other studies in relevant fields. According to the literature, three types of factors have been proved to influence human mobility, including urban form (e.g. land use characteristics), transportation system characteristics (e.g. accessibility, convenience and service quality), and socioeconomic attributes of individual or family. In this study, all these three aspects are addressed and the spatially heterogeneous impact of each element in urban form system is measured explicitly by controlling other factors in the statistical models.

Urban form is not only the immediate outcome of spatial plans but also the core elements affecting urban sustainability. As an important determinant for energy conservation and low carbon economy, it can impact the sustainable development of an urban system at the very beginning. Many existing research have empirically indicated that the urban form featuring poly-centers, higher density and mixed use corresponds to lower average transportation energy consumption. For example, Anderson et al. (1996) concluded that urban form has strong relationship with transportation energy consumption including passengers and cargo. Newman and Kenworthy (1989) used many cities as samples and found that average transportation energy consumption apiece decreases with population density. Holden and Norland (2005) found significant relationships between urban form and household & transportation energy consumption via analyzing eight neighborhoods in the greater Oslo region, indicating that the compact city policy corresponds with a sustainable urban form. Alford and Whiteman (2009) evaluated various types of urban form in different sub-regions of the Melbourne area in Australia and found

Y. Long, Z. Shen, *Geospatial Analysis to Support Urban Planning in Beijing*, GeoJournal Library 116, DOI 10.1007/978-3-319-19342-7_7

that areas with higher residential and employment density are more likely to consume transport energy efficiently.

One important linkage between urban form and energy conservation is traveler's commuting behavior and total commuting distance. To measure the impact quantitatively, the activity-based modeling approach is widely applied using travel diaries as the basic dataset. Urban forms at housing and job places, respectively, are used as variables for quantitative evaluation. These empirical studies range from the impact of urban form on travel behavior, mobile travel behavior, and children travel behavior, to pedestrian travel behavior and to non-work travel behavior (Dieleman et al. 2002; Giuliano and Narayan 2003; Horner 2007; Maat and Timmermans 2009; McMillan 2007; Pan et al. 2009; Schlossberg et al. 2006; Zhang 2005). Moreover, Krizek (2003) indicated that the traveling behavior of a family would vary from their living neighborhood. As for land-use characteristics, another essential component of urban form, Lin and Yang (2009) found mixed land use reduces trip generation and indirectly increases the share of private mode ridership. Accordingly, the impacts of urbanization process can be statistically assessed by the relationship between the spatial distribution of commuting trips and those mobility factors associated with other urban problems in question.

Aside from urban form, socioeconomic attributes and transportation system characteristics also influence travel behavior to various extents. Some research believed that travelers' socioeconomic and demographic characteristics exert small or insignificant influence on travel demand (Lin and Yang 2009; Pan et al. 2009). However, other research argued that socio-economic characteristics have an almost equal or even greater influence on travel behavior than urban form. Giuliano and Narayan (2003) indicated that females travel shorter distance than males; children and older people make less daily trips than that by the 'middle' age-groups; and trips increase with income and employment. Research also found that car ownership is an important variable to explain travel choice (Dieleman et al. 2002) and in-home/out-of-home recreation patterns which will affect trip generation (Arun and Ram 2001). For transportation system characteristics, Dieleman et al. (2002) looked into at least 70,000 households and more than 150,000 people through Netherlands National Travel Survey and found that the supply of good public transport clearly reduces car use. Consequently, socioeconomic factors and features of transportation system should be well considered together with the elements of urban form in one model, so as to truly count the roles of urban form in shaping mobility pattern.

The spatial nature of urban form and commuting data implies highly possible spatial dependence with the variables and spatial heterogeneity of associations between the variables. However, most of previous studies using Ordinary Least Square (OLS), a conventional global regression method, failed to consider the presence of spatial dependence and spatial heterogeneity. It is suggested that urban system is far more complex than we expected, which essentially requires the inherent spatial properties to be addressed in urban modeling. Spatial regression models can take two threads of strategies, namely the global models and local models. The global models mainly consider and account for the spatial autocorrelation. One typical model is the Spatial Auto-Regressive model (SAR) introduced by Anselin in 1988. In contrast, the local model focuses on the spatial heterogeneous, or non-stationary, relationships in the data. Geographically Weighted Regression is such an approach dealing with the issue of local variations of spatial associations (Fotheringham et al. 2002). Until now, although most urban mobility models have been established to explore the global relationships between mobility patterns and their determinants, there were some studies in which human mobility models are fitted using local regression approaches (Goetzke 2008; Kawabata et al. 2007; Mulley 2013). Yet, recent studies have recognized that urban data distribution is affected by the global fixed effect and local effect simultaneously. As an extended version of Geographically Weighted Regression (GWR), Mixed Geographically Weighted Regression (MGWR) was proposed to prevent the limitation of capturing purely local or global model by modeling the urban distributions by incorporating the spatial stationary and non-stationary in the same model. One such example is the modeling of urban hedonic price pattern (Wei and Qi 2012). However, very few studies adopted this model to explore the multi-scaled relationships between urban commuting and its determinants.

The focus of this work is to explore spatial heterogeneity of urban form's impact on mobility. We choose the mixed-scaled regression techniques in order to unfold the spatial complexity in mobility modeling that is usually hidden when applying the global regression method. We are particularly interested in exploring the answers to the following questions:

1. What kinds of urban form factors are suitable predictive variables for modeling the spatial heterogeneity of urban mobility?
2. Is MGWR model a better-specified model with proper technical corrections in comparison with other standard models?
3. To what extent does urban form influence human mobility in consideration of spatial variation of the influence?

The rest of this chapter is organized as follows. Section 7.2 discusses the methods to account for spatial heterogeneity and introduces the proposed empirical model. Section 7.3 reports the datasets used for evaluating urban form and human mobility. At the next step, empirical results in Beijing Metropolitan Area are presented and summarized in Sect. 7.4, before Sect. 7.5 highlights the final remarks.

7.2 Methodology

7.2.1 Modeling Spatial Effects in Urban Mobility

Most of mobility models are expressed in a traditionally standard linear regression model, in which the traits of mobility, e.g. the travel distance, are regressed on a series of structural, socioeconomic and traffic characteristics. However, these traditional econometrics have largely ignored spatial dependence and spatial heterogeneity that violate the traditional Gauss-Markov assumptions used in regression modeling (Anselin 1988). Due to the nature of spatial dependence which is widely observed in urban data, spatial regression techniques are preferred over the traditional OLS model. Spatial regression modeling is developed to address spatial autocorrelation and/or heterogeneity at the same time. Anselin (1988) introduced the most widely applied spatial regression model called Spatial Auto-Regressive model (SAR), in which spatial dependence can be incorporated in two distinct ways: as an additional operator in the form of a spatially lagged dependent variable, or in the error structure. The former is referred to as a Spatial Lag Model (SLM) and is appropriate when the focus of interest is the assessment of the existence and strength of spatial interaction. Formally, a spatial lag model, or a spatial autoregressive model is expressed as

$$y = X\beta + \rho Wy + \varepsilon \tag{7.1}$$

where ρ is a spatial autoregressive coefficient, β is common regression coefficient and ε is a vector of error terms. A spatial lag for y is expressed as Wy, where W is a spatial weights matrix arbitrarily defined by the modeler reflecting the geographical continuity throughout the study landscape.

Spatial dependence in the regression disturbance term, or a spatial error model, referred to as Spatial Error Model (SEM), is appropriate when the concern is correcting for the potentially biasing influence of the spatial autocorrelation. Hence,

$$y = X\beta + \varepsilon \tag{7.2}$$

$$\varepsilon = \sigma W\varepsilon + \mu \tag{7.3}$$

is equivalent to

$$y = \sigma Wy + X\beta - \sigma WX\beta + \varepsilon \tag{7.4}$$

which is a SLM with an additional set of spatially lagged exogenous variables (WX) and a set of k nonlinear (common factor) constraints on the coefficients (the product of the spatial autoregressive coefficient with the regression coefficients β should equal the negative of the coefficients of WX). In this sense, the SLM model deals with the overall spatial spillover effect with defined factors, whereas the SEM model measures the extent that the predictability is interfered by other undefined key factors. In other words, if SEM is more significant, it means that there are some critical variables missed in the proposed model. Clearly, despite the fact that SAR models have taken into account the fixed regional effect with a fixed lagged operator, they are still global models. These global models are overtaken based on an assumption that all the parameters are homogenous over the geographical environment, which may be problematic to reflect the local bias in reality. Consequently, locally weighted models are hardly demanded if the locally explicit results are aimed.

Based on the observed spatial non-stationery relationships, various approaches haven been developed in order to deal with spatially varying coefficients. Among which, GWR is the most widely adopted method in the literature with relevant context. It provides an elegant and easily grasped means of modeling such relationships. In fact, GWR could be considered as a local version of spatial regression that generates parameters disaggregated by the spatial units of analysis, which allows assessment of the spatial heterogeneity in the estimated relationships between the independent and a set of dependent variables (Fotheringham et al. 2002). For a location i, GWR model is formally defined as:

$$y_i = \beta_0 (u_i, v_i) + \sum_{j=1}^{k_b} \beta_j (u_i, v_i) x_{ij} + \varepsilon_i \tag{7.5}$$

where (ui,vi) denotes the coordinates of the ith location in space, $\beta_0(u, v)$ denotes to the local constant for the place i, and $\beta_j(u_i, v_i)$ is a realization of the continuous function $\beta_j(u, v)$ at regression point i, which is a parameter to be estimated. ε_i is a random error term, assumed to be normally distributed. In this model, observations located closer to the regression point are weighted more heavily than the observations located far away in the study area on the basis of a distance decay function, for example, a Gaussian function. After local estimates are derived at all locations in the study area, a continuous surface of local parameters will be automatically generated. This map presents various information about each estimation which includes not only its magnitude, but also the signs which could be positive and negative according to the specific location. In a GWR model, both positive and negative values can be observed with different degrees of significance for a coefficient. In this regard, it is obvious that employing the local GWR model to estimate human mobility patterns can provide far more valuable information on the spatial variations of related variables than conventional regression model.

7.2.2 *Mixed-GWR: Modeling Mobility on Multi-levels*

A recognized issue with GWR is that not all factors will present significant spatial variability across space. The reality is that both global effects and local effects may be in place. Therefore, the purely local regression model might not always be the best option to explore the relationship between the response and the explanatory variables. Some studies have discovered that socioeconomic attributes are more suitable to be treated as global variables whereas the structural features are more likely to be the local factors with significant geographical variations (Fotheringham et al. 2002). One proper solution to this issue is applying a regression model where both local and global effects are properly defined and placed. Mixed GWR (MGWR) is such a model introduced by the developer of GWR. In the MGWR model, those coefficients that were proved to be non-fluctuant across locations will be kept constant thereby improving the efficiency of prediction. Therefore, the pure local GWR model is extended to a multi-scaled one which can reflect the real spatial complexity in urban system. MGWR model can be formulated by Eq. 7.5 as follows:

$$y_i = \sum_{g=1}^{k_a} \beta_g x_{ig}(a) + \sum_{j=1}^{k_b} \beta_j(u_i, v_i) x_{ij}(b) + \varepsilon_i \tag{7.6}$$

where k_a and k_b denote the total account of global and local parameters for the variables respectively; $x_{ig}(a)$ refers to the global variables and $x_{ij}(b)$ stands for the local variables; β_g is the g th parameter associated with the global explanatory variables at all locations.

The adoptability of MGWR relies on a calibration procedure by a multiple stepwise regression algorithm to test the geographical variability for each variable. This is conducted by model comparisons between all pairs of the fitted GWR model, say the purely local GWR, and a modified model where only kth coefficient is fixed globally. By comparing with the difference of criterion measured by AICc, we can further decide which local factor should be assumed as global. Therefore, MGWR is more promising than standard GWR model when non-stationary and spatial stationary are detected.

Local t-value generated in GWR model helps us to investigate the spatial significance of local coefficient estimates. We use a significance level of 0.05 to determine the significance of the local coefficients.

7.3 Data

7.3.1 Study Area and Sample Data

We select the metropolitan area of Beijing as the study area and intend to measure the influence of selected factors on residents' mobility (Fig. 7.1). Beijing household travel survey data in 2005 are employed to evaluate mobility in Beijing. The 2005 survey covers the whole administrative areas including all 18 districts with 1118 TAZs as the basic geographical survey unit (Beijing Municipal Commission of Transport and Beijing Transportation Research Centre 2007). The sampling size is 81,760 households and 174,957 people, with a sampling ratio of 1.36 %. This survey adopts a travel diary form. For each trip, the survey records the departure time/location, arriving time/location, trip purpose and mode, as well as other important information including trip distance, destination building type, and transit line numbers. The household and personal information are also included in this survey. The household information consists of household size, Hukou (official residence registration) status and residence location, while the personal information includes gender, age, household role, job type and location, and whether having driving license

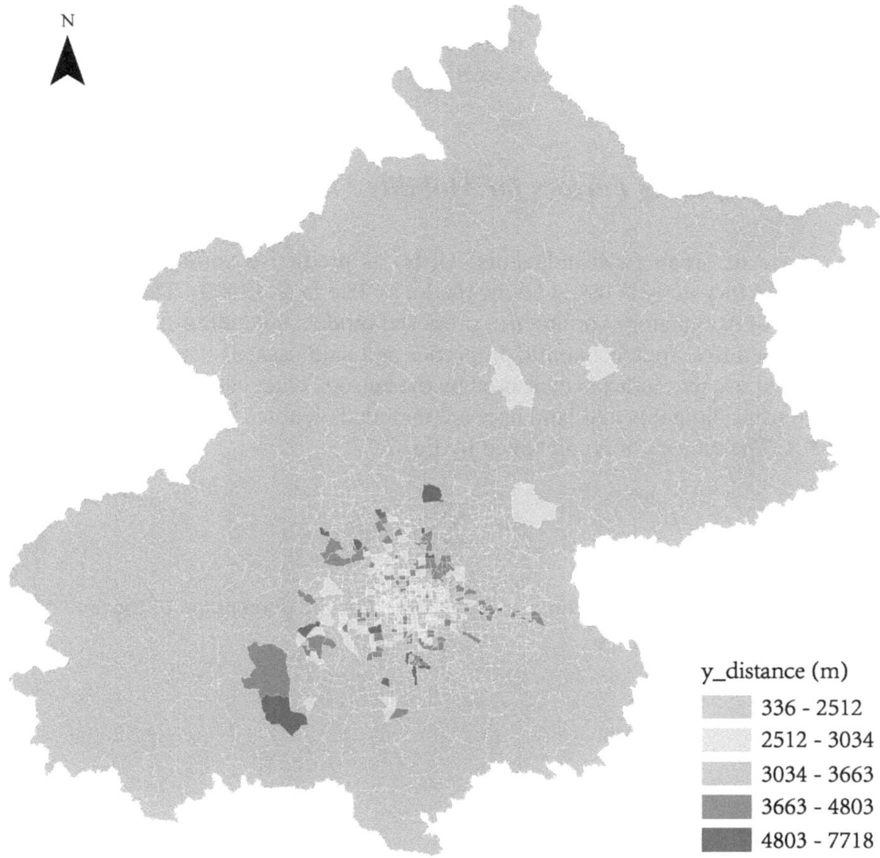

Fig. 7.1 Spatial distribution of mobility in terms of average travel distance

or transit month pass. The trip purposes in this survey include: (1) work, (2) school, (3) back to home, (4) back trip, (5) shopping, (6) entertainment, (7) daily life (such as dining, medical, social visiting, leisure/fitness, and picking up/ delivery), (8) business, (9) other. Job types include: (1) worker, (2) researcher, (3) office employee/ public employee, (4) teacher, (5) student, (6) self-employed, (7) attendant, (8) retiree, (9) specialized staff (such as medical staff, professional driver, bus/metro/ taxi driver, and soldier/police), (10) farmer, (11) unemployed, (12) other. Trip modes are: (1) walk, (2) bicycle, (3) electric bicycle, (4) motor, (5) bus, (6) mini bus, (7) metro, (8) employer-provided bus, (9) private car, (10) employer-provided car, (11) legal and illegal taxi. Among all transportation modes, the share of bus ridership is 13.81 % in the Beijing Metropolitan Area according to this survey.

In the following analysis, we aim to analyze the stimulus–response relationship between urban form and human mobility in both TAZ and individual scales. The analyses are performed at the TAZ level and individual level respectively.

7.3.2 Computing Factors for Mobility Modelling

The choice of urban form indicators (UFIs) as predictive variables for mobility modeling in this study is based on the model by Pan et al. (2009). There are mainly five types of explanatory variables in proposed models, including geometry, accessibility, amenities, socioeconomic properties and land uses, as listed in Table 7.1.

The land use mix index is measured by the entropy calculated from areas of various land uses. Nine types of land uses are included, denoted as C, D, F, M, R, S, U, W and X. The entropy S is calculated by Eq. 7.7:

$$S = -\sum_{i=1}^{n} P_i log_{10} P_i \tag{7.7}$$

in which n is the number of land use types, and pi is the percentage of the area of the ith land use type in the TAZ

Table 7.1 The various types of factors in mobility modeling in the TAZ scale

Main type	Abbreviation	Description
Geometry	A_PS	Average parcel/block size
Accessibilities	D_TAM	Distance to Tian'anmen square
	D_CBD	Distance to CBD
	D_ZGC	Distance to Zhongguancun
	D_NSC	Distance to the nearest sub-city center
	D_NSS	Distance to the nearest subway stations
Amenities	ST_D	Road/street density
	BS_D	Bus stops density
	PF_D	Density of public facilities
Socioeconomic properties	POP_D	Population density in 2005
	JOB_D	Job density in 2010
	H_INC	Average household income (if no date in the survey, then use 3400 average value)
	H_CAR	Average car ownership in households
	A_AGE	Average age
Land uses	L_MIX	Land use mix index

Thus, in the theoretical regression model at the TAZ scale, a vector y_{dis} is regressed on a series of determinates as Eq. 7.8.

$$\begin{aligned}\ln(y_{dis}) = \beta_0 + \beta_1 \ln(A_{PS}) + \beta_2 \ln(D_{TAM}) + \beta_3 \ln(D_{CBD}) \\ + \beta_4 \ln(D\_ZGC) + \beta_5 \ln(D\_NSC) + \beta_6 \ln(D\_NSS) \\ + \beta_7 \ln(ST\_D) + \beta_8 \ln(BS\_D) + \beta_9 \ln(PF\_D) \\ + \beta_{10} \ln(POP\_D) + \beta_{11} \ln(JOB\_D) \\ + \beta_{12} \ln(H\_INC) + \beta_{13} \ln(H\_CAR) \\ + \beta_{14} \ln(A\_AGE) \\ + \beta_{15} \ln(L\_MIX)\end{aligned} \tag{7.8}$$

The data for deriving these indicators are Land use parcels in 2005, and locational GIS data in 2005, bus stops in 2005, public facilities in 2005, roads in 2005, population for parcels in 2005, and the 2005 household travel survey in 2005 (the 2005 survey) for demographic properties. The descriptive statistics are summarized in Table 7.2.

At the individual level, we add some factors regarding the personal elements including gender, age, career status (student or not) measured by variable S_STU, and the home-to-work trip distance measured by variable D_RJ. In addition, some aggregated variables at the TAZ level are replaced by the individual level data. These include income and car ownership. The descriptive statistics of all variables are presented in Table 7.3.

Table 7.2 Descriptive statistics for independent variables in the TAZ scale (N = 1118)

Variables	Minimum	Maximum	Mean	Std. deviation
A_PS	0.18041	44.18365	3.38	3.719
D_TAM	550	115,481	19,900	17148.244
D_CBD	335	113,185	20,400	16854.603
D_ZGC	230	115,498	21,900	16403.987
D_NSC	375	68,265	19,500	9217.940
D_NSS	247	102,121	10,300	14581.321
ST_D	0	213.25	48.93	38.916
BS_D	0	2.675573	0.23	0.321
PF_D	0	1.079499	0.11	0.169
POP_D	0	938	82.57	123.588
JOB_D	0	21.86201	0.37	1.246
H_INC	1000	25,000	3510.02	1521.604
H_CAR	0	1	0.13	0.181
A_AGE	33	60	42.86	2.076
L_MIX	0	0.85066	0.41	0.181

Table 7.3 Statistical descriptive table for variables of the person level data (N = 174,957)

Variables	Minimum	Maximum	Mean	Std. deviation
SEX	0	1	0.50	0.500
AGE	0	95	41.62	17.397
S_STU	0	1	0.13	0.334
D_RJ	0	111	3.67	6.132
A_PS	0.18041	23.90190	2.0402337	1.67606557
D_TAM	851.00	72516.0	12400.131	12440.692
D_CBD	642.00	75166.0	14188.847	12114.0551
D_ZGC	790.00	69804.0	15156.941	11565.8723
D_NSC	743.00	36259.0	18483.116	7711.76117
D_NSS	276.00	53717.0	5084.9656	9271.08157
ST_D	3.82524	208.025	77.726260	44.09671165
BS_D	0.00000	1.79886	0.4718081	0.34736583
PF_D	0.00000	1.07950	0.2469767	0.21420801
POP_D	0.00000	554.00	199.80923	134.036610
JOB_D	0.00000	21.86201	0.8413297	1.93712864
INC	1000	40,000	3541.80	2733.996
CAR	0	3	0.34	0.491
L_MIX	0.00000	0.79741	0.4157790	0.13105488
y_{dis}	1.95	12.47	7.5358	1.02938

The theoretical model for regression analysis of personal mobility can be formally expressed as Eq. 7.9.

$$\begin{aligned}\ln\left(y_{dis}\right) = {} & \beta_0 + \beta_1 SEX + \beta_2 \ln\left(AGE\right) + \beta_3 S\_STU + \beta_4 \ln\left(D\_RJ\right) + \beta_5 \ln\left(A_{PS}\right) \\ & + \beta_6 \ln\left(D_{TAM}\right) + \beta_7 \ln\left(D_{CBD}\right) + \beta_8 \ln\left(D\_ZGC\right) + \beta_9 \ln\left(D\_NSC\right) \\ & + \beta_{10} \ln\left(D\_NSS\right) + \beta_{11} \ln\left(ST\_D\right) + \beta_{12} \ln\left(BS\_D\right) + \beta_{13} \ln\left(PF\_D\right) \\ & + \beta_{14} \ln\left(POP\_D\right) + \beta_{15} \ln\left(JOB\_D\right) + \beta_{16} \ln\left(INC\right) + \beta_{17} \ln\left(CAR\right) \\ & + \beta_{18} \ln\left(L\_MIX\right)\end{aligned} \tag{7.9}$$

7.3.3 *Calibration of OLS, SAR and Mixed-GWR*

7.3.3.1 Calibration of Theoretical Model

Before conducting regression models, the problem of multicollinearity is examined and resolved. It happens when two or more of the variables in the model are highly correlated, which will result in an over counting bias and an unstable/unreliable model. One method to detect multicollinearity is to use the Pearson product–moment correlation coefficients, or Pearson's correlation in short. Generally speaking, two variables with Pearson's correlation above 0.8 suggest a high degree of multicollinearity.

Another way to judge the degree of multicollinearity is to examine the so-called Variance Inflation Factor (VIF), which is a formal detection for multicollinearity. If one variable has a VIF value bigger than 7, it means that it has to be dropped from the model.

In the proposed regression model at the TAZ level, four variables of accessibilities including D_TAM, D_CBD, D_ZGC and D_NSS are found to be highly correlated, which is also proved by their large values of VIF. Therefore, only one of these four variables can be kept in the theoretical model formulated by Eq. 7.8. Likewise, those four variables are also found to cause multicollinearity in the regression model based on individual data. Thus three of the distance variables are dropped in both models proposed in this study, while only D_NSC is taken into account as the measure of accessibility generated by urban form.

7.3.3.2 Spatial Autocorrelation Detection

As concisely expressed by Tobler's First Law of Geography, spatial dependence is the fundamental character of geographically distributed phenomena. When significant spatial dependence is identified, spatial autocorrelation is said to be present in the variable at issue. As a result, the classic statistical regression methods are inadequate to account for the spatial autocorrelation in the variables. Therefore, spatial regression methods have been developed in response to such inadequacy. In this study, we first performed Moran's I analysis to examine the presence of spatial autocorrelation in the data. Using the toolkit named GeoDa developed by Anselin, we calculated the value of global Moran's I and generated the map of local Moran's I by Local Indicators of Spatial Association (LISA) analysis. According to the results in Fig. 7.2, there is significant spatial heterogeneity discovered in human mobility pattern at the TAZ level. The global Moran's I value for all the TAZs in Beijing Metropolitan Area is 0.14 (p-value equals to 0.001 by running permutation test 999 times), suggesting that there is slightly clustered when all TAZs are considered. The local analysis results in Fig. 7.2d reveal significant clusters for some TAZs. Low-Low clusters distribute both in the downtown areas and part of the suburbs. High-High clusters distribute in places which have mostly residential functions and few job opportunities. The evidence of significant spatial autocorrelation and particularly local variations suggest the possibility of spatial heterogeneity in the spatial relationships. Thus SAR and MGWR are adopted to study travel patterns in Beijing. SAR is used to account for spatial autocorrelation, while MGWR is adopted to consider spatial heterogeneity.

7.3.3.3 Calibration of the MGWR Model

An advantage of MGWR is that it enables modelers to incorporate the fixed effects as a subset of all explanatory variables based on their prior judgment. However, it is not always easy to objectively define which factors should be fixed. We adopt the geographical variability test to solve this problem. The difference is examined

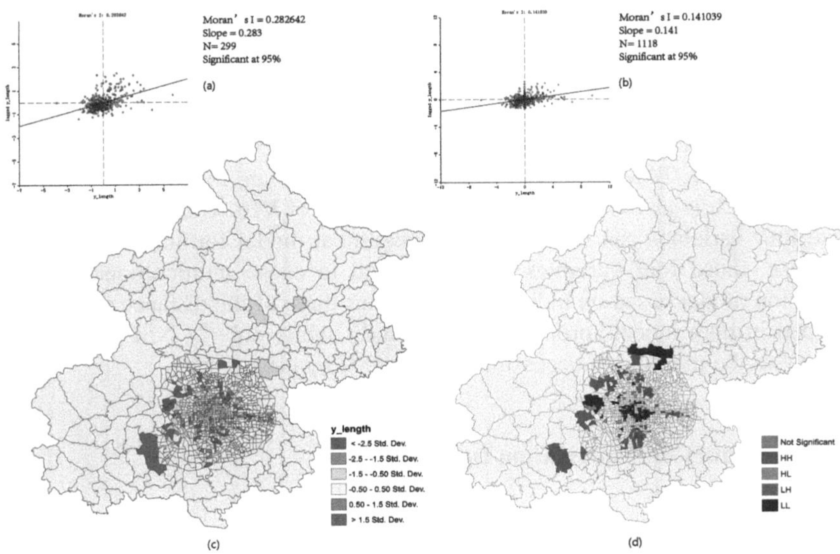

Fig. 7.2 Spatial autocorrelation analysis for average travel distance: (**a**) Moran scatter plot for special TAZs (<–0.5 SD and >0.5 SD); (**b**) Moran scatter plot for all TAZs; (**c**) Hotspots TAZs (<–0.5 SD and >0.5 SD); (**d**) LISA analysis map

Table 7.4 Geographical variability test for variables

Variables	F-statistics	DIFF of criterion	Type of parameter
Intercept	13.92	−162.39	Local
A_PS	0.57	67.57	Global
D_NSC	1.60	17.64	Global
ST_D	1.97	21.09	Global
BS_D	0.36	69.22	Global
PF_D	1.00	43.88	Global
POP_D	0.67	58.31	Global
JOB_D	1.51	37.64	Global
H_INC	15.51	−246.61	Local
H_CAR	12.36	−200.61	Local
A_AGE	5.56	−56.73	Local
L_MIX	1.11	53.09	Global

between a fitted GWR model and a counterpart model in which only the kth coefficient is fixed globally. A positive difference indicates that kth variable should be identified as the global one. The critical value here is 2 in terms of AICc, which means that the switched model achieve better fitting results if it can reduce at least 2 in AIC. The results are reported in Table 7.4, showing that only three variables

(H_INC, H_CAR, A_AGE) replicating socioeconomic characteristics are defined as local parameters, whereas others are modeled globally in GWR.

Another issue for MGWR model is the best bandwidth selection. Some exiting methods can help to select the most promising bandwidth towards the best fitting model. For instance, cross-validation (CV) procedure is such a way introduced by Cleveland in 1979 for local regression models. In the following analysis, we follow the procedure suggested by Fotheringham et al. (2002) using a Gaussian Kernel function to select the golden bandwidth for minimizing the AICc value. This procedure is conducted by iterations in which only the bandwidth changes but other model settings are kept constant. The generated bandwidth in this model is 104 for the proposed MGWR model.

7.4 Empirical Results

7.4.1 Primary Findings on the TAZs Level

7.4.1.1 OLS Analysis for TAZs

Table 7.5 shows the coefficients generated for a stepwise OLS regression analysis. The second column from the left is the results of OLS analysis when all travel modes are combined. In addition, we conducted the OLS analysis in the TAZ scale for each of the specific traffic modes, including car, bus, bike, moto and walk. Here we regard taxi as car, metro as bus. It seems that socioeconomic features matters more significantly for all modes, while the accessibilities affect the mobility of certain mode more significantly. Furthermore, it implies that people tend to travel shorter distance when they live in the area where the urban density is higher, which is on the basis of the evidence that all the urban density features show a negative linkage to the mobility length. Meanwhile, a person is more likely to travel longer if he/she lives in the richer area with higher income and car ownership. Yet, the adjusted R square of each mode is generally low implying that the global OLS models can hardly inform a good fitted model as we have predicted. We also conducted OLS analysis for the dependent variables y_count and y_time. For y_count, $R2=0.06$. For y_time, $R2=0.03$. This suggests that the explanation power of the OLS model for trip frequency and trip time is quite limited.

7.4.1.2 SAR

We analyze regression diagnostics for trend surface regression models, which includes a spatial weights matrix for use in a SAR model. A linear trend surface is considered here, meaning that only explanatory variables are included but no cross-terms. Two kinds of contiguity based spatial weights matrices, Rook and Queen, are

Table 7.5 Stepwise OLS results for various traffic modes (in terms of average travelling distance)

Variables	All modes	Car	Bus	Bike	Moto	Walk
Intercept	3209.524***	15570.931	12622.127	5627.796	6179.490	531.878
	(33.726)	(18.2***)	(18.3***)	(23.2***)	(10.456***)	(9.4***)
A_PS	–	–	–	–	–	–
D_NSC	–	0.262	0.163	–	0.105	–0.003
		(7.5***)	(5.6***)		(3.983***)	(–2.3**)
ST_D	–	–63.594	–64.969	–11.096	–	–
		(–6.2***)	(–8.4***)	(–3.0***)		
BS_D	–	–3847.367	–	–	–	–
		(–3.0***)				
PF_D	–567.579***	–	–	–	–	–
	(110.33)					
POP_D	–0.543***	–	–6.777	–	–	–
	(0.162)		(–2.6***)			
JOB_D	–46.789***	–	–	–	–	–
	(11.455)					
H_INC	0.02**	–	–	–	–	–
	(0.009)					
H_CAR	822.728***	–8082.251	–4505.999	–4717.215	–5577.429	265.816
	(92.987)	(–4.0***)	(–2.7***)	(–6.2***)	(–5.030***)	(3.6***)
A_AGE	–	–	–	–	–	–
L_MIX	–	–	–	–	–	–195.742
						(–2.5**)
N	1118	1015	982	848	597	1045
R square	0.111					
Adjusted R square		0.195	0.195	0.077	0.063	0.028
AIC	16,797					

Note: Standard errors in parentheses; *, **, and *** represent for the confident level at 90 %, 95 % and 99 %, respectively

attempted in our regressions, which are provided by GeoDa. The statistics of Queen Matrix tend to be more significant. Therefore, in the following analysis the Queen Continuity Matrix is employed in SAR models.

In the next step, SLM, SEM and OLS models are compared and the results are presented in Table 7.6. It suggests that spatial regression models improve the fitting results (bigger R square, smaller AICc value), reduce the standard errors of almost all variables and prove the significance of spatial lag. In the comparison between SLM and SEM, we find that SLM is more appropriate for modelling spatial autocorrelation due to the robustness significance. The outputs of SAR models confirm the significant variables in OLS. However, just like OLS, the SAR models hardly have a satisfactory prediction.

Table 7.6 Comparison among SLM, SEM and OLS with queen spatial weight matrix

Variables	SLM	SEM	OLS (enter)
W_Y_distance	0.231*** (0.043)	–	–
Intercept	2827.520*** (324.588)	3555.605*** (292.735)	3649.306 (295.053)
A_PS	1.899 (3.786)	1.904 (3.865)	1.839 (3.873)
D_NSC	0.001 (0.002)	0.002 (0.002)	0.001 (0.001)
ST_D	−0.577 (0.485)	−0.638 (0.529)	−0.714 (0.496)
BS_D	−11.817 (58.460)	1.148 (60.085)	−11.365 (59.772)
PF_D	−369.794*** (123.854)	−410.389*** (133.823)	−462.801*** (125.820)
POP_D	−0.426*** (0.165)	−0.439*** (0.169)	−0.439*** (0.169)
JOB_D	−44.995*** (11.259)	−44.298*** (11.213)	−45.597*** (11.517)
H_INC	0.031*** (0.008)	0.019*** (0.009)	0.022*** (0.009)
H_CAR	780.518*** (96.689)	793.189*** (98.111)	806.582*** (98.745)
A_AGE	−9.794 (6.604)	−8.784 (6.691)	−11.416* (6.753)
L_MIX	45.631 (75.348)	28.584 (78.265)	73.715 (77.050)
LAMBDA	–	0.249*** (0.047)	–
N	1118	1118	1118
R square	0.146	0.146	0.116
Log likelihood	−8374.74	−8375.88	−8389.05
AIC	16840.7	166775.8	16,802
Robustness	3.994 **	0.300	–

Note: Standard errors in parentheses; *, **, and *** represent for the confident level at 90 %, 95 % and 99 %, respectively

7.4.1.3 MGWR

The MGWR provides both global and local estimations (see Table 7.7). The ranges of local estimates show that the H_INC, H_CAR and A_AGE variables are very geographically varying. All these local variables present negative and positive effects in different parts of our study area, illustrating that a constant coefficient in global models would have ignored the important particularity with locations. Moreover, it is easy to be recognised that the explanatory ability of MGWR model is much better than the OLS and SAR models based on the increased R square and the decreased AIC, suggesting that GWR models explain more about mobility patterns. Other global coefficients are recognised to be similar to the ones in OLS and SAR regressions. Consequently, we paid more attention to the spatial variability of local parameter estimates and the corresponding t-value distributions for H_INC, H_CAR and A_AGE.

Average Household Income (H_INC)

The spatial distribution of H_INC parameter estimates and corresponding t-value map are depicted in Fig. 7.3. Figure 7.3a shows that the spatial relationship between TAZ-based average income and the travel length. It reveals that residents live in

Table 7.7 The summary of MGWR results

Fixed (global) coefficients		Geographically varying (local) coefficients			
Variables	Estimates (standardized)	Variables	Estimates (mean)	Estimates (min)	Estimates (max)
A_PS	3.164	Intercept	3348.313	2478.510	4512.987
D_NSC	−18.049	H_INC	2.224	−4468.907	511.494
ST_D	−6.049	H_CAR	184.962	−1113.669	1861.479
BS_D	7.832	A_AGE	−10.293	−2223.535	2932.294
PF_D	−26.263				
POP_D	−8.438				
JOB_D	−41.840				
L_MIX	7.121				
Classic AIC	16331.2				
AICc	16353.3				
R square	0.509				
Adjusted R Square	0.441				

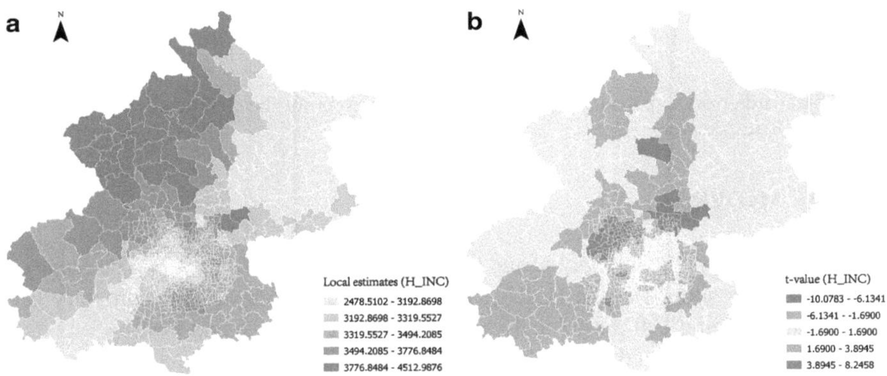

Fig. 7.3 Spatial variation of H_INC local estimates (**a**) and t-value map (**b**)

most places in the inner city of Beijing, the area within the fifth ring road, would travel to closer places if they have more income. It should be noted that about half of the TAZs show negative impacts of people's income on their travel length, which seems to be a reversed result in comparison with the OLS result. The negative peak appears in the areas in proximity to Beijing Capital International Airport in Shunyi District. This trend spreads in most of areas in Beijing except for three clusters where the income standard exerts positive effects on travel distance. Those three clusters are generally suburban areas including the area around Tongzhou District in the south-east of Beijing, the places in Changping District in the north-west and the rural areas in the very north of Beijing. The statistical reliability of this analysis is supported by the t-value pattern shown in Fig. 7.3b. Almost all the positively related to hot-pots and the negative ones are statistically significant at the confidence level of 95 %. It suggests that high accuracy of local estimates can be explained in our case study by the significant correlation between household income and people's mobility. The reasons why the relationships in suburban areas vary differently should be explored more in further detailed studies.

Average Household Car Ownership (H_CAR)

Figure 7.4 illustrates the local variation of the average car-ownership. It shows negative and very small values of coefficients for the inner city of Beijing, the south-west and north-east Beijing; however, these results are not significant. Other areas (darker coloring), particularly the broad areas in the northwest and the south of Beijing City exhibit higher positive local estimates, which demonstrates that people living in these areas are willing to travel further if they have more cars. It confirms the common sense that cars will encourage the traveling length to some extent, though more empirical research is to be conducted. The t-value map (Fig. 7.4b) indicates that the significant relationship between car-ownership and mobility pattern is particularly obvious in the town of Majuqiao (the dark cluster in the south-east of Beijing) and the areas around Yanqin (the clusters on the northwest in Beijing). Thus, combining the pattern of local estimates and the statistical significance distribution, the result implies a significant trend in Beijing that car ownership presents a significant factor for long-distance job opportunities.

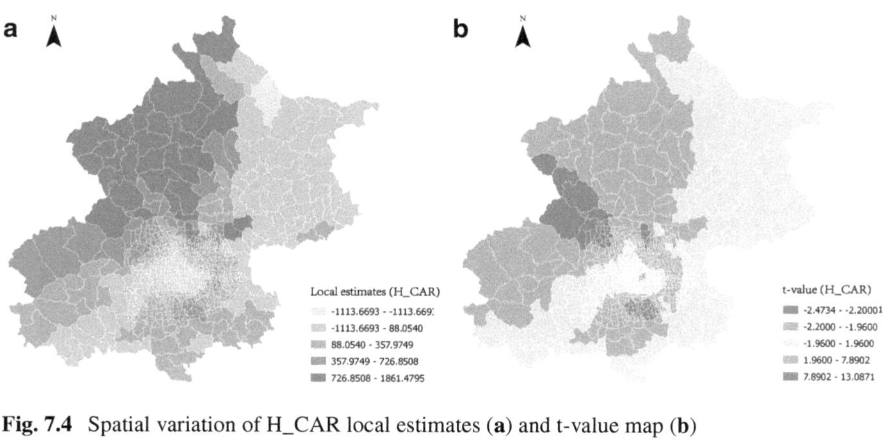

Fig. 7.4 Spatial variation of H_CAR local estimates (**a**) and t-value map (**b**)

Average Age (A_AGE)

Figure 7.5 reveals that the inner city of Beijing and the broad east part of Beijing Metropolitan Area are found to present a negative relationship between average age and travel length. This is easy to be understood because younger people tend to be more likely and able to commute further for work and there are more old or retired people living in the inner city than the ones in the outer suburban or rural areas.

Key findings from the MGWR model are discussed as below. Firstly, it is confirmed that spatial heterogeneity exists in the spatial association between trip length and the explanatory factors and can be properly modeled by MGWR. Secondly, the inner city and the suburban areas are clearly distinguished by the signs of localized coefficients. Thirdly, the western part of Beijing tends to exhibit a positive significant relationship between household socioeconomic characteristics and mobility distance, whereas the relationship in the eastern parts is generally insignificant from a statistical perspective. Finally, the empirical results indicate that socioeconomic factors impact the mobility pattern in Beijing vary spatially in the metropolitan area. Thus, more efforts should be made in understanding the mechanism between policy delivery and socioeconomic reaction, thereby making proper political and planning decisions based on socioeconomic situations to reduce traffic congestions, air pollutions and other issues caused by lengthy commuting distance. That is to say, our findings here inform that proper design of socio-economic structures can help to optimize urban mobility.

7.4.2 *Primary Findings on the Individual Level*

Ordinary least square regression (OLS) is conducted for the person level data. The dependent variable is the natural logarithm of average trip distance of a person in a day. The regression results are listed in Table 7.8 in comparison with that of the

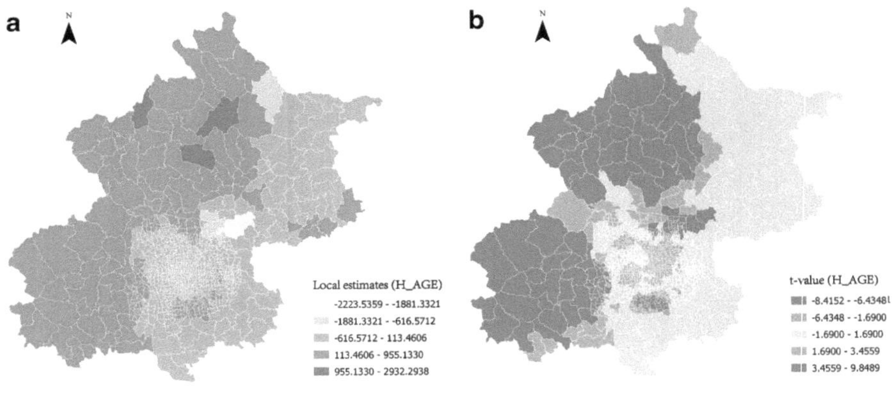

Fig. 7.5 Spatial variation of A_AGE local estimates (**a**) and t-value map (**b**)

Table 7.8 The summary for comparing OLS for persons and MGWR for TAZs

Names	OLS (y_distance)	GWR (y_distance)
Intercept	1552.59***	3348.313 (mean)***
SEX	428.686***	–
AGE	–0.481	–
S_STU	–527.620***	–
D_RJ	283.184***	–
A_PS	28.725***	3.164
D_NSC	0.008***	–18.049
ST_D	–0.731**	–6.049
BS_D	–26.778	–7.832
PF_D	–17.658	–26.263***
POP_D	.024	–8.438***
JOB_D	–41.640***	–41.840***
INC	.027***	–
CAR	558.672***	–
L_MIX	269.188**	7.121
A_AGE		–10.293 (mean)***
H_INC		2.224 (mean)**
H_CAR		184.962 (mean)***
R square	0.146	0.509
Adjusted R square	0.146	0.441

Note: Standard errors in parentheses; *, **, and *** represent for the confident level at 90 %, 95 % and 99 %, respectively

GWR model. The results show that personal statues (SEX, S_STU and A_PS), local accessibilities (D_NSC, ST_D, L_MIX) and socioeconomic features (INC and CAR) are positively related to an individual's commuting length. It suggests that individual decisions about the travel distance in our observed dataset are made by considering various characteristics of the urban form rather than simply minimizing the length of travel. However, the MGWR on TAZ level tells a different story in which urban density (i.e. PF_D, POP_D, JOB_D) and local socio-economic factors play more important roles in influencing mobility on a group scale. In short, the explanatory factors function differently on various scales. For individuals, personal status, homes' accessibilities are more important, whereas for the groups of people in TAZs, the socioeconomic characteristics and urban densities of the urban form function more significantly. This also informs the urban planners that urban policies should pay specific attention to different aspects of urban form at various levels.

7.5 Conclusions

This chapter analyses the impact of urban form on human mobility in Beijing using a large-scale travel survey and GIS datasets. Totally 15 indicators including geometry, accessibility, amenity, demographic and land use composition types were derived to measure the urban form at various parts of Beijing quantitatively. We aggregated urban form indicators and human mobility indicators for the 1118 TAZs in Beijing and performed analysis at the TAZ scale. The dependent variable is the average trip distance of each person for each TAZ. Classic linear regression, spatial autoregressive models, and the MGWR were adopted to examine the impact. MGWR enables us to understand the spatial heterogeneity of the impacts on human mobility. By comparing all three types of models, we find that MGWR model achieves the best performance with much higher explanatory capacity. In the MGWR experiment, three socioeconomic indicators of urban form are found to have significant and spatially varying impacts on human mobility. They are the average household income, average car ownership and average age in households. Meanwhile, we also conducted OLS analysis using the raw survey data on the individual level and compared it with the MGWR model. The results highlight that people's mobility are related to their personal statuses and various urban form factors. However, the aggregated group mobility on the TAZ level is impacted by the local elements of urban density and socioeconomic characteristics to a larger extent. The findings suggest strong spatial heterogeneity in the influence of predictive variables on residents' travel behavior in Beijing.

The contributions of this study lie in the following three aspects. First, although extensive previous studies are available on identifying the impact of urban form on human mobility, few were focused on Beijing, a mega city in the urbanizing China. The research findings of study can provide useful information for future urban planning in Beijing. Secondly, our study is based on fine-grained, best available data. We use a large-scale survey data covering the entire Beijing Metropolitan Area with a sampling ratio of about 2 %, rather than small-scale survey with limited samples in only selected neighborhoods. In this research, all TAZs representing all types of urban forms in the Beijing Metropolitan Area are taken into account. Furthermore, fine-scale urban GIS datasets were used to calculate urban form indicators. Thirdly, in contrast to the conventional OLS regression, our work paid special attention to the spatial heterogeneity of the impact. Indeed it is found that the impact of each urban form indicator varies geographically, which was not addressed by previous studies. Finally, the large-scale, fine-grained survey enabled us to detect the differences between the individual level OLS results and the TAZ level regression results.

References

Alford, G., & Whiteman, J. (2009). Macro-urban form and transport energy outcomes: Investigations for Melbourne. *Road & Transport Research, 18*, 53–67. doi:10.1061/(ASCE)0733-9488(2006)132:3(132).

Anderson, W. P., Kanaroglou, P. S., & Miller, E. J. (1996). Urban form, energy and the environment: A review of issues, evidence and policy. *Urban Studies, 33*, 7–35. doi:10.1080/00420989650012095.

Anselin, L. (1988). *Spatial econometrics: Methods and models*. Dordrecht: Kluwer Academic Publishers.

Arun, R. K., & Ram, M. P. (2001). A structural equations analysis of commuters' activity and travel patterns. *Transportation, 29*(1), 33–54. doi:10.1023/A:1005253813277.

Beijing Municipal Commission of Transport, Beijing Transportation Research Centre. (2007). *The 3rd transportation comprehensive survey report of Beijing*. Internal working report (In Chinese).

Dieleman, F. M., Dijst, M., & Burghouwt, G. (2002). Urban form and travel behaviour: Micro-level household attributes and residential context. *Urban Studies, 39*, 507–527. doi:10.1080/00420980220112801.

Fotheringham, A. S., Brunsdon, C., & Charlton, M. (2002). *Geographically weighted regression: The analysis of spatially varying relationships*. Chichester: John Wiley & Sons Ltd.

Giuliano, G., & Narayan, D. (2003). Another look at travel patterns and urban form: The US and Great Britain. *Urban Studies, 40*(11), 2295–2312.

Goetzke, F. (2008). Network effects in public transit use: Evidence from a spatially autoregressive mode choice model for New York. *Urban Studies, 45*, 407–417. doi:10.1177/0042098007085970.

Holden, E., & Norland, I. T. (2005). Three challenges for the compact city as a sustainable urban form: Household consumption of energy and transport in eight residential areas in the greater Oslo region. *Urban Studies, 42*, 2145–2166. doi:10.1080/00420980500332064.

Horner, M. W. (2007). A multi-scale analysis of urban form and commuting change in a small metropolitan area (1990–2000). *Annals of Regional Science, 41*, 315–332. doi:10.1007/s00168-006-0098-y.

Kawabata, M., & Shen, Q. (2007). Commuting inequality between cars and public transit: The case of the San Francisco Bay area, 1990–2000. *Urban Studies, 44*, 1759–1780. doi:10.1080/00420980701426616.

Krizek, K. J. (2003). Residential relocation and changes in urban travel – Does neighborhood-scale urban form matter? *Journal of the American Planning Association, 69*, 265–281. doi:10.1080/01944360308978019.

Lin, J., & Yang, A. (2009). Structural Analysis of How Urban Form Impacts Travel Demand: Evidence from Taipei. *Urban Studies, 46*(9), 1951–1967. doi: 10.1177/0042098009106017.

Maat, K., & Timmermans, H. J. P. (2009). A causal model relating urban form with daily travel distance through activity/travel decisions. *Transportation Planning and Technology, 32*, 115–134. doi:10.1080/03081060902861285.

McMillan, T. E. (2007). The relative influence of urban form on a child's travel mode to school. *Transportation Research Part A: Policy and Practice, 41*, 69–79. doi:10.1016/j.tra.2006.05.011.

Mulley, C. (2013). Accessibility and residential land value uplift: Identifying spatial variations in the accessibility impacts of a bus transitway. *Urban Studies*. doi:10.1177/0042098013499082.

Newman, P. W. G., & Kenworthy, J. R. (1989). Gasoline consumption and cities: A comparison of US cities with a global survey. *Journal of American Planning Association, 55*, 24–37. doi:10.1080/01944368908975398.

Pan, H. X., Shen, Q., & Zhang, M. (2009). Influence of urban form on travel behaviour in four neighbourhoods of Shanghai. *Urban Studies, 46*, 275–294. doi:10.1177/0042098008099355.

Schlossberg, M., Greene, J., Phillips, P. P., Johnson, B., & Parker, B. (2006). School trips – effects of urban form and distance on travel mode. *Journal of the American Planning Association, 72*, 337–346. doi:10.1080/01944360608976755.

Wei, C., & Qi, F. (2012). On the estimation and testing of mixed geographically weighted regression models. *Economic Modelling, 29*(6), 2615–2620.

Zhang, M. (2005). Exploring the relationship between urban form and nonwork travel through time use analysis. *Landscape and Urban Planning, 73*, 244–261. doi:10.1016/j.landurbplan.2004.11.008.

Chapter 8
Finding Public Transportation Community Structure Based on Large-Scale Smart Card Records in Beijing

8.1 Introduction

Public transportation in big cities is a crucial part of urban transportation infrastructures. Exploring the spatiotemporal patterns of public trips can help us to understand dynamic human movements, transportation patterns and the complex urban systems thus supporting better urban planning and design. The availability of large-scale smart card data (SCD), which is one type of urban Big Data collected from public transportation operations and management institutions, offers new opportunities to study the intra-urban structure and spatial interaction dynamics by zooming into individual-based public trips. Previous research has investigated the jobs-housing relationships and commuting patterns using such data and demonstrated comparisons with traditional high-cost travel survey approach (Long et al. 2012; Long and Thill 2013). The study of spatial interactions is one of the traditional researches in geography and regional science. For regional studies, the functional region is defined by regional geographers based on interactions between its distinctive land-use zones (Johnston et al. 1981). Representation forms of spatial interactions between different zones include human movement, commodity flow, resource allocation, information communication and so on. For the past several decades, studies of spatial interaction processes have mainly been based on the census datasets (Rae 2009; Jang and Yao 2011). Recent fast development in information and communications technology (ICT) and the availability of big geospatial data (such as mobile phone records, GPS-enabled taxi/cab traces, location-based social network check-ins) has supported several frontier researches on spatial interactions and networks (Ratti et al. 2010; Gao et al. 2013; Kang et al. 2013; Liu et al. 2014), identifying functional urban regions (Manley 2014), as well as to reveal spatiotemporal intra-urban land use variations from travel patterns (Liu et al. 2012).

In this research, we are interested in extracting origin-destination (OD) flow matrices in the aggregation scale of traffic analysis zones (TAZs) and analyzing the intra-urban spatial interaction patterns revealed by human movements among TAZs

Y. Long, Z. Shen, *Geospatial Analysis to Support Urban Planning in Beijing*,
GeoJournal Library 116, DOI 10.1007/978-3-319-19342-7_8

using public transportation. Traditional spatial clustering approaches, which group similar spatial objects into classes, are not sufficient to explore the network structure of spatial interactions between different regions. Thus we applied the novel community detection methods from the study of complex networks to examine the dynamic spatial structures of public transportation communities in the Beijing Metropolitan Area (16,410 km^2). It can help to find the ground-truth community structure of strongly connected TAZs by public transportation, which may yield insights for urban planners on land use patterns or for transportation engineers on traffic congestion.

This chapter is organized as follows. Sections 8.2 and 8.3 describe the data and methodology used for this study. We elaborate the detailed results in Sect. 8.4. We conclude this work and propose next-step plans in Sect. 8.5.

8.2 Data

In Beijing, most bus/metro passengers use smart cards when getting on and off buses and metros to pay their fares. Thus, individual OD trips which connect bus stops (or metro stations) can be extracted directly from the detailed records of SCD. The collected SCD consists of 97.9 million trips from anonymized 10.9 million smart card users during a one-week period from April 5 to April 11, 2010. More details on public transportation and SCD in Beijing are available in the chapter entitled "Profiling underprivileged residents with mid-term public transit smartcard data of Beijing" in this book. In order to create the public transportation OD flow matrices in the TAZ level, we first georeferenced all bus stops and metros stations with latitude/longitude coordinates, and then spatially joined them into the total 1911 Beijing TAZ boundaries (see Fig. 8.1). A directed-weighted linkage between two TAZs represents the total number of public trips from the origin-TAZ to the destination-TAZ in a given time interval. Regarding the temporal dynamics, we aggregated the data into different hourly and daily periods to study the spatiotemporal patterns in public transportation, as well as variations between weekdays and weekends.

8.3 Methodology

In the study of complex networks, a community is defined as a subset (group) of the whole network and the nodes in the same community are densely connected internally and grouped together. The identification of such densely connected nodes in networks is called community detection. Popular community detection methods can be classified into two groups: graph partitioning and hierarchical clustering. Graph partitioning divides a network graph into a set of non-overlapping groups,

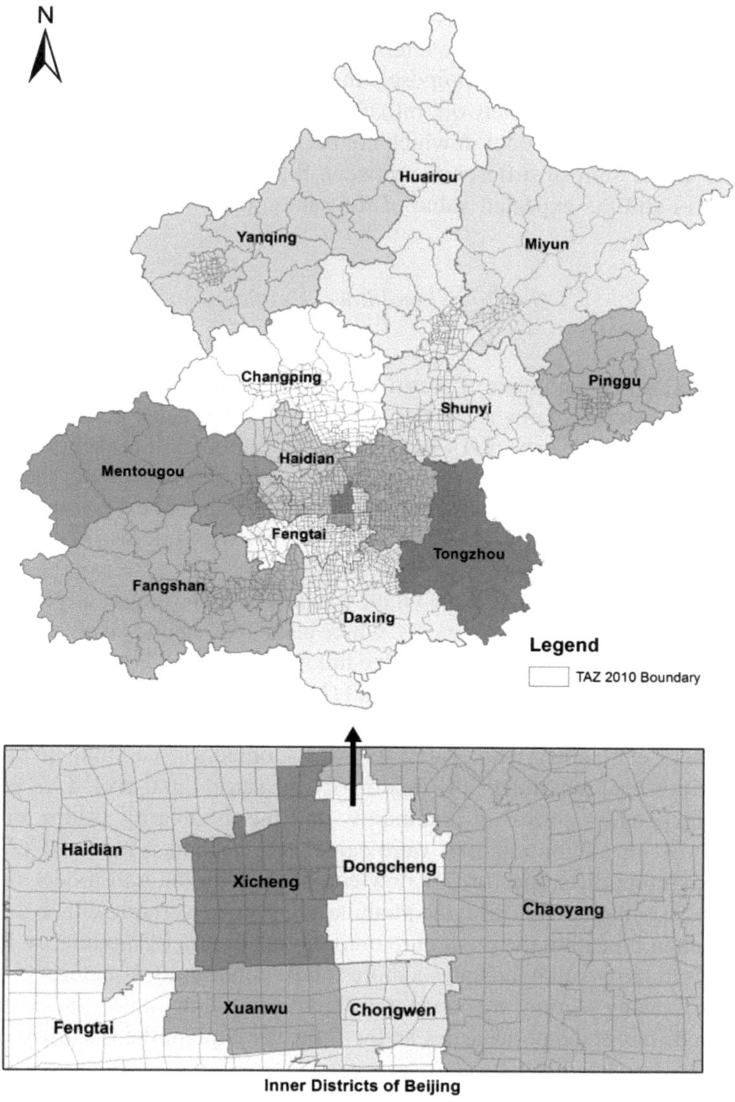

Fig. 8.1 The study area in administrative districts (*different gray levels*) of Beijing. It is noteworthy that the basic spatial unit is a TAZ, and the district divisions showed here was the 2010 version without the merges of Xuanwu and Chongwen districts in order to keep consistent with the SCD data collection period

while hierarchical clustering seeks to build a hierarchy of clusters of nodes, such that for each cluster there are more internal than external connections.

Newman and Girvan (2004) propose a modularity metric to evaluate the quality of a particular division of a network into communities. Modularity compares a proposed division to a null model in which connections between nodes are random. It is defined as the sum of differences between the fraction of edges falling within communities and the expected value of the same quantity under the random null model.

$$Q = \sum_{k} \sum_{ij \in C} \left(realflow_{ijk} - estflow_{ijk} \right) \tag{8.1}$$

In Eq. 8.1, k is the number of partition communities, $realflow_{ijk}$ gives the actual fraction of interactions between nodes i and j within the same community C, and $estflow_{ijk}$ represents the expected values under the random null model or other theoretical models. If the fraction of edges within communities is no better than the null model the modularity $Q=0$, while $Q=1$ indicates the most robust community structure. In practice, modularity values of different real world networks with varying sizes fall into the range 0.3–0.7.

We first converted the TAZ-scale ODflow matrices in the consecutive seven days into seven undirected-weighted graphs, where each TAZ can be taken as a node and each OD-flow interaction as a weight edge linking two TAZs. Then, the widely used Newman-modularity-maximization method (Newman 2004) was applied to find the daily public transportation communities. In practice, a bottom-up fast greedy algorithm (Clauset et al. 2004; Gao et al. 2013) was adopted for searching an optimized graph partition that maximizes the modularity measure. First, each TAZ started in its own independent cluster of community and the modularity values among all pairs of TAZs for all communities were calculated. Second, a pair of TAZs which has the maximum difference of OD flow compared with the null model should be merged into a community. Third, the modularity of the new graph will be calculated again and then repeating the procedure until the maximum of modularity is found. A larger modularity value indicates a more robust community structure.

In the following section, we conduct different community detection experiments on SCD in different temporal scales and will further explain the identified community structures in detail.

8.4 Results

8.4.1 Identification of Communities on Weekdays and Weekends

We first examine the daily public transportation community structure. Table 8.1 shows the detailed network information of daily community detection results of public transportation OD trips during a week. We find the community consistent

Table 8.1 Daily community detection results of public transportation OD trips in a week

Day of week	# of nodes	# of edges	# of groups	Mean of community size	MAX Q
Monday	1214	55,205	6	202	0.467
Tuesday	1214	55,598	6	202	0.461
Wednesday	1216	55,510	6	203	0.462
Thursday	1213	55,641	6	202	0.475
Friday	1213	55,614	6	202	0.457
Saturday	1212	55,490	6	202	0.470
Sunday	1213	53,273	6	202	0.471

pattern in terms of the number of divided groups (6), the average size of each community (202) and the maximum value of modularity (0.457–0.475) in the detection processes. The community on Thursday has the largest modularity value, which indicates a more stable network community structure than other days.

Although the network statistics are similar in the seven days of a week, the spatial distributions of these detected communities lie in slightly different. As shown in Fig. 8.2, the daily community detection results demonstrate that in general geographically cohesive regions that correspond well with administrative districts (such as Tongzhou) or merged boundaries (such as Fengtai and Daxin) in Beijing were identified by weekday public transportation patterns, while some unexpected spatial structures might uncover hidden urban structure that needs further investigation. The suburb public transportation communities usually contain more TAZs than urban central TAZs. There exist strong public transit connections among TAZs which locate along the middle west-to-east corridor including the Chang'an Avenue in Beijing, where the Beijing Subway line 1 also runs through the street. Note that the passengers can use smart card when they travelled on metro lines. Surprisingly, most of the southern TAZs in Daxing, Fengtai and Fangshan districts were aggregated into a large transportation community. It indicates there are more frequent intra-public trips within its own community in the southern region than the inter-community trips across other sub-regions of the Beijing Metropolitan Area. The same giant community pattern lies in the northwest TAZs in Yanqing district and the majority of TAZs in Tongzhou, although there is several connected TAZs from inner districts to Tongzhou through the Beijing Subway Batong line. Also, it is remarkable to see an enclave in the southern part of Fangshan district has been aggregated into a spatially separated large community north/northeast parts of Beijing (covering a large portion of Chaoyang, Shunyi and Changping districts) in all seven consecutive days, which indicates a strong public transportation connection pattern. The integrated analysis of geographical contexts, land-use types, housing prices, job opportunities, and the prominent points of interest in these regions might offer better explanation about the patterns identified in the community detection results. In addition, a northwest TAZ in Changping district was aggregated into a large number of spatially separated TAZs which belong to inner districts of Beijing (Dongcheng, Xicheng, Chongwen and Xuanwu) only in weekends not in weekdays. It reveals a

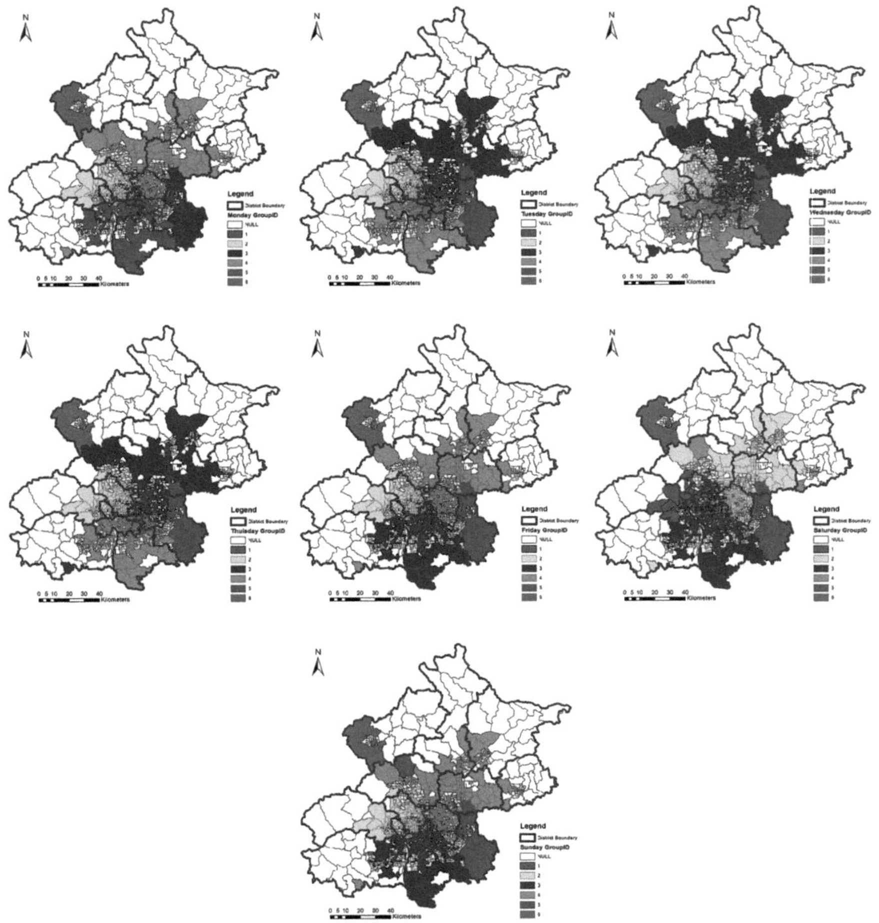

Fig. 8.2 The spatial distributions of daily community detection results of public transportation OD trips using SCD in a week

recreation place of interests in the northwest TAZ and attracts a large portion of public travel trips. Potentially, this pattern could help local transportation agency to identify the needs to provide temporal services for increasing public transportation demands in these connected regions.

We also created an interactive web map for exploring the public transportation community detection results in the geographical context (Fig. 8.3). The online geovisualization of communities for comments and validation using local knowledge can be accessed at http://www.beijingcitylab.com/projects-1/3-bus-landscapes/. We have invited online browsers to propose comments on the communities detected.

Fig. 8.3 The interactive web map for exploring the public transportation communities in Beijing

8.4.2 *Comparison with Household Survey Data*

Household travel survey is a traditional data-collection approach for acquiring information about residents' travel behaviors and estimating transportation demands (Beijing Transportation Research Center 2011). The survey tracks travelers' socio-economic attributes, as well as trip origin and destination, time and duration, purpose and mode. We used the survey in 2010 for comparing with results from SCD. The sample size of the survey is 46,900 households (116,142 residents) in the whole Beijing Metropolitan Area. The 2010 Survey provides the one-day travel diary of all respondents covering all travel modes. We applied the same data processing and community detection procedure introduced above to one-day household survey data. We found that the daily community detection results using the household travel surveys are different from that using SCD.

As shown in Fig. 8.4, there are 13 communities identified when maximizing the modularity of the TAZ network connected by survey OD trips. It is clear to see that most of the TAZs in suburbs have been grouped into the corresponding outer districts. The community boundaries generally match well with administrative boundaries of Beijing Districts. For those places that didn't match, especially for the spatial separated communities, it usually indicates some interesting travel patterns, land-use or urban structure, which could be identified through geographical contexts analysis (Gao et al. 2013).

By comparing the community detection results of SCD and household survey data, we also find that the SCD results match better with the Beijing planned

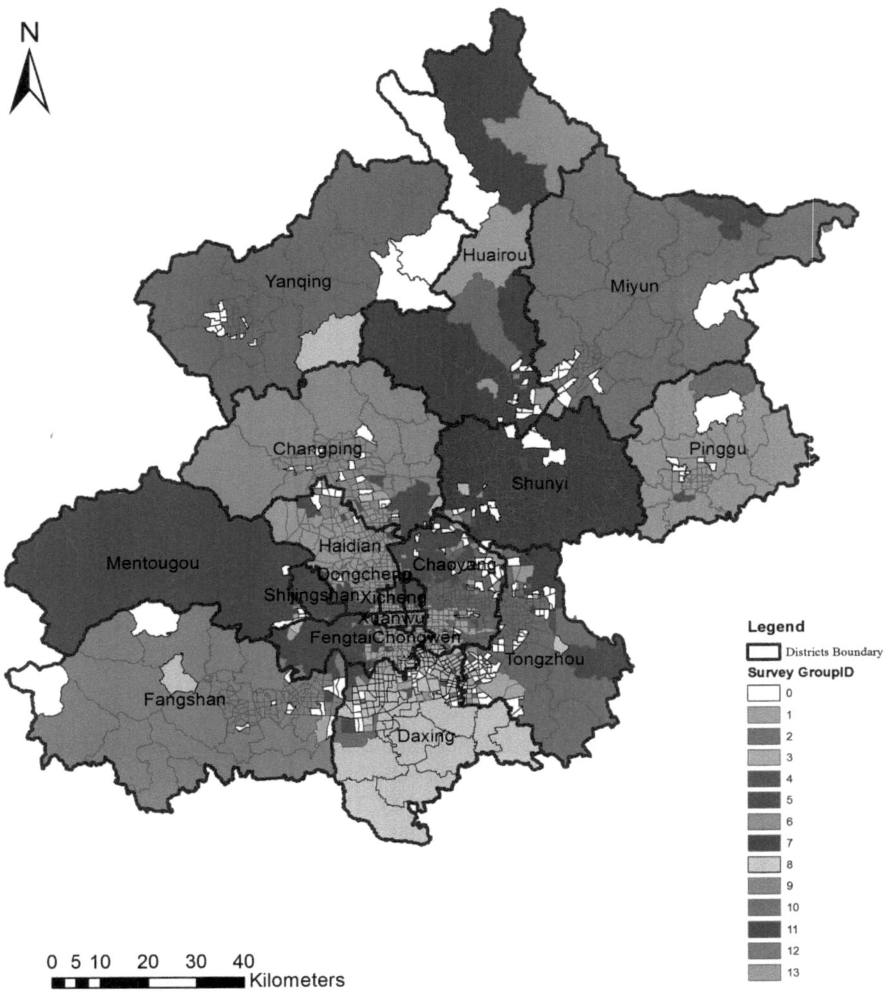

Fig. 8.4 The spatial distribution of community detection results of one-day household survey data

urban area boundary (see Fig. 8.5), which means that the actual operation data of publication transportation might be a good source to validate the urban planning and development.

8.4.3 Hourly Patterns

Coming back to the SCD, it is also valuable to study the spatiotemporal patterns in a micro-time scale. The community detection results of three-hourly aggregated trips, especially the commuting trips at peak hours yield insights on the overall

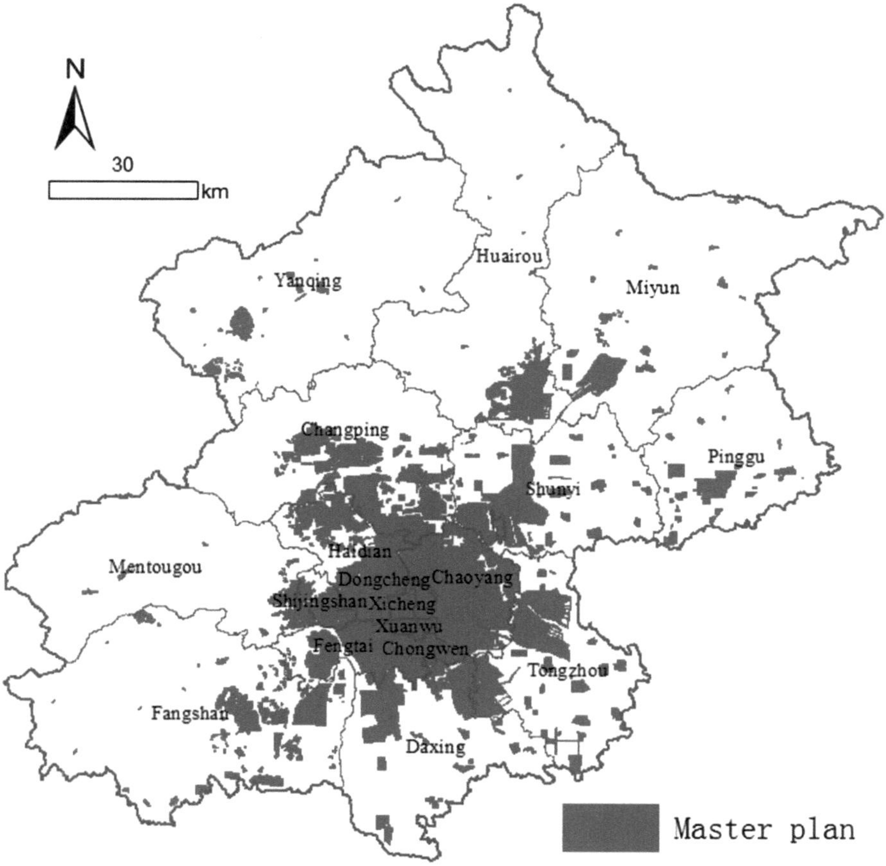

Fig. 8.5 Planned urban area in Beijing

job-related mobility patterns and intra-TAZ spatial interactions using public transportation. Table 8.2 shows the detailed network information of three-hourly community detection results of SCD during a weekday. We find that the hourly network structures change more (nodes and edges) than those of daily networks. The maximum modularity in hour 18–21 (0.473) and 09–12 (0.470) has the largest values and thus indicates a more robust community structure. For the spatial distribution patterns (see Fig. 8.6), the northern TAZs change more frequently than the southern parts, especially in the Changping District. In addition, similar to the daily patterns, there exist strong public transit connections through the whole day in TAZs that are located along the central west-to-east corridor including the Chang'an Avenue in Beijing, where the Beijing Subway line 1 (west-east) run through, as well as those TAZs along the northern part of the subway line 5 (north-south corridor). The interactive web map could also help us to identity underlying patterns by overlaying the detection results on the geographical contexts.

Table 8.2 Three-hourly community detection results of public transportation OD trips using SCD in a weekday

Hours	# of nodes	# of edges	# of groups	Mean of community size	Max Q
06–09	1205	45,359	7	172	0.459
09–12	1207	45,607	6	201	0.470
12–15	1206	55,510	6	203	0.462
15–18	1212	45,519	6	202	0.452
18–21	1208	46,378	5	242	0.473
21–24	1145	31,064	7	164	0.451

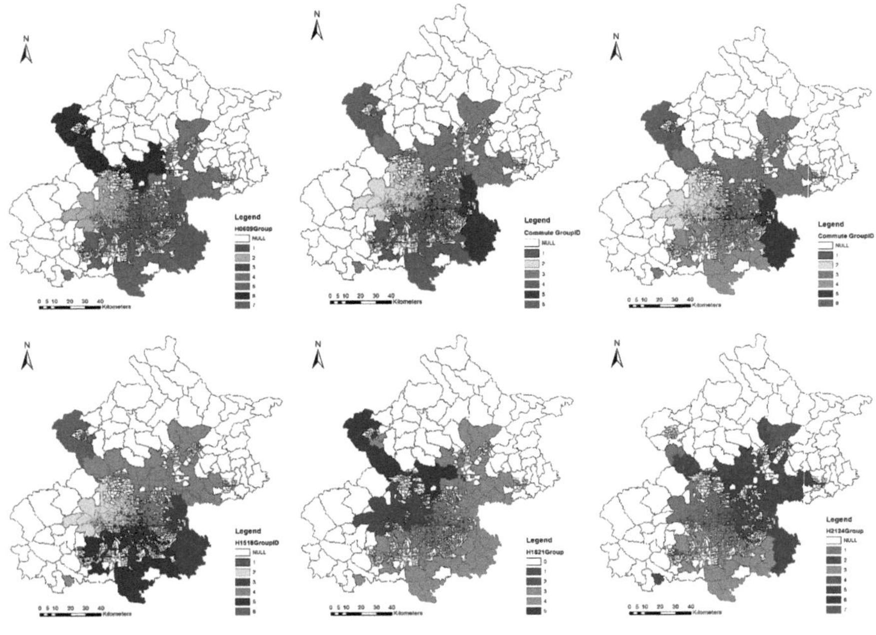

Fig. 8.6 The spatial distributions of three-hourly community detection results of public transportation OD trips using SCD in a weekday

8.4.4 Identification of Community Structure on Commuting Trips

In big cities, commuting trips usually contribute the largest share of daily public transportation. The commuting behaviors rely on the spatial distribution of regional job opportunities and housing markets. Long et al. (2012) introduced an algorithm to identify the residential zones and job places of the smart card users and further extract their commuting trips in Beijing. We extracted more than 700, 000

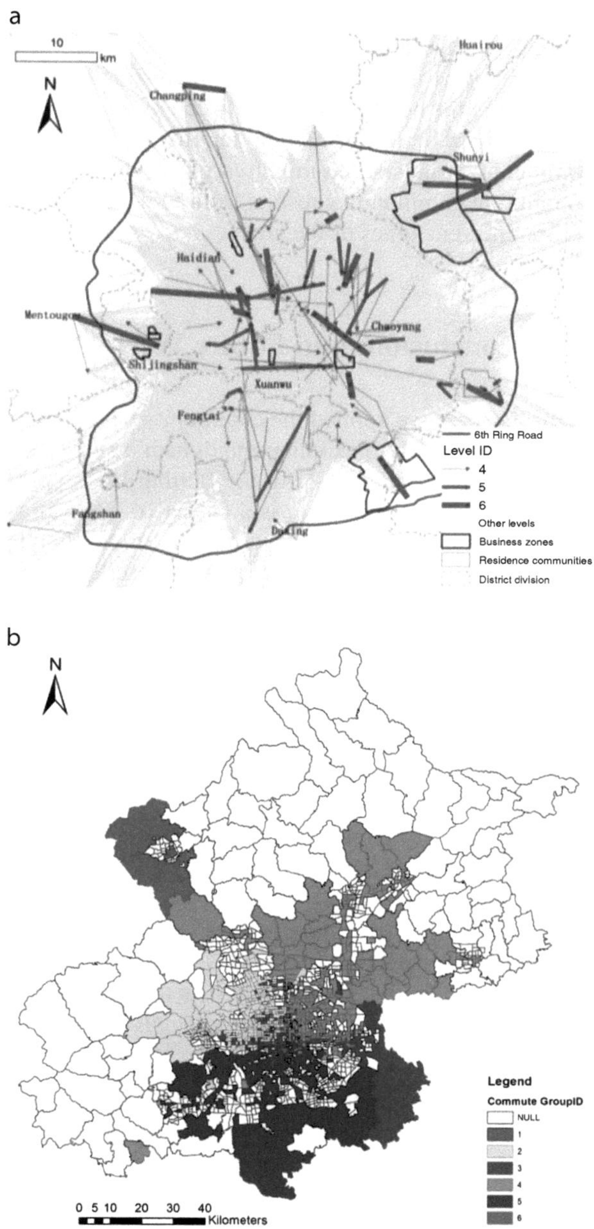

Fig. 8.7 (**a**) The *arrow-links* in the TAZ scale illustrates the commuting patterns (from home-region to job-region) in Beijing; (**b**) the spatial distributions of community detection results on commuting trips

commuting trips based on the 2010 SCD using this algorithm. Figure 8.7(a) shows the prominent home-to-job flows that represent heavy commuting traffic by a head/tail division. The spatial distribution of identified six commuting network communities is shown in Fig. 8.7(b). The largest commuting community contains 375

TAZs, while the smallest one has only 15 TAZs and the mean size is about 197. It tends to have more heavy commuting trips within the same community zones compared with inter-communities. Interestingly, the commuting community result is similar to the hourly patterns between 9 AM and 12 PM instead of that of 6 AM–9 AM (in Fig. 8.6), but a small modularity value 0.349 indicates that it is not a robust community structure and might vary over different time periods. The formation of commuting communities results from the influence of where people live and work across different zones of the city.

8.5 Conclusions and Future Work

In this chapter, we applied the community detection methods based on the study of complex networks to examine the dynamic spatial interaction structures of public transportation communities in the Beijing Metropolitan Area using SCD. A community represents that there is a subset of TAZs in which the smart card holders have more intra-community trips than inter-community trips via public transportation (i.e., bus and subway/metro). It also reflects the spatial heterogeneity of OD trips. There are several findings based on our experiment results:

First, the community detection results help to identify the functional connected traffic analysis zones by public transportation and most of them are consistent in both weekdays and weekends. Some detected spatially separated TAZs which belong to the same community indicate strong public travel demands in these regions, either for commuting trips on weekdays or for recreational trips on weekends.

Second, the daily community detection results using SCD are different from that using household travel surveys and the SCD community boundaries match better with Beijing urban planned area than the household travel survey.

Third, the hourly network structures change more than those of daily networks; the community detection results also have more variances in spatial distribution.

Fourth, the identified community structure on commuting trips shields insights on where and which residential- and job-related TAZs are connected by public transportation.

This research applies a network-analysis approach to investigate the ground-truth community structure of strongly connected TAZs via public transportation, which yields insights on urban structure in Beijing from the public transportation functional zone perspective. In further research, we would like to conduct more detailed analysis by integrating land-use data, points-of-interest (POI) database with human activities from household surveys or social media to give a more holistic view of public transportation using emerging urban big data and computing techniques. In addition, the map matching of these OD trips to actual streets and further analysis could be beneficial for the reliability analysis of street networks and emergency transportation management.

References

Beijing Transportation Research Center. (2011). Beijing transportation annual report 2011 (In Chinese).

Clauset, A., Newman, M. E., & Moore, C. (2004). Finding community structure in very large networks. *Physical Review E, 70*(6), 066111.

Gao, S., Liu, Y., Wang, Y., & Ma, X. (2013). Discovering spatial interaction communities from mobile phone data. *Transactions in GIS, 17*(3), 463–481.

Jang, W., & Yao, X. (2011). Interpolating spatial interaction data. *Transactions in GIS, 15*(4), 541–555.

Johnston, R., Gregory, D., & Smith, D. (1981). *The dictionary of human geography*. Oxford: Blackwell Reference.

Kang, C., Zhang, Y., Ma, X., & Liu, Y. (2013). Inferring properties and revealing geographical impacts of intercity mobile communication network of China using a subnet data set. *International Journal of Geographical Information Science, 27*(3), 431–448.

Liu, Y., Wang, F., Xiao, Y., & Gao, S. (2012). Urban land uses and traffic 'source-sink areas': Evidence from GPS-enabled taxi data in Shanghai. *Landscape and Urban Planning, 106*(1), 73–87.

Liu, Y., Sui, Z., Kang, C., & Gao, Y. (2014). Uncovering patterns of inter-urban trip and spatial interaction from social media check-in data. *PloS One, 9*(1), e86026.

Long, Y., & Thill, J. C. (2013). Combining smart card data and household travel survey to analyze jobs-housing relationships in Beijing. arXiv preprint. arXiv:1309.5993.

Long, Y., Zhang, Y., & Cui, C. Y. (2012). Identifying commuting pattern of Beijing using bus smart card data. *Acta Geographica Sinica, 67*(10), 1339–1352.

Manley, E. (2014). Identifying functional urban regions within traffic flow. *Regional Studies, Regional Science, 1*(1), 40–42.

Newman, M. E. (2004). Fast algorithm for detecting community structure in networks. *Physical Review E, 69*(6), 066133.

Newman, M. E., & Girvan, M. (2004). Finding and evaluating community structure in networks. *Physical Review E, 69*(2), 026113.

Rae, A. (2009). From spatial interaction data to spatial interaction information? Geovisualisation and spatial structures of migration from the 2001 UK census. *Computers, Environment and Urban Systems, 33*(3), 161–178.

Ratti, C., Sobolevsky, S., Calabrese, F., Andris, C., Reades, J., Martino, M., Claxton, R., & Strogatz, S. H. (2010). Redrawing the map of Great Britain from a network of human interactions. *PloS One, 5*, e14248.

Chapter 9
Profiling Underprivileged Residents with Mid-term Public Transit Smartcard Data of Beijing

9.1 Introduction

China has witnessed rapid economic development in recent decades since the country's reforming and opening-up in the 1980s. Despite the overall economic development, urban poverty has increasingly plagued Chinese cities, among which the post-Olympic Beijing would not be the exception (Song et al. 2009). A better understanding of identities and social-spatial mobilities of underprivileged residents with relatively low income is crucial for designing public policies and social programs, such as e.g. affordable housing and public transit system. In this chapter, we aim to provide a method for identifying socially and spatially underprivileged residents in the Chinese capital Beijing and detailing their social and spatial mobility based on increasingly available smartcard data from public transit system.

It is challenging to identify these underprivileged residents by conventional methods such as questionnaires (online or offline), face-to-face interview and telephone interview. One major obstacle remains that personal fortune – wealth or poverty – is still a taboo subject in the Chinese culture. Consequently, people's revealed attitudes and/or conditions about poverty is oftentimes not a robust measurement. For instance, the Household Travel Survey of Beijing suggests the overall annual household income in 2010 is around 30,000 CNY (less than 5000 USD), which is a far cry from numbers coming from other sources, such as other academic studies, statistical yearbooks, and local tales. In addition, poverty conditions assessed via surveys and interviews are often over-stated, as these surveys and interviews are often used to gather information for welfare programs. Still, conventional methods such as survey suffer from issues such as limited sample size and coarse spatial resolution. For these reasons, relevant official reports are prone to unrobust measurement and deliberate manipulation. To summarize, the identification of underprivileged residents and their social and spatial mobility is of significant social value, however the operationalization by conventional methods is difficult.

Y. Long, Z. Shen, *Geospatial Analysis to Support Urban Planning in Beijing*,
GeoJournal Library 116, DOI 10.1007/978-3-319-19342-7_9

With the development of information and communication technologies (ICT), voluminous, longitudinal, and ubiquitous digital measurements – often dubbed as "big data" emerge as an alternative solution to figure out our research questions through sensing and diagnosing cities (Batty 2012). Human mobility and activity could be captured by big data, providing detailed spatial and social information in a fine scale for academic research (Goodchild 2007). Among all types of big data such as GPS trajectories, social networks and phone call records, smartcard data (SCD) have been widely used in public transit systems in China.

Such public transit SCD is a promising candidate for charting the social and economic geography of underprivileged residents for the following reasons. **First**, bus/metro riders in Beijing are generally the low-income group. Most residents who could afford private cars would commute by driving, instead of by taking oftentimes over-crowed public transit. Economic-wise, car ownership is both a positional good as well as a life necessity in the increasingly car-based living in Beijing. After all, those who spend hours every day on the city's overcrowd transit system (Fig. 9.1) oftentimes have no other options. **Second**, over 90 % of all public transit passengers (mode share in 2010 is 40 %) possess identifiable smartcards, as regular card users receive a 60 % discount on the fares. Given the large size of public transit riders and the high percentage of smartcard ownership, SCD would indeed cover a large group of population. **Third**, a person who rides bus/metro for a relatively long period (e.g. two years) would have a higher probability to be an underprivileged resident in Beijing: long-time public transit riders often have limited social mobility, as in the case of Beijing. To sum up, public transit smart cardholders would overlap with the most underprivileged residents in Beijing.

With the advantages of SCD in mind, we should be cautious about some potential bias incurred from treating all frequent cardholders as underprivileged residents. For instance, several colleagues of the first author would commute everyday by bus, in order to live in the downtown and stay close to their jobs. Therefore, only a portion of cardholders meets the definition of socially and spatially underprivileged residents. Additional rules are needed to separate these residents from the general

Fig. 9.1 Crowd for riding bus (*left*, Line 930) and metro (*right*, Line 4) in afternoon peak of Beijing (Source: www.caing.com)

pool of cardholders. In addition, long-term observations would help to circumvent issues arising from short term fluctuations in ridership. This is consistent with the claim in Batty (2013) that it is possible to identify long-term dynamics using intermediate big data.

In this chapter, we identify cardholders' places of residence and work using SCD in 2008 and 2010, respectively. Their residential and job location changes could be derived via comparing the results of both years, since every smartcard has a unique ID which remains the same during the study period. The cardholders with identified housing place in both 2008 and 2010 are taken as underprivileged residents in Beijing, which were then profiled into various groups based on travel frequency, commute time, as well as jobs-housing location and their change. The human mobility of final identified cardholders associated with lower social and spatial mobility is then depicted in fine spatial scale based on which corresponding policy implications are drawn. This chapter is structured as follows. Section 9.2 reviews existing research in both urban poverty and SCD application. Study area, data, and underprivileged residents in Beijing are explained in Sect. 9.3. The methods and their results are described in Sects. 9.4 and 9.5, respectively. We draw conclusions and propose potential applications of the study in Sect. 9.6.

9.2 Related Research

9.2.1 *Urban Poverty of Chinese Cities*

Underprivileged residents have attracted extensive attention from scholars, in the form of research on urban poverty, urban migration, urban villages etc. Almost all existing studies were conducted using small-scale surveys or focus groups (e.g. several-hundred interviewers in several typical neighborhoods) (Chen et al. 2006; Fan et al. 2011; Wu et al. 2010; Zheng et al. 2009). Most of these studies focus on interviewers' living condition, traffic expenditure, public service requirements as well as commuting conditions, rather than residential and workplace variation and detailed mobility status, which are not easy to measure by conventional surveys. However, to the best of our knowledge, there is no study that identifies large-scale underprivileged residents, analyzes their jobs-housing variation during a middle term, and visualizes their mobility in a very detailed manner.

9.2.2 *Social-Economic Level Identification Using Trajectories*

There are few studies on automatically identifying social-economic levels of travelers (SELs) using trajectories like mobile traces, SCD, and floating car trajectories. Most studies focus on correlating SELs with large-scale-trajectory-extracted human mobility. For instance, Amini et al. (2014) analyzed the impact of developing level on human mobility using 150-day mobile call records. Frias-Martinez et al. (2012) studied the relationship between mobility variables and SELs using cell phone traces over six months, and found that populations with higher social-economic levels are strongly linked to larger mobility ranges than populations from lower socio-economic status (Aggregated correlation, not identification from individual mobility). One exception is the research of Kang et al. (2010), which used mobile call records of nine consecutive days, found those users with fewer but frequently-visited anchor points are associated with stable jobs, while those with more but less frequently-visited anchor points with unstable jobs. This is proven by common sense of Chinese residents as discussed in the introduction section. Therefore, identifying social-economic levels according to the jobs-housing place variation provides an alternative solution.

There are considerable publications on inferring housing and job places from trajectories like mobile phone call data records and location-based social networks (LBSN). For housing place identification, Lu et al. (2013) regarded the location of the last mobile signal of the day as the housing place of a mobile user. The most frequently visited point-of-interest (POI) (Scellato et al. 2011) or grid (Cheng et al. 2011; Cho et al. 2011) was regarded as an LBSN user's housing place. It is not easy to infer housing places form LBSN with a high spatial resolution. Comparing housing place identification, there are fewer studies on identifying job places using trajectories, with the exceptions of Cho et al. (2011) using LBSN and Isaacman et al. (2011) using cellular network data. It should be mentioned that taxi trajectories are not suitable for identifying a passenger's housing and job places considering the passenger-sharing nature of taxis. However, less attention has been paid to using SCD to identify housing and job places as well as their variation across time in a metropolitan city, which is the main focus of this work.

9.2.3 Smartcard Data Mining

The smartcard recording all cardholders' bus trip information is an alternative form of location-acquisition technology. Smartcard-automated fare collection systems are increasingly applied in public transit systems. Simultaneously with collecting fares, such systems can produce intensive travel patterns of cardholders, data that are useful for analyzing urban dynamics. Since 1990 the use of smartcards has become significant owing to the development of the Internet and the increased complexity of mobile communication technologies (Blythe 2004). Intelligent Transportation Systems (ITS) that incorporate smartcard-automated fare systems either existed or had been being established in over 100 Chinese cities as of 2007.

The data generated by smartcard systems track the detailed onboard transactions of each cardholder. We argue that smartcard technology can provide valuable information because it is a continuous data collection technique that provides a complete and real-time bus travel diary for all bus travelers.

Previous studies advocate using SCD to make decisions on the planning and design of public transit systems (see Pelletier et al. (2011) for a review) as well as to analyze urban structure. In South Korea, Joh and Hwang (2010) analyzed cardholder trip trajectories using bus smartcard data from ten million trips off our million individuals, and correlated this data with land use characteristics in the Seoul Metropolitan Area. Jang (2010) estimated travel time and transfer information using data of more than 100 million trips in Seoul from the same system. Roth et al. (2011) used a real-time "Oyster" card database of individual traveler movements in the London subway to reveal the polycentric urban structure of London. Gong et al. (2012) explored spatiotemporal characteristics of intra-city trips using metro SCD of five million trips in Shenzhen, China. Sun et al. (2012) used subway SCD in Singapore to extract passengers' spatio-temporal density and train's trajectory. Sun et al. (2013) analyzed encounter patterns of cardholders using around 20 million bus SCD from 2.8 million anonymous users over one week in Singapore. Long and Thill (2015) mapped commuting pattern of Beijing using one-week SCD in 2008 via identifying cardholders' housing and job places. Zhou and Long (2014) used the over 200 thousand identified commuting trips in Long and Thill (2015) to analyze bus commuters' jobs-housing balance in Beijing. In sum, considering urban dynamics process in terms of different response time and duration proposed by Wegener et al. (1986) and strengthened by Simmonds et al. (2011), existing research focuses on using a short period SCD for instant city mapping or revealing instant urban dynamics, but few of them is related with identifying medium- or long-term urban dynamics (e.g. land use change, residential construction or economic change) using a long-period SCD, with an exception of Sun et al. (2014).

> …… the power of big data is that if collected for long enough then the longer term will emerge from the short term. At the moment these data are about what happens in the short term, but over ten years or longer we will have a unique focus on the longer term – in fact we will have a snapshot of urban dynamics which is unprecedented. (Batty 2013)

It is possible to identify medium- or long-term (medium or low-speed) dynamics using intermediate big data. In this chapter, housing or job place change for a person belongs to medium-term urban dynamics, which is not easy to analyze using short-term big data. To our knowledge, this manuscript would be the first exploration in this field, in the context that all studies on big data focus on instant dynamics. We argue that, long term dynamics is possible to be identified using big data expanding several years.

9.3 Study Area, Data and Local Background

9.3.1 Study Area

As the capital of China, the Beijing Metropolitan Area (BMA) with an area of 16,410 km^2 has over 20 million residents in 2010 and is becoming one of the most populous cities in the world. The BMA lies in northern China, to the east of the Shanxi altiplano and south of the Inner Mongolian altiplano. The southeastern part of the BMA is a flatland, extending east for 150 km to the coast of the Bohai Sea. Mountains cover an area of 10,072 km^2, 61 % of the whole study area (Fig. 9.2). See Yang et al. (2013) for more background information on Beijing.

As of 2010, there were 184 km of commuter subway lines (excluding the airport express rail) in Beijing. Beijing Metro manages and maintains the subway system, which was built and financed by the BMG. Beijing Public Transportation Company,

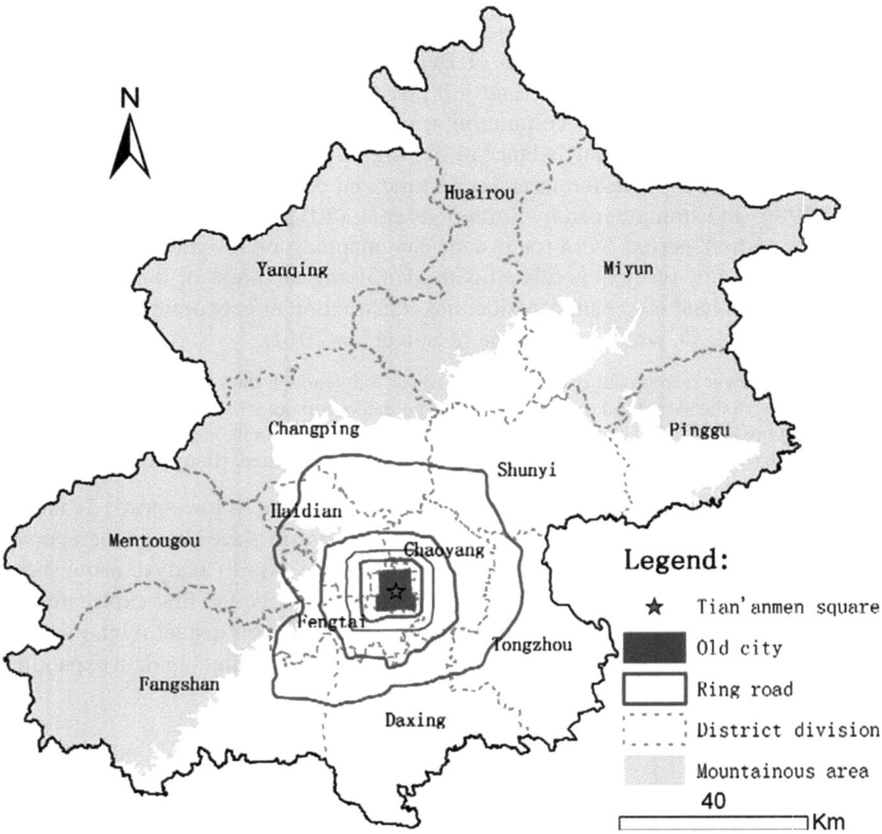

Fig. 9.2 The Beijing metropolitan area map

Table 9.1 Mode share of Beijing residents

Mode	2008 share (%)	2010 share (%)
Bus	28.8	28.9
Metro	8.0	10.0
Taxi	7.4	7.1
Car	33.6	34.0
Bike and walking	20.3	18.1
Company shuttle	1.9	1.9
Total	100	100

Sources: Beijing Transportation Research Center (2011)

a state-owned company provides public bus services in the Beijing metropolitan area. In 2011 alone, these buses produced 1.7 billion of vehicle kilometers traveled and transported total passenger of 4.9 billion.[1] Bus trips accounts for a significant share of all the trips by public transit (subway, bus and company shuttle). Thanks to the continuous expansion of the subway lines in Beijing, the share of subway trips have gradually increased in the city. Table 9.1 shows the mode share of the local residents of Beijing in 2008 and in 2010.

9.3.2 Data

9.3.2.1 Smartcard Data in 2008 and 2010

Since 2005, over 90 % of bus riders in Beijing have swiped an anonymous smartcard (images see Fig. 9.3) when boarding and alighting (for suburb routes) or when boarding (for inner-city routes) to pay for their fare. The high rate of smartcard usage among bus riders is largely because of the subsidy the government gives to riders paying the bus fare by a smartcard. Those riders enjoy 60 % discounts on any routes in the local bus system (80 % for students). Smartcard also enables owners to pay for other services such as taxi, electricity and sewage that are offered by the local government or companies it subsidizes.

When cardholders use their smartcard for paying bus services, the card reader installed on the bus automatically generates the following information, Bus trip origin and/or destination stops, boarding and/or alighting time, as well as the unique card number and the card type (student card at a discount vs. regular card). Two bus fare types exist. The first is fixed-fare, which is associated with short routes, and the other is distance-fare, which is associated with long routes. For the first type, 0.4 CNY is charged for each single bus trip, and the corresponding SCD contains only the departure time and stop ID and no arrival time or stop ID. Cardholders' spatio-temporal information is incomplete for this kind of route. For the latter type, the fare depends on the route ID and trip distance, and the SCD contains full information.

[1] Information based on: http://www.bjbus.com/home/view_content.php?uSec=00000002&uSub=00000012, accessed July 01, 2012.

Fig. 9.3 A public transit smartcard in Beijing (*Left*: front cover, *Right*: back cover)

Table 9.2 The summary of SCD in 2008 and 2010

Year	2008	2010
Date	7–13 April	5–11 April
Cardholder (m)	8.5	10.9
SCD (m)	78.0	97.9
1 Bus (m)	78.0	82.7
1.1 Flexible fare (m)	27.0 (34.7 %)	23.4 (28.3 %)
1.2 Fixed fare (m)	51.0 (65.3 %)	59.3 (71.7 %)
2 Subway (m)	0	15.2

Both types of bus SCD records are used in this chapter for identifying cardholders' housing and job places as well as residents that are socially and spatially underprivileged. Our strategy for tackling incomplete information issue of fixed-fare records have been elaborated in Long and Thill (2015), and the strategy proposed by Ma et al. (2012) could be used in future for solving the incomplete information of Beijing bus SCD. The records of subways have complete spatiotemporal information, the same as those of flexible-fare bus records.

We collected one-week SCD in 2008 and 2010 from Beijing Municipal Administration & Communications Card Co. (BMAC). They are the key data of this chapter. Details are in Table 9.2. Subway records are missing for 2008 SCD and included in 2010 SCD. SCD have been geo-coded using bus route and stops/stations in Fig. 9.4, which is a time consuming process.

9.3.2.2 Bus Routes/Stops, Subway Lines/Stations, and Traffic Analysis Zones (TAZs) of Beijing

GIS layers of bus routes and stops and subway lines and stops are essential for geocoding and mapping SCD. Since SCD in two years are included in this chapter, and bus routes and stops change a lot during 2008–2010, we used different spatial layers to map SCD in 2008 and 2010, respectively. In 2008, there were 1287 bus routes (Fig. 9.4a) and 8691 bus stops (see Fig. 9.4b) in the BMA. In 2010, there were 1928 bus routes (Fig. 9.4c) and 21,372 bus stops (Fig. 9.4d) in the BMA. Note that a bus route in this chapter has direction, e.g., the bus No. 113 has two routes, one from

Fig. 9.4 (**a**) Bus routes in 2008, (**b**) bus stops in 2008, (**c**) bus routes in 2010, (**d**) bus stops in 2010, (**e**) subway lines and stations in 2010, (**f**) traffic analysis zones in 2010 (TAZs) of the BMA. Note: All maps are from the Beijing Institute of City Planning. Some bus routes and stops are outside the BMA, as shown in (**a**–**d**) since some residents live outside the BMA and in adjacent towns in Hebei province

Dabeiyao to Qijiahuozi and the other from Qijiahuozi to Dabeiyao. They are counted separately in this chapter. The average distance between each bus stop and its nearest neighbor in 2010 is 231 m. Since SCD in 2010 include subway information, subway lines and stations are used to map SCD2010. In 2010, there are nine subway lines including the airport express line and 147 subway stations associated with the 2010 SCD (see Fig. 9.4e). We use Beijing TAZ data to aggregate the analytical results for better visualization. Totally 1911 TAZs are defined (see Fig. 9.4f) according to the administrative boundaries, main roads, and the planning layout in the BMA.

9.3.2.3 Household Travel Survey in 2010

The 2010 Beijing Household Travel Survey data (the 2010 survey hereafter) was used for getting underprivileged resident samples from all surveyed residents in Beijing. This survey adopts a multistage sampling strategy with a target of 1 % sampling rate. 1085 out of 1911 TAZs in the whole BMA are selected (few people reside in the other TAZs). In each TAZ, 10–50 households are selected to take a face-to-face interview. The final sample size is 46,900 households (116,142 residents) in the BMA. The 2010 Survey provides one-day travel diaries of all respondents, which gives commuting time for each employee (distance unavailable except the TAZs of Origin-Destination). The commuting mode is identified as the mode of the trip segment with the longest duration. We choose the mode of the highest mobility when there are two or more segments that have the same, longest duration. The 2010 Survey presents also household information including household structure, income, and residential location at the TAZ level, as well as personal information including gender, age, occupation, industry of employment, etc.

We identify underprivileged residents with only metro/bus trips from the 2010 survey to reveal their spatial mobility. These residents are limited to bus/metro users to link with SCD. We selected 4432 underprivileged residents (around 4 %) by three steps. Note that since the 2010 survey income information is not convinced and we therefore only select underprivileged residents with extreme conditions (thus 4432 selected residents are with extreme underprivileged in terms of the selection standards we used). First, we select any residents in households with no car, annual income lower than 50,000 CNY, housing area smaller than 60 m^2, and with no Moto or e-bicycle. Second, from those residents selected by the first step, we select residents whose age is between 18 and 80 (including 18 and 80), and who are not students or foreigners, and have smartcard. Third, we limited the residents selected by the second step to those only having bus/metro trips and over one ride per day.

We then analyze their profile to gain knowledge of the underprivileged residents of Beijing via the 2010 survey. Out of all 3479 households with identified underprivileged residents, 699 households do not have at least an employed person, and average housing area of the 699 households is 44.1 m^2 (6–60). Regarding their housing condition, 68 live in affordable housing (should be 16 % according to official report, Liao 2013), 935 as tenants, and 94 in relatives' or friends' housing, which shows that affordable housing was still not common for underprivileged residents in 2010. Among all identified underprivileged residents (1913 male and 988 with driving license), 3102 (70.0 %) residents have jobs, in which 2958 are full time, and others part-time. For others, there are 929 retired, 418 housewives, and 362 unemployed. For their education, seven hold postgraduate degrees, 121 hold bachelor degrees, and 2223 occupational education. They are 40 years old on average, 3348 (75.5 %) have local *Hukou* (namely residence permit). There are 128 staying in the current house for less than six months. Note that the 2010 survey only documents one-day travel diary and their socioeconomic attributes, but lacks housing and job change information, which is only available in long-term records.

Their one-day travel diaries are then extracted and explorative analyzed in detail. The trajectories of identified underprivileged residents are regarded as representative samples for depicting spatial mobility of underprivileged residents using metro/bus. There are totally 9367 trips (1757 walking trips not included) with 1337 metro trips (14.3 %). We found that 4123 (93.0 %) residents have two trips in a day, while others have more than two trips. We mapped all residents' housing and job places in space and found no significant characteristics of their jobs-housing places (see Fig. 9.5), which means that it is not easy to define underprivileged residents simply with their housing and job places. In addition, there are 2827 (30.1 %) commuting trips and 4405 (47.0 %) back-home trips among all trips, denoting that they have limited recreational and social activities in their daily life.

9.3.3 Underprivileged Residents in Beijing and Their Mobility

Better understanding the conditions of underprivileged residents in Beijing is necessary for answering our research questions with SCD. Generally, there are negligible underprivileged residents in Beijing. In 2010, the average household income for the

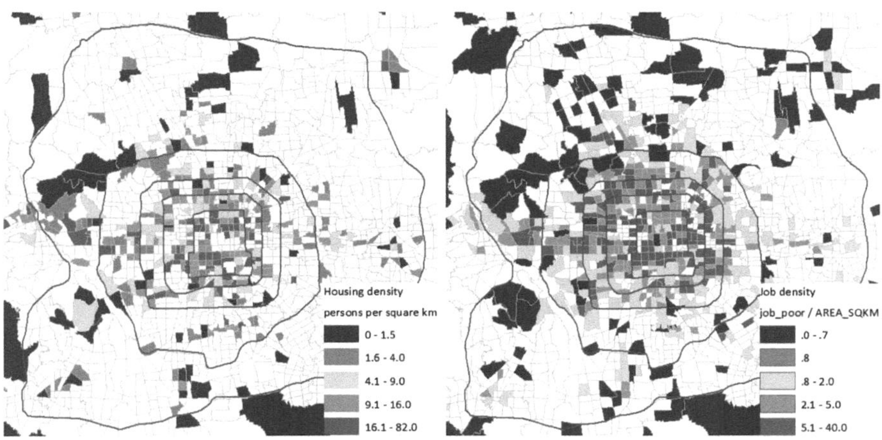

Fig. 9.5 Housing and job density in the TAZ level

bottom 20 % households in terms of income was 16,988 CNY in Beijing (the Chinese poverty line is 2300 CNY in 2011), while the average household income for all households is 33,360 CNY (Beijing Municipal Statistics Bureau and NBS Survey Office in Beijing, 2011). Among them, there are 137,063 residents (23 million for the whole country) receiving pension from the local government (http://files2.mca.gov.cn/cws/201107/20110711152146727.htm), whose average monthly income was 366 CNY in Dec. 2010. A negligible number of them stay in urban villages in Beijing (Zheng et al. 2009). They could be classified into four social groups, the traditionally poor, fresh graduates (Yizu), unemployed adults, and rural migrants. The traditionally poor residents live in the old city and have low income due to the lack of work skill. Fresh graduates generally are from middle to low class universities, most of which are outside Beijing. The fresh graduates work in small private firms and cannot afford to rent a flat in downtown. Unemployed adults lose their jobs thus have no income or only have limited income from part-time jobs. Rural migrants, peasants from other provinces in China, come to Beijing for low-skilled jobs, like construction and household/business services. It should be noted that a small portion of the traditionally poor and unemployed adults can receive pension from the local government. The very limited pension, however, can only afford very basic living requirements in the capital city Beijing.

We developed several methods to gain knowledge on each type of underprivileged residents in Beijing, on who they are, how they are, and how their mobility is. First, we used literature review, planner interview, common sense of the first author of this manuscript as a local resident, as well as online small-scale survey via Sina Weibo, the Chinese Twitter widely used in mainland China. We analyzed each social group in terms of their conditions of housing, job, transportation and leisure, as highlighted by The Athens Charter.

1. **Housing.** With the booming supply of constructed affordable housing projects by the government mainly since 2008 (construction starts during 2006–2007), more and more of them would assemble in these housing projects, 85 % of which are located between the fourth and the sixth ring roads of Beijing. The application requirement for affordable housing is the size of current housing (less than 15 m^2 per person), annual household income (below 88,000 CNY), and *Hukou* conditions (limited to local urban residents). For the traditionally poor living in the center city, they have to live further away from central locations after the demolition of urban villages or with the gentrification process. For those who cannot afford buying a house, to keep close to their jobs and to save transportation expense, they tend to group-rent in downtown that several tenants squeeze into a room for low rent per person (a latest act in Beijing forbids this behavior). Some tenants living in small properties in suburb suffer from long distance commute by metro or bus. In addition, due to the informal rental market in Beijing, the tenants frequently change their housing places.
2. **Job.** Most of them work for private enterprises rather than state-owned enterprises and do not have stable jobs, which is a common sense in Beijing (Gu et al. 2013).

Since their jobs are not stable, a considerable proportion of them frequently change their jobs.

3. **Transportation**. Due to the vehicle plate restriction, the increased taxi fare and parking fee, most of them have no car and travel by bus/metro/(e-)bicycle. Sheer cost of cars and petro is a more important factor to restrict their mobility. Most of them commute a long-distance to work, as revealed by Long and Thill (2015). For some extreme underprivileged residents like rural migrants, they still prefer buses and bicycles/electric bicycles than metro (although the metro fare in Beijing is very cheap as 2 CNY). They would only choose metro if they need to go somewhere so far that metro is the only sensible method. But for some fresh graduates, metro is very attractive.
4. **Leisure.** Most of the underprivileged residents' activities are around their housing places. Fresh graduates tend to travel around metro stations with good accessibility, and they are the heavy users of online shopping, which has competitive price comparing with high-end shopping centers.

The above analysis has been summarized in Table 9.3, revealing that the places of residence and job of the underprivileged residents are generally not stable and change a lot over time (e.g. two years). Housing change, job change, and commute are essential for profiling and differentiating underprivileged residents in Beijing. We expect to find these aspects by using SCD.

Table 9.3 The profile of underprivileged residents in Beijing

Type		Housing	Job	Transport	Leisure
Traditional poor	Spatial	Downtown or in suburb after demolition	Local or far		Local
	Non-spatial	Affordable housing	Small business/ unstable	(Electric) bicycle/bus	
Fresh graduates	Spatial	Downtown group rent or suburb rent	Business zones/ downtown	Long commuting	Around subway stations
	Non-spatial	Affordable housing	Private firms/ overload work/ frequently change job	Metro/bus	Online shopping/ buy sale
Unemployed adults	Spatial				Local
	Non-spatial	Affordable housing	No job	(Electric) bicycle/bus	
Rural migrants	Spatial	Close to job			Local
	Non-spatial	Small property housing	Service/industry	(Electric) bicycle/bus	

9.3.4 Most of Frequent Bus/Metro Riders in Beijing Are Underprivileged Residents

Almost all public transit smartcards are associated with only one cardholder during 2008–2010, which is our elementary assumption for the following analysis on SCD 2008 and 2010.[2] That is, the user of a smartcard does not change during 2008–2010 in most cases, which especially applies to those frequent bus/metro riders. This is confirmed by Beijing Transportation Commission, BMAC, and our small-scale interview in Beijing. There are a considerable amount of public transit passengers in Beijing having several smartcards, some of which might be used by their relatives who visit Beijing for business or leisure. Further, it is not common for several passengers to share one smartcard in Beijing. In addition, some cardholders get their smartcards refunds when they do not need the smartcards or the smartcards do not work. Totally 20 CNY would be refund for each smartcard. As of 8 Jan 2014, there are 44 refund locations in Beijing, the few amounts of which cause a long queue at each location (generally over one hour according to the interviews). This problem leads to some passengers throwing the broken smartcards rather than refund in Beijing. The IDs of refunded smartcards would not be used anymore according to our inquiries with Beijing Planning Commission and BMAC.

With the proven assumption of one-card-for-one-passenger, we further argue that most of frequent bus/metro riders in Beijing were and are representatives of underprivileged residents in Beijing from the following aspects, although this has been discussed in the introduction section (the third paragraph). This argument needs further explanation since it is the elementary assumption of our chapter to regard (most of) frequent bus/metro riders as underprivileged residents in Beijing and then to profile their typology and spatial mobility. This is supported by (1) a local household travel survey in 2010 (N = 116,142); (2) a small-scale survey of transit riders in 2012 (N = 709); (3) our interview with local residents in Beijing (N = 46), and (4) the first author's local knowledge of Beijing.

Evidence 1, by the 2010 Survey

We have described the spatial mobility of extremely underprivileged residents in Beijing from the 2010 survey. That analysis is for revealing the mobility profile of them. To avoid potential bias in the revealed profile, we used very strict condition in selecting urban underprivileged residents in Beijing. In addition to the analysis, we released several standards to select frequent bus/metro riders, who are defined as not students, having a smartcard, and having over two (>=3) riding bus/metro ridings per day. Among all 1946 persons from 618 households identified as frequent riders, 1601 have Beijing Hukou, 1386 have no driving license, 913 have full time job, 47 have part-time job, 798 are retired, and other shave no job. For the income of the 618 households, 417 are in the level 1, 166 are in the level 2, 191 are in the level 3, and

[2] The lottery policy for getting a permit to buy a vehicle by individuals was issued on 1 Jan 2011. Therefore these middle-class residents who have no permit to buy a private car are not included in our 2008 and 2010 SCD.

8 are in level 4. Moreover, 81 households have a car. According the aforementioned criteria of frequent bus/metro riders, they are economically underprivileged in terms of income, car ownership, as well as driving license ownership.

Evidence 2, by a Small-Scale Survey
In addition to the 2010 survey, we also conducted a small-scale survey on socio-economic attributes of cardholders in Shangdi and Qinghe sub-districts, Haidian district during September to December 2012. The survey was conducted by means of location sensing devices, an interactive survey website, questionnaires in person, and telephone. Totally 709 valid samples were retrieved (133 with valid smartcard ID) and each was associated with a valid card ID, socioeconomic attributes and corresponding travel diaries (one week). Among all 709 samples, there are 125 persons with at least one bus/metro trip per day. They are regarded as frequent bus/metro riders. Among 125 identified frequent riders, 80.8 % residents' monthly income are lower than 6000 CNY, and most of them (50 % of all) only in the 2001–4000 CNY level. Furthermore, among all 126 riders, 68 have no Beijing Hukou (a precondition for a stable and decent job in Beijing), only 20 have driving licenses, and 31 have a car. Therefore, frequent bus/metro riders are also economically underprivileged.

Evidence 3, from Interview with Local Residents in Beijing
For getting further knowledge of frequent bus/metro riders in Beijing, we also interviewed local residents in Beijing. Most of them were friends and colleagues of the first author. We asked their, their relatives', friends' and colleagues' ideas on or experience with frequent bus/metro riders. Totally 34 residents among all 40 agree with the argument.

Evidence 4, from Common Sense of Local Planners Who Stayed in Beijing
Finally, according to common sense of local planners on bus/metro frequent riders, those who ride bus/metro frequently for a long time, especially those with several transfers or a long distance, tend to be economically underprivileged.

Therefore, according to the evidences, most of frequent bus/metro cardholders (FCs) are underprivileged residents in Beijing. Some of FCs with a short commuting distance might be not underprivileged. We would consider this in the following method section via classifying FCs.

9.4 Method

Overall, the process for identifying and profiling underprivileged residents from SCD includes three steps. First, we identify housing and job places of cardholders using SCD in 2008 and 2010, respectively. The variations of housing and job places of cardholders are analyzed in detail. Second, those cardholders with housing places in both 2008 and 2010 are regarded as FRs. Third, we profile all FRs identified into 20 groups in terms of their housing place variation (change, no change),

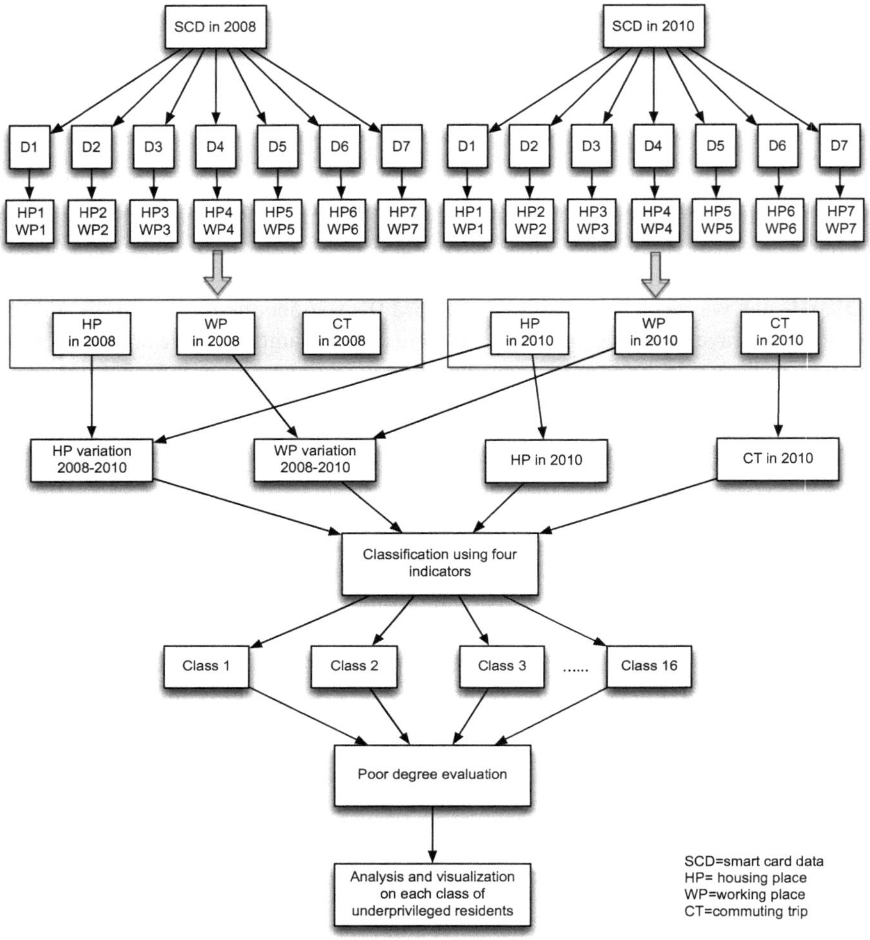

Fig. 9.6 Flowchart for identifying and profiling underprivileged residents using SCD in 2008 and 2010

job place variation (change, no change, find a job, lose a job, and jobless all the time) during 2008–2010 and housing place in 2010 (within the fourth ring road or not) for deriving policy implications for practitioners. Last, the underprivileged degree of each FR is then evaluated (Fig. 9.6).

9.4.1 *Housing and Job Place Identification of All Cardholders*

To identify a cardholder's workplace, we queried one-day data on a MS SQL Server and repeated the work for seven days based on these rules (see Long et al. 2013 for more details):

- The card type is not a student card;
- Dj>=6 h, where Dj is the duration that a cardholder stays at place j, which is associated with all bus stops within 500 m of one another;
- j <>1, which means that j is not the first place in a weekday that the server records.
- The place where a cardholder visited most frequently in five weekdays will be defined as the final workplace of the cardholder in this study.

Similarly, we deduced from the data queries that a place would be a cardholder's housing place if the data meet these conditions:

- The cardholder has an identified workplace;
- The card type is not a student card;
- Dh>=6 h, where Dh is the duration that a cardholder stays at place h, which is associated with all bus stops within 500 m of one another;
- Fh>= Fj, where Fh is the first and the most frequent place a cardholder starts a bus trip of a day within the week, Fj is the trip frequency to or from j that the cardholder has.

We applied the methods for identifying housing and job places for SCD in 2008 and 2010, respectively. It should be mentioned that in the 2008 SCD identification, to ensure that we singled out commuters solely by bus, we only selected cardholders that had continuous bus swipes. That is, our study does not consider commuters who are multimodal and take bus and metro in identification of the 2008 SCD. This problem is absent from the 2010 SCD identification. Since the card ID keeps the same in 2008 and 2010, we could reveal a portion of cardholders' residential place change and job place change, which would be used in the following step. In this chapter, we used the benchmark 2 km for variation in housing and working places. E.g., if a cardholder's housing place in 2008 is over 2 km desired distance to that in 2010, the cardholder's housing place would be assumed to change.

9.4.2 *Underprivileged Residents Identification and Classification*

Considering our discussion in the introduction section, Sect. 9.3.3 and Sect. 9.3.4, we regard cardholders with identified housing places in both the 2008 SCD and the 2010 SCD as underprivileged residents due to the following reasons. First, bus/metro riders in Beijing are already less privileged. Second, a cardholder with identified housing place could guarantee that he/she used bus/metro frequently and had a stable housing place in the week. Third, continuing frequently riding bus/metro since 2008 to 2010 would be more underprivileged among all cardholders. We admit that small portions of identified underprivileged residents are not really poor. For these, we would figure it out via the following classification.

Table 9.4 The profile of all types of underprivileged residents in Beijing (R4=The fourth ring road of Beijing)

Group	Housing change	Job change	Housing location	Cardholders
1	Yes	Yes	Out of R4	6044
2	Yes	Yes	Within R4	4055
3	Yes	Find a job	Out of R4	13,007
4	Yes	Find a job	Within R4	8556
5	Yes	Jobless	Out of R4	25,129
6	Yes	Jobless	Within R4	16,760
7	Yes	Lose the job	Out of R4	7635
8	Yes	Lose the job	Within R4	4803
9	Yes	No	Out of R4	672
10	Yes	No	Within R4	421
11	No	Yes	Out of R4	1728
12	No	Yes	Within R4	886
13	No	Find a job	Out of R4	3585
14	No	Find a job	Within R4	1771
15	No	Jobless	Out of R4	9023
16	No	Jobless	Within R4	3918
17	No	Lose the job	Out of R4	2374
18	No	Lose the job	Within R4	1097
19	No	No	Out of R4	761
20	No	No	Within R4	349

In accordance with the four urban functions by the Athens Charter, we classify all identified underprivileged residents into several groups in terms of three indicators, housing (changed or not), job (no change, changed, lose a job, find a job, jobless all the time), and housing location (within the fourth ring road or not). There would be 2×5×2=20 groups for all identified underprivileged residents as shown in Table 9.4. The size/ratio of each group could be calculated, and their trajectories in the 2010 SCD could be analyzed to profile their spatial mobility. Their social conditions could be profiled with the three indicators used for classification. Therefore, each group represents living styles of more underprivileged residents in Beijing.

9.5 Results

9.5.1 Identified FRs and Their Dynamics During 2008–2010

The SCD identified results in 2008 and 2010 are listed in Table 9.5. There are totally 7.3 m cardholders having recorded trips in both 2008 and 2010 SCD. There are far more cardholders with housing and job places in 2010 than those in 2008, which attributes to the absent of metro records in the 2008 SCD.

Table 9.5 The profile of SCD identification results

Indicator		SCD2008 ($\times 10^3$)	SCD2010 ($\times 10^3$)	Both ($\times 10^3$)
Total cardholders		8.5 m	10.9 m	7.3 m
Cardholders with	Housing place	1046	2121	113
	Job place	363	986	15
	Commuting trip	222	703	15
	Commuting distance	222	703	15
	Commuting time	222	420	9
Average commuting distance (km)		8.2	10.6	N/A
Average commuting time (min)		36.0	35.4	

Note: Not all commuting trips in 2010 have the commuting time, since we only have the boarding time for distance-fare bus lines in the 2010 SCD

Table 9.6 Housing place dynamics of FRs during 2008–2010

Housing place			Cardholders	Ratio (%)
Not changed			25,492	22.6
Changed		Total	87,082	77.4
	Inward (km)	All	42,013	37.3
		2–5	9211	8.2
		5–10	9651	8.6
		10–20	13,150	11.7
		>=20	10,001	8.9
	Outward (km)	All	45,069	40.1
		2–5	7990	7.1
		5–10	10,139	9.0
		10–20	16,400	14.6
		>=20	10,540	9.4
Sum			**112,574**	**100.0**

There are 112,574 cardholders with identified housing places in both 2008 and 2010, which are defined as FRs and underprivileged residents in Beijing. Their housing place, workplace, and commuting dynamics during 2008–2010 are analyzed as shown in Tables 9.6, 9.7, and 9.8.

There are 77.4 % FRs' housing places were changed during 2008–2010, and more FRs moved outward (most of them with over 5 km outward distance).

The changes of FRs' workplaces are also significant during 2008–2010. Only 13.3 % FRs kept working in 2008 and 2010. There are 14.1 % FRs lost their jobs during 2008–2010. That is, each of them had an identified workspace in 2008 but not in 2010. We also found that 23.9 % FRs got a job in 2010. Surprisingly there are 48.7 % FRs kept jobless in 2008 and 2010. We assume most of them are retired residents, part-time workers and jobless residents (students not excluded from housing and job identification process).

Table 9.7 Workplace dynamics of FRs during 2008–2010

Workplace				Cardholders	Ratio (%)
Working				14,916	13.3
	Not changed			2203	2.0
	Changed		Total	12,713	11.3
		Inward (km)	All	6142	5.5
			2–5	1444	1.3
			5–10	1893	1.7
			10–20	2071	1.8
			>=20	734	0.7
		Outward (km)	All	6571	5.8
			2–5	1371	1.2
			5–10	2018	1.8
			10–20	2156	1.9
			>=20	1026	0.9
Losing job				15,909	14.1
Finding a job				26,919	23.9
Jobless				54,830	48.7
Sum				**112,574**	**100.00**

Table 9.8 Commuting distance variation of FRs (with commuting trips both in 2008 and 2010)

Commuting distance in 2010 – that in 2008 (km)	Cardholders
>=20	436
10–20	1885
5–10	2266
2–5	2419
0–2	2647
(−2)–0	1984
−5–(−2)	1416
−10–(−5)	1069
−20–(−10)	622
<=−20	172
Sum	**14,916**

For those FRs with commuting trip identified in both 2008 and 2010, we analyzed the variation of their commuting distance. Totally 78.0 % FRs' commuting distance increased from 2008 to 2010, indicating the commuting situation by bus/metro got worse during the period.

We classified all these 112,574 FRs into 20 types according to the method discussed in Sect 9.4. The total number of each type is shown in Table 9.4.

9.5.2 Evaluation on Underprivileged Degree

There are several separate approaches for evaluating the underprivileged degree for each group of underprivileged residents identified as follows. (1) Trajectory comparison. The trajectories of identified results of each type can be compared with those of urban poor only by bus/metro documented in the 2010 survey (4432 extremely poor), in terms of the similarity of trajectories. The more similar, the more underprivileged a type would be. (2) Residence context of housing places. This approach calculates the decent level of each stop/station (e.g. the spatial context of a stop/station like housing price or affordable housing projects or high-end amenities or high-income TAZs) in trajectories of identified cardholders by each model, and sum for each cardholder. The model with the least average decent level by cardholders would be selected as the best model for identifying urban poor using SCD. (3) Double check using the 2014 SCD. Average trip count in 2014 of all cardholders appearing in both the identified urban poor and the 2014 SCD (also one week) could be calculated. The model with a greater trip count total would be selected as the best model used in this chapter. In this chapter, we adopt the last approach for evaluating the underprivileged residents we identified.

Among all 112,574 FRs identified, 29,189 (25.9 %) have at least one trip in a week in 2014. Their average trip count is 13.0 in 2014 and the standard deviation is 9.7. Those 29,189 residents are regarded as the most underprivileged FRs, and their underprivileged degree increases with the trip count in 2014. For those majorities not appearing in the 2014 SCD, there are several reasons, (1) their economic condition has improved and become private car drivers; (2) they do not use the previous cards but still ride bus by other cards. This condition is not dominating in Beijing according to our local knowledge; (3) they moved out of Beijing, which is common with increasing living cost in the city. In sum, the dynamics of identified FRs was huge during 2010–2014, which is normal when comparing with those middle-class residents in Beijing.

9.6 Conclusions and Discussion

This chapter explores on using public transit smartcard data during 2008–2010 for identifying and profiling economically underprivileged residents in Beijing. The ground truth of underprivileged residents in Beijing has been elaborated from their housing, job, transport and leisure aspects. The mobility of them has been disclosed by analyzing economically underprivileged residents documented in the 2010 Beijing household travel survey. Considering Chinese situation, we then regarded these frequent public transit passengers as underprivileged residents, which have been proven by several avenues. This empirical study includes the following aspects. First, we extracted rules from the 2010 survey for identifying housing and job places for cardholders in the 2008 and 2010 SCD, respectively. The inherent smartcard IDs during 2008–1010 enabled us to detect shared cardholders' housing, job and commuting dynamics in the period. Second, we regarded these cardholders with identified housing place in both 2008 and 2010 as underprivileged residents in Beijing. We then profile them into 20 groups considering their housing place variation (change, no change), job place variation (change, no change, find a job, lose a job, and jobless all the time) during 2008–2010 and housing location in 2010 (within the fourth ring road or not). The mobility pattern of each group has been profiled in detail. Last, the underprivileged degree of cardholders in each group were estimated and compared with each other.

The methodological contributions of this chapter are as follows. **First**, we use two-year-spanning immediate big data for understanding long-term urban dynamics, which is not possible by short-term data and only available when big data are accumulated (Batty 2013). **Second**, we propose a promising solution on extracting rules from conventional travel surveys and using them for identifying interested information from big data. The straightforward method has its potential in applying to other cities with access to SCD. **Third**, a large number of underprivileged residents in Beijing are for the first time profiled in terms of their housing, job and spatial mobility, which is not easy by conventional surveys in the background that Chinese people dislike disclosing their social condition and detailed spatial mobility (the gap of average household income in the 2010 survey and actual is an evidence). In sum, the urban poverty analysis is not easy by other types of big data, and SCD have advantage for this issue in China considering the special situation.

In addition to our methodological contributions, we see opportunities on its application in practices. **First**, the revealed profile of underprivileged residents in terms of their housing, job and mobility can be regarded as references to the location choice of poor-aware facilities/amenities, like affordable housing and related job markets. **Second**, the identification results can be referred in the process of social welfare application like affordable houses and monetary pensions. That is, whether an applicant qualifies the social welfare standard can be double-checked by his/her trips using SCD, in addition to stated or reported indicators. **Third**, the 60 % public transit fare discount on using smartcard has been applied to all passengers since the use of smartcards in 2005. There are appeals on raising the fare due to unaffordable operation cost by the local government. In this situation, one possible solution is that public transit fare subsidization can go to these identified underprivileged residents, who are in great need of fare discount, comparing to others. **Fourth**, the profiled mobility patterns of identified underprivileged residents can be addressed in revising public transit planning and design.

While admitting the merits of our study, there are several potential biases to be addressed in our future research. **First**, metro records are not included in the 2008 SCD, resulting in those cardholders commuting by metro from home are absent from identified underprivileged residents. **Second**, how the underprivileged residents identified from SCD are representative of all is an unavoidable question, as discussed by Liu et al. (2014) for social network users and Sun et al. (2012) for train passengers. For instance, underprivileged residents with no bus/metro trips are not accounted in this study. These residents by e-bicycles are more underprivileged by our common sense in Beijing, which could be complemented by the 2010 survey. In addition, we are extending our data analysis from 2008–2010 to 2008–2013 for understanding a longer term urban dynamics like residential and job location choice.

References

Amini, A., Kung, K., Kang, C., Sobolevsky, S., & Ratti, C. (2014). The impact of social segregation on human mobility in developing and urbanized regions. arXiv preprint. arXiv:1401.5743.

Batty, M. (2012). Editorial. *Environment and Planning B: Planning and Design, 39*, 191–193.

Batty, M. (2013). Urban modelling: A progress report. In Symposium on applied urban modelling (AUM 2013), University of Cambridge.

Beijing Municipal Statistics Bureau, NBS Survey Office in Beijing. (2011). *Beijing statistical yearbook 2011*. Beijing: China Statistics Press.

Beijing Transportation Research Center. (2011). Beijing transportation annual report 2009. Unpublished official report. (trans: Chinese). http://www.bjtrc.org.cn/InfoCenter/NewsAttach// aeb7c878-d31e-4f08-982f-3c17c717c87b.pdf. Accessed 21 Sept 2013.

Blythe, P. (2004). Improving public transport ticketing through smart cards. *Proceedings of the Institute of Civil Engineers, Municipal Engineers, 157*, 47–54.

Chen, G., Gu, C., & Wu, F. (2006). Urban poverty in the transitional economy: A case of Nanjing, China. *Habitat International, 30*(1), 1–26.

Cheng, Z., Caverlee, J., Lee, K., & Sui, D. Z. (2011). Exploring millions of footprints in location sharing services. *ICWSM, 2011*, 81–88.

Cho, E., Myers, S. A., & Leskovec, J. (2011). Friendship and mobility: User movement in location-based social networks. In Proceedings of the 17th ACM SIGKDD international conference on knowledge discovery and data mining (pp. 1082–1090), ACM.

Fan, C. C., Sun, M., & Zheng, S. (2011). Migration and split households: A comparison of sole, couple, and family migrants in Beijing, China. *Environment and Planning - Part A, 43*(9), 2164.

Frias-Martinez, V., Virseda-Jerez, J., & Frias-Martinez, E. (2012). On the relation between socio-economic status and physical mobility. *Information Technology for Development, 18*(2), 91–106.

Gong, Y., Liu, Y., Lin, Y., Yang, J., Duan, Z., & Li, G. (2012). Exploring spatiotemporal characteristics of intra-urban trips using metro smartcard records. In Proceedings of geoinformatics, Hong Kong.

Goodchild, M. F. (2007). Citizens as sensors: The world of volunteered geography. *GeoJournal, 69*(4), 211–221.

Gu, Y., Xu, Y., & Zheng, S. (2013). Job uncertainty, residential location choices and commuting patterns in Beijing. Working paper.

Isaacman, S., Becker, R., Cáceres, R., Kobourov, S., Martonosi, M., Rowland, J., & Varshavsky, A. (2011). Identifying important places in people's lives from cellular network data. In K. Lyons, J. Hightower, & E. M. Huang (Eds.), *Pervasive computing* (pp. 133–151). Berlin/Heidelberg: Springer.

Jang, W. (2010). Travel time and transfer analysis using transit smart card data. *Transportation Research Record: Journal of the Transportation Research Board, 2144*, 142–149.

Joh, C-H., & Hwang, C. (2010). A time-geographic analysis of trip trajectories and land use characteristics in Seoul metropolitan area by using multidimensional sequence alignment and spatial analysis. AAG 2010 annual meeting, Washington, DC.

Kang, C, Gao, S., Lin, X., Xiao, Y., Yuan, Y., Liu, Y., & Ma, X. (2010). Analyzing and geo-visualizing individual human mobility patterns using mobile call records. In Geoinformatics, 2010 18th international conference on. IEEE (pp. 1–7).

Liao, Z. (2013). Beijing affordable housing planning. Zhuqu, submitted (in Chinese).

Liu, Y., Sui, Z., Kang, C., & Gao, Y. (2014). Uncovering patterns of inter-urban trips and spatial interactions from check-in data. arXiv preprint. arXiv:1310.0282.

Long, Y., & Thill, J.-C. (2015). Combining smart card data, household travel survey and land use pattern for identifying housing-jobs relationships in Beijing. arXiv preprint. arXiv: 1309.5993.

Lu, X., Wetter, E., Bharti, N., Tatem, A. J., & Bengtsson, L. (2013). Approaching the limit of predictability in human mobility. *Scientific Reports, 3*, 2923.

Ma, X. L., Wang, Y. H., Chen, F., & Liu, J. F. (2012). Transit smart card data mining for passenger origin information extraction. *Journal of Zhejiang University SCIENCE C, 13*(10), 750–760.

Pelletier, M.-P., Trepanier, M., & Morency, C. (2011). Smart card data use in public transit: A literature review. *Transportation Research Part C, 19*, 557–568.

Roth, C., Kang, S. M., Batty, M., & Barthélemy, M. (2011). Structure of urban movements: Polycentric activity and entangled hierarchical flows. *PloS One, 6*(1), e15923.

Scellato, S., Noulas, A., Lambiotte, R., & Mascolo, C. (2011). Socio-spatial properties of online location-based social networks. *ICWSM, 11*, 329–336.

Simmonds, D., Waddell, P., & Wegener, M. (2011). Equilibrium vs dynamics in urban modelling. In Symposium on applied urban modelling (AUM 2011) (pp. 23–24), University of Cambridge.

Song, S., Zhu, E., & Mukhopadhyay, S. (2009). Urban poor in China. *Chinese Economy, 42*(4), 44–62.

Sun, L., Lee, D. H., Erath, A., & Huang, X. (2012). Using smart card data to extract passenger's spatio-temporal density and train's trajectory of MRT system. In Proceedings of the ACM SIGKDD international workshop on urban computing (pp. 142–148). ACM.

Sun, L., Axhausen, K. W., Lee, D. H., & Huang, X. (2013). Understanding metropolitan patterns of daily encounters. *Proceedings of the National Academy of Sciences, 110*(34), 13774–13779.

Sun, L., Jin, J. G., Axhausen, K. W., Lee, D. H., & Cebrian, M. (2014). Quantifying long-term evolution of intra-urban spatial interactions. arXiv preprint. arXiv:1407.0145.

Wegener, M., Gnad, F., & Vannahme, M. (1986). The time scale of urban change. In B. Hutchinson & M. Batty (Eds.), *Advances in urban systems modelling* (pp. 145–197). Amsterdam: North Holland.

Wu, F., He, S., & Webster, C. (2010). Path dependency and the neighbourhood effect: Urban poverty in impoverished neighbourhoods in Chinese cities. *Environment and Planning A, 42*(1), 134.

Yang, Z., Cai, J., Henk, F., Ottens, L., & Sliuzas, R. (2013). *Beijing. Cities, 31*, 491–506.

Zheng, S., Long, F., Fan, C. C., & Gu, Y. (2009). Urban villages in China: A 2008 survey of migrant settlements in Beijing. *Eurasian Geography and Economics, 50*(4), 425–446.

Zhou, J., & Long, Y. (2014). Bus commuters' jobs-housing balance in Beijing: An exploration using large-scale synthesized smart card data. *Journal of the Transportation Research Board* (Forthcoming).

Chapter 10
Discovering Functional Zones Using Bus Smart Card Data and Points of Interest in Beijing

10.1 Introduction

A city, whether from a material or a social perspective, is a complex space system (Lai and Han 2009; Batty 2009), and the study based on the microscopic objects is an important way to understand the operation of this complex system. Due to the limitation of data, the classic study of urban elements, organization and structure is often limited in large scale (such as townships or traffic analysis zones). But with the emerging of big data, Ubiquitous volunteered data have provided a new channel to describe and understand the structure of urban space (Batty 1990).

Big data, which have been a major focus in some research areas, mainly consist of bus smart card records, flight records, credit card records, twitter records, mobile phone records and so on. These data can be formed into a log of citizens' trip behavior (Neuhaus 2009), and can be used to real-time monitor the urban activities (Calabrese and Ratti 2006), analyze the intensity and temporal distribution of urban activities (Ratti et al. 2006) and study human mobility pattern in urban areas (Gonzalez et al. 2008).

Human mobility has a close relationship with urban spatial structure (Goodchild and Janelle 1984; Goodchild et al. 1993). In existing research, urban spatial structure is often used by researchers to study human activity, analyze citizens' trip behavior and explore the impact of urban spatial structure on people' traveling (Jiang et al. 2009). For example, by studying the structure of urban land-use, urban commuter model was studied (Hamilton 1982), and the impact of the spatial structure on residents' commuting behavior was analyzed (Liu and Wang 2011; Wang and Chai 2009). However, few studies have discussed how to use the existing data of urban human activity for urban functional zoning. Because along with the development of the city, urban land use and spatial structure are changing rapidly and the city is developing from the previous mono-center model to a polycentric model (Anas et al. 1998; McMillen and McDonald 1997). The immediate and clear division of urban functional areas can give inspiration to city planners on the future planning,

Y. Long, Z. Shen, *Geospatial Analysis to Support Urban Planning in Beijing*, GeoJournal Library 116, DOI 10.1007/978-3-319-19342-7_10

and validate the land use planning of the past. However, traditional studies on land use and urban spatial structure are mainly based on remote sensing data (Lu and Weng 2005; Xiao et al. 2006) which is expensive and lacks timely updates, thus cannot meet the needs of urban planners and scholars. Therefore, it will become prevailing to use the vast amounts of data provided by LBS to analyze human activity and carry out research of urban spatial structure.

In existing researches of urban spatial structure, GSM and GPS data have been extensively used. For example, Qi et al. used GPS information of driving taxis in Hangzhou to analyze the relationship between passengers' patterns of getting on and off and the corresponding social function of urban areas (Qi et al. 2011). Based on one week's GPS data from more than 6600 driving taxis in Shanghai, Liu et al. used the "source-sink" model proposed by Pulliam to characterize the daily traffic model, and then to analyze the present situation of land use in Shanghai (Liu et al. 2012; Pulliam 1988). Yuan used GPS information of taxi and urban POIs (Point of Interests) data to establish a semantic model and study the functional partition of different regions of the city with the help of data mining method (Yuan et al. 2012).

In recent years, SCD (Smart Card Data) have gradually been used in urban studies as a kind of large-scale data with spatiotemporal labels. Sun analyzed passengers' spatio-temporal density and activity tracks based on the SCD of Singapore (Sun et al. 2012). Joh and Hwang used 10 million pieces of SCD records of Seoul metropolitan area to analyze the feature of cardholders' routes and urban land use (CH and Hwang 2010). Long et al. analyzed the relationship between working and living locations as well as the feature of commuting direction using SCD of Beijing (Long et al. 2012). In addition, some researchers have begun to use POIs in the study of urban space. POIs are some basic locations of the city and mainly include buildings with landmark function in the local area. POIs have been widely applied to the study of urban spatial structure. According to the significant degree of difference, Zhao et al. extracted hierarchical landmarks from POIs, and obtained the hierarchical knowledge space that can be used in intelligent route guidance (Zhao et al. 2011). Since LBS technology is featured with high positioning accuracy, interactive feature and huge amount of data while POIs have obvious advantages in identifying geographical pattern, its id of great significance to integrate the two in the study of urban spatial structure.

Studies using LBS technologies are still in their starting stage. Because the amount of data is too large, it is usually hard to extract useful information using the traditional data analysis method and technology (Witten and Frank 1999). In recent years, with the rapid development of computer techniques, database management system and artificial intelligence technologies gradually became mature and their combination has contributed to the effective characterization and analysis of large data (Tan et al. 2006). Scholars also have begun to use the classification, association analysis and cluster analysis to discovery potentially and useful information from vast amounts of GIS data. For the past few years, cluster analysis has been widely used, to analyze daily activities of urban people like working, living, attending school, travelling and shopping based on GPS, GSM, SCD and other data (Jiang et al. 2012a; Sun et al. 2011), and thereby to identify the space-time structure of the city (Jiang et al. 2012b) as well as the immediate and detailed information of urban land use (Pan et al. 2013).

In actual studies, the data provided by LBS technology is often continuous in time and such data are called time-series data (Agrawal et al. 1993). Time-series data are usually massive data that may include a lot of noise and have poor efficiency or even impossibility in direct cluster analysis of raw data. Therefore, dimensionality reduction and feature transformation for multi-dimensional time series data is needed. Discrete Fourier Transform (DFT) (Agrawal et al. 1993), Principal Component Analysis (PCA) (Sun and Chen 1994), Singular Value Decomposition (SVD) (Korn et al. 1997) are some commonly-used methods for this. Cluster analysis of time series data can be similarity-based, feature-based, model-based and segmentation-based. And the choice mainly depends on the type of application data and the purpose of cluster analysis (Jiang et al. 2005). Traditional cluster analysis is mostly vector-based which cannot well solve the problem of time series clustering. In recent years, model-based clustering was more used in the study on clustering analysis of time series.

This study is to establish the model of Discovering Zones of different Functions (DZoF) based on the SCD and POIs. In this model, the bus platform level traffic data model was developed. We have used the model-based algorithm – expectation maximization (EM) in cluster analysis of 8,691 bus platforms in Beijing, as well as the pattern recognition rules of SCD data mining based on the traditional studies of residents' commuting behavior, general cognition and POIs model, and conducted a functional interpretation of clusters resulted from previous clustering analysis. According to DZoF model, this study ultimately determined the function of each bus platform in Beijing, and made a summary on a scale of TAZ to achieve recognition of the function of different regions. In order to verify the validity of DZoF (Discovering Zones of different Functions) model in identifying the result, the chapter also made a comparison analysis between the land-use map of the overall urban planning of Beijing (2004–2020) and Google map of the area.

The method section of this chapter gives a definition of DZoF model used for urban functional identification, and introduces the specific method to use DZoF model in the cluster analysis of multi-dimensional time series. In the application section, Beijing is taken for an example, a continuous week's bus SCD of Beijing in April 2008 and POIs are used to identify each functional area of the city while experimental result is also tested. The final part is about the summary and discussion of the entire study.

10.2 Overview of Study Area and Explanation of Data

10.2.1 Overview of Study Area

The study area of this chapter is Beijing city area, with a total area of 16,410 km^2 and a permanent population of 20,693,000.[1] Beijing has a modern and tridimensional transportation network which ramifies all over the city. Till 2008, there are 17,000 km bus-lines and 20,600 operating vehicles, eight lines of rail transit with 200 km of operating mileage, and 66,000 operating taxis.[2]

10.2.2 Data

10.2.2.1 Lines and Platforms of Buses in Beijing

The data of this study is mainly based on one continuous week (April 7th–13th)'s bus smart card records at Beijing in 2008 (smart card records of rail transit not included), which covers more than 600 bus lines (a total of 1,287 inbound and outbound records which contains 566 records of one-ticket lines and 721 of segmented-pricing lines), and about 37,000 bus stops[3] and 8,691 bus platforms. Figure 10.1 shows the distribution of bus platforms in Beijing.

10.2.2.2 Smart Card Data

Due to the non-technical factors, the smart card data applied in this study do not include those of bus and rail transit lines operated by Xianglong Company, which is a state-owned bus company at Beijing. The basic information of each record includes: card-swiping time and location of each card holder (the location is represented by the number of the line and the stop), type of card (ordinary, student or staff card), and transaction sequence number (representing the cumulative number of a cardholder's swiping card), driver number and vehicle number. There are altogether 77,976,010 card-swiping records in that week.

Beijing bus lines have two modes of tolling: (1) one-ticket lines with short distance, which are mostly located in the city centre, and the SCD of this mode only record passengers' pick-up time instead of drop-off time; (2) segmented-pricing lines, usually with long route, and both terminal and starting station located outside the Fifth Beltway. The SCD of this mode record the complete spatial-temporal information of cardholders' behavior of card swiping. In order to collect both pick-up

[1] The data is from the 2012 Statistical Yearbook of Beijing (http://www.bjstats.gov.cn).

[2] The data is from the statistics on Beijing Transportation website (http://www.bjbus.com/).

[3] It is not the number of but platforms, but the sum of bus stops of all bus lines. A bus stop is the name of a platform in a bus line.

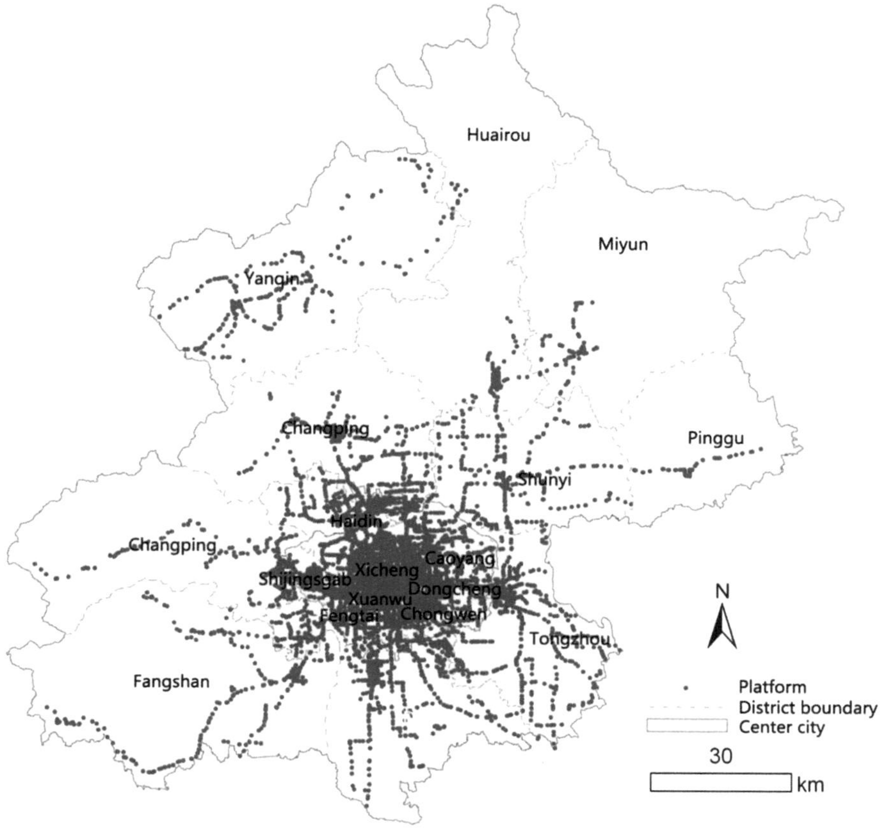

Fig. 10.1 The bus platforms of the Beijing Metropolitan Area (BMA)

and drop-off flows, most of the data used in this study are those of segmented-pricing lines, which include 37,649,207 records in total.

10.2.2.3 Points of Interest

POIs used in this study were collected in 2010 at Beijing and have 113,810 records, obtained from Sina Micro-blog Geographic Service Platform[4] (as shown in Fig. 10.2).

The classification and explanation of POIs is shown in Table 10.1.

The amount of each category of POIs is shown in Fig. 10.3.

[4] Sina Micro-blog LBS Platform, officially opened in 2012 April, provides third party developers with free access to Sina location service. Its most outstanding part are the two functions that base on user and POI. Related interface based on user can allow users to obtain individual's dynamic time line. and POI interface is based on a specific location (Http://open.weibo.com/).

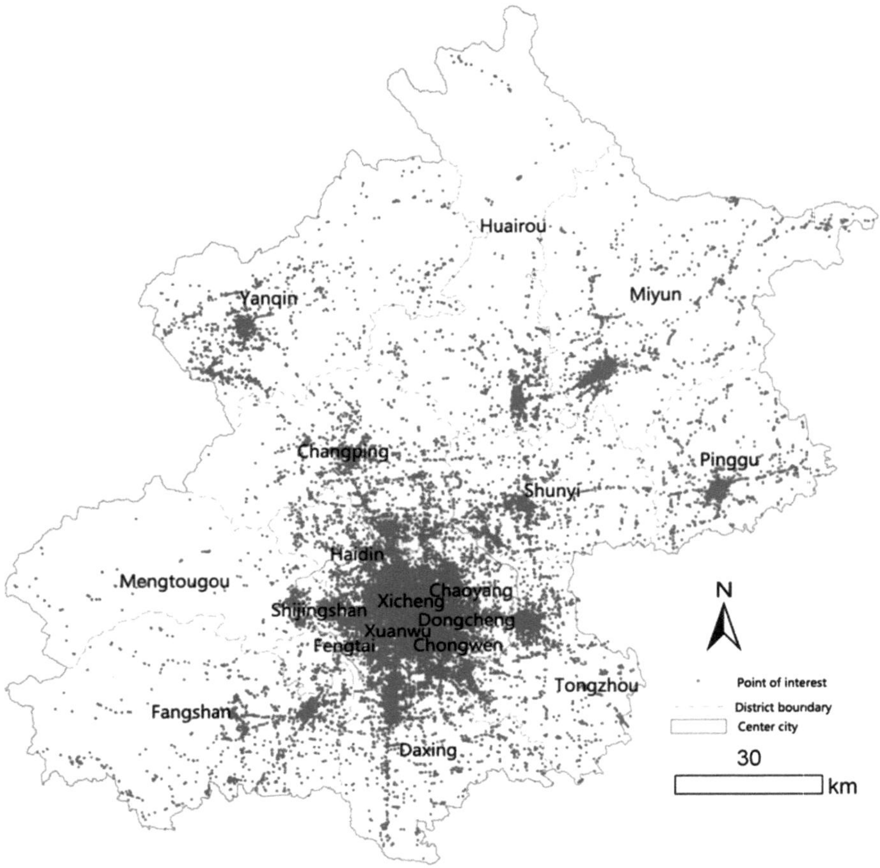

Fig. 10.2 The POIs of the Beijing Metropolitan Area (BMA)

10.3 Method

In this study, the identification of city functional regions was achieved mainly through the construction of DZoF (Discovery Zones of different Functions) model.

First, by using SQL Server, acquire and pre-process SCD and POIs data, Platform Flow model (PF model) and POIs data model were built. Second, by using the cluster analysis techniques of multi-dimensional data, we designed the feature of one week's traffic data, used the EM algorithm for clustering analysis and acquired clusters. Third, the function of the result clusters was interpreted in three aspects: POI model, feature of residents' commuting behavior and residents' general cognition. The function of each bus platform was then defined, including public management, science, educational and cultural service, residential, commercial and entertainment function, and scenic spots. The analysis diagram is as shown in Fig. 10.4.

Table 10.1 Codes, categories and description of POIs

First level code	Classification of POIs	Explanation	First level code	Classification of POIs	Explanation
01	Automobile service	Oil station, gas station, automobile care, car-washing station, car renting, etc.	11	Place of interest	Park plaza, scenic spot
02	Vehicle sales	VW, Toyota, Honda, GM and BMW sales	12	Business and residential area	Industrial park, residential area, etc.
03	Automobile maintenance	Automobile integrated maintenance, maintenance of VM, Honda, etc.	13	Government agency and social organization	Government agency, foreign institution, social organization, etc.
04	Motorcycle service	Motorcycle sales and maintenance	14	Science, educational and cultural service	Museum, library, cultural center, school, research institution, etc.
05	Catering service	Chinese and foreign restaurants, fast- food restaurant, coffee house, etc.	15	Transport facility	Airport, train station, harbor and wharf, subway station, etc.
06	Shopping service	Shopping mall, convenience shop, household appliance store, supermarket, home furnishing store, etc.	16	Finance and insurance service	Bank, insurance company, securities company, finance corporation
07	Life service	Travel agency, post office, logistic express, talent market, electric power office, beauty salon, etc.	17	Corporation	Well-known enterprise, company, factory, agricultural base, etc.
08	Sport and leisure service	Stadium, entertainment venue, leisure facility, movie theater, etc.	18	Road affiliated facility	Toll station, service center of gas station.
09	Health care service	General and specific hospital, clinic, etc.	19	Address information	Transport-related place name, city center
10	Accommodation service	Hotel, guest house	20	Public facility	Newsstand, public toilet, shelter

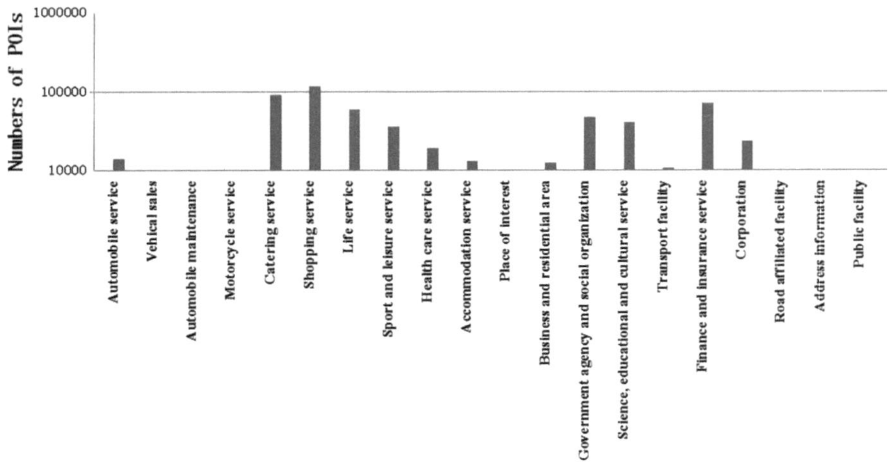

Fig. 10.3 POIs counts for each category in 2010

10.3.1 Blind Clustering of Bus Platforms

10.3.1.1 Pre-processing of SCD

Flow statistics collected according to bus lines compose the raw data of the study, which have a problem of containing many bus lines going through the same platform. By using SQL Server, statistics of different bus lines that go through the same platform were collected, and we acquired the flow statistic of each bus platform, fx, y, z (x is Platform ID, x=1,2,…,8691; y represents date, y=7,8,…,13; z represents time, z=0,1…,23).

10.3.1.2 PF (Platform Flows) Data Model

For each bus platform, an inflow vector was built (X7,0, X7,1, …, Xi, j, …, X13,23). Xi, j represents the number of passengers that have been picked up on the platform in the jth hour on April ith 2008 (i=7, 8, …, 13; j=0, 1…, 23). While at the same time, an outflow vector was built (Y7,0, Y7,1,…,Yi,j, …, Y13,23). Yi, j represents the number of passengers that have been dropped off on the platform in the jth hour on April ith 2008 (i=7,8, …, 13, j=0, 1…, 23).

By transforming the data into two-dimensional time-series data, making dimensionality reduction and constricting a linear function, the study uses a ratio of the number of pick-up passengers to the number of drop-off ones at different time as an index for comparing the similarity of platform flows:

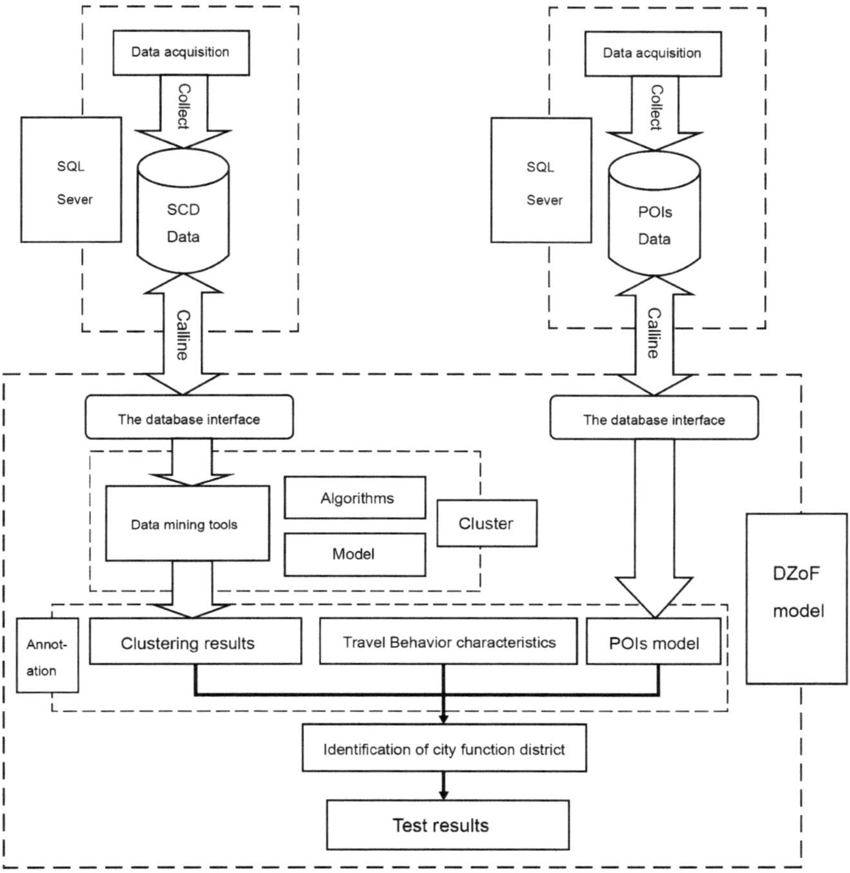

Fig. 10.4 The analysis diagram

$$Z_{ij} = \frac{X_{i,j}}{Y_{i,j}} \tag{10.1}$$

(Xi, j represents the number of passengers that have been picked up on the platform in the jth hour on April ith; Yi, j represents the number of passengers that have been dropped off on the platform in the jth hour on April ith)

Thus, for each platform, a PF model was built (Z7,0, Z7,1,…, Zi, j,…,Z13,23). Zi, j represents the ratio of passengers that have been picked up on the platform in the jth hour on April ith.

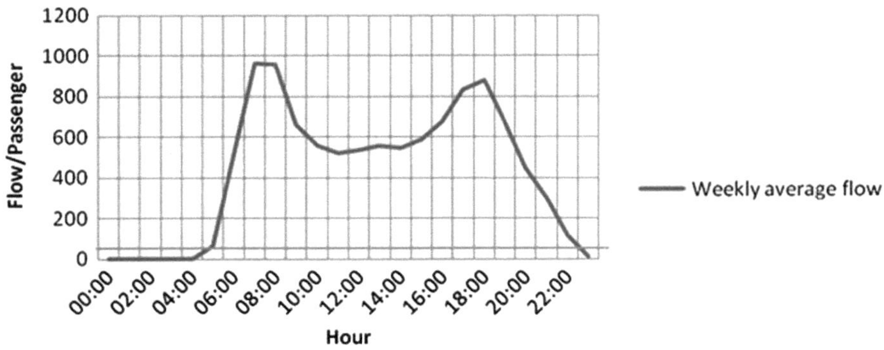

Fig. 10.5 The average passenger flows in various departure hours of a week

10.3.1.3 Data Dimensionality Reduction

From the statistics of average inflows and outflows in various departure hours of a week (as shown in Fig. 10.5), it illustrated people's trip hour converges on 5:00–23:00 (above the red line in Fig. 10.5). After removing redundant features of the raw data, PF data decreased from 168d to 126d.

Next, we drew a graphics of the inflows in various departure hours of one week for the No. 1934 platform (Figs. 10.6 and 10.7). The graphics suggest that the statistics of weekdays and those of weekends each have strong consistency seperately.

Then, correlation analysis of statistics of platform flows on weekdays and weekends was made and Pearson correlation coefficient between each two attributes was calculated. The results showed that the flow data in the same hour of different weekdays were significantly correlated at 0.01 (bilateral) levels Therefore, we designed the feature of flow data and calculated the arithmetic mean of statistics of weekdays and weekends.

10.3.1.4 Expectation-Maximization (EM) Algorithm

In this chapter, expectation maximization (EM) algorithm was chosen as the method for clustering on bus platforms. In this algorithm, for each object, its probability of each distribution is calculated, which is equivalent to a procedure in K-means algorithm that assigns each object to a cluster; and maximum likelihood estimating in EM algorithm is equivalent to cluster's centered calculating in K-means algorithm. But compared with the K-means algorithm, EM algorithm is more general and can be applied to different classes and find clusters of different sizes. At the same time, EM algorithm is model-based, which can eliminate the complexity of data association.

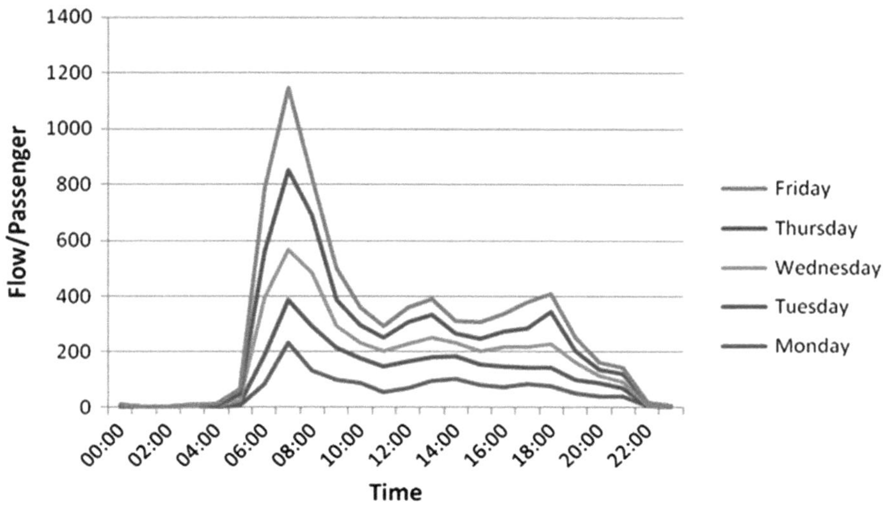

Fig. 10.6 The inflows in various departure hours on weekdays for No. 1934 platform

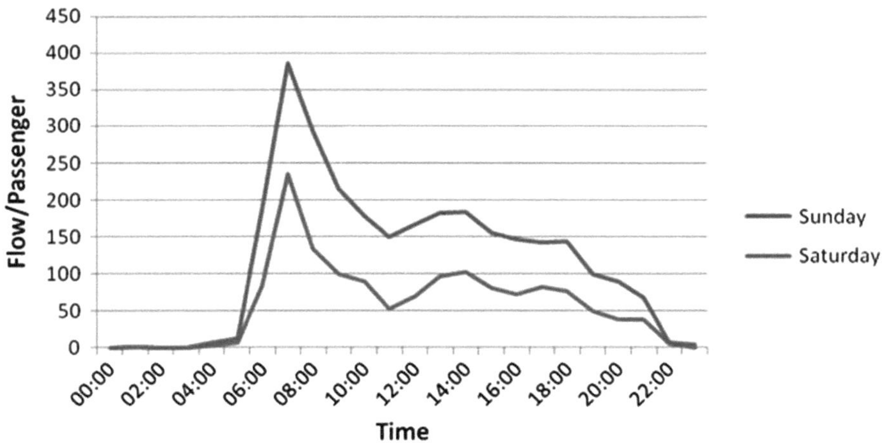

Fig. 10.7 The inflows in various departure hours on the weekend for No. 1934 platform

EM algorithm is a clustering method that uses hybrid model. The principle of model-based clustering is to assume that data is obtained by a statistical process which can be classified with a statistical model. Therefore, a statistical model that best fits the data can be found and parameters of the model can be estimated from data. The basic process of EM algorithm can be summarized as: first, do the initial guess of parameters; then, iteratively refine the estimation. Estimation of parameters in the algorithm is carried out by using the maximum likelihood method. Probability density of points generated from one-dimensional Gaussian distribution is:

$$\text{prob}(\chi \mid \Theta) = \Pi \frac{1}{\sigma\sqrt{2\pi}} e^{-\frac{(x_i-\mu)^2}{2\sigma^2}}. \tag{10.2}$$

If the value of σ and μ is unknown, then it requires a process to estimate them, which means choosing σ and μ that can maximize the formula above. This way of estimating model parameters is called maximum likelihood estimation in statistics.

10.3.2 Identification of Urban Functional Areas

10.3.2.1 Collection of POI Data of Bus Service Area

In this study, service area of bus platform is defined the area within a radius of 500 m (Huang 2006) around the bus platform. For each bus platform, the number of POI data of different categories is calculated, indicated as .i is platform ID (i = 1, 2, ..., 8691); j is 1st level code of POI (j = 1,2,...,20).

10.3.2.2 Data Standardization

Among POIs data of Beijing in 2010, there are 90,819 records of catering service POI, 116,499 shopping service POIs and only 4,575 POIs of place of interest. In the process of statistical analysis, function identification will be affected by difference in POIs' magnitude.

Therefore, original POIs data need to through Z-Score standardization according to the following formula:

$$x_{ij}^{*} = \begin{cases} \dfrac{x_{i,j} - \overline{x_j}}{S_j} & (S_j \neq 0) \\ 0 & (S_j = 0) \end{cases} \tag{10.3}$$
$$(i = 1,2,\ldots,n;\, j = 1,2,\ldots,m).$$

10.3.2.3 POI Data Model

For each platform, a POIs feature vector, FD (Frequency Density) model was built, denoted as fd1, fd2, ..., fd20. Fdi denotes the frequency density of the ist category of POI in platform service area R:

$$fd_i = \frac{\textbf{The standardized number of the ist category of POI of the platform}}{\textbf{The area of platform service arear r}}. \quad (10.4)$$

Similarly, for each platform, another POI feature vector, CR (Category Ratio) model was built, denoted as cr1, cr2, …, cr20.

cri denotes the percentage of the ist category of POI in all POI of the area:

$$cr_i = \frac{\textbf{The standardized number of the ist category of POI of the platform}}{|\,\textbf{The number of POI in platform service arear r}\,|}. \quad (10.5)$$

10.3.2.4 Urban Functional Identification

The study used traditional rules of data acquisition, and then used these rules for pattern recognition of large data, which was to recognize the function of the clusters obtained from blind clustering of bus platforms by using the correlation between residents' trip time and purpose, residents' general cognition and POI model.

Function recognition is a comprehensive application of the following three methods:

1. Calculate each cluster's FD (Frequency Density) model, and sort the results (achieving internal ranking); second, calculate each cluster's CR (Category Ratio) model and sort the results (achieving external ranking).
2. Identify the feature of flows in trip hours of each cluster.
3. Residents' general cognition. People usually are aware of the function of some well-known places, such as the Imperial Palace, Chinese Silicon Valley and Beihai Park.

10.4 Results

10.4.1 Clustering Results of Bus Platforms and Summary at TAZ Scale

After clustering of bus stops based on flow data by using EM algorithm in the software Weka, six different clusters were obtained (each bus stop only belongs to one cluster, C0–C5). Then, according to the spatial subordination between platforms and traffic analysis zones (TAZ), statistical work was carried out for each traffic analysis zone, and the cluster that exists most in a TAZ represents the category of the TAZ. The clustering results were summarized at TAZs scale (sparse for unclassified area), as shown in Fig. 10.8.

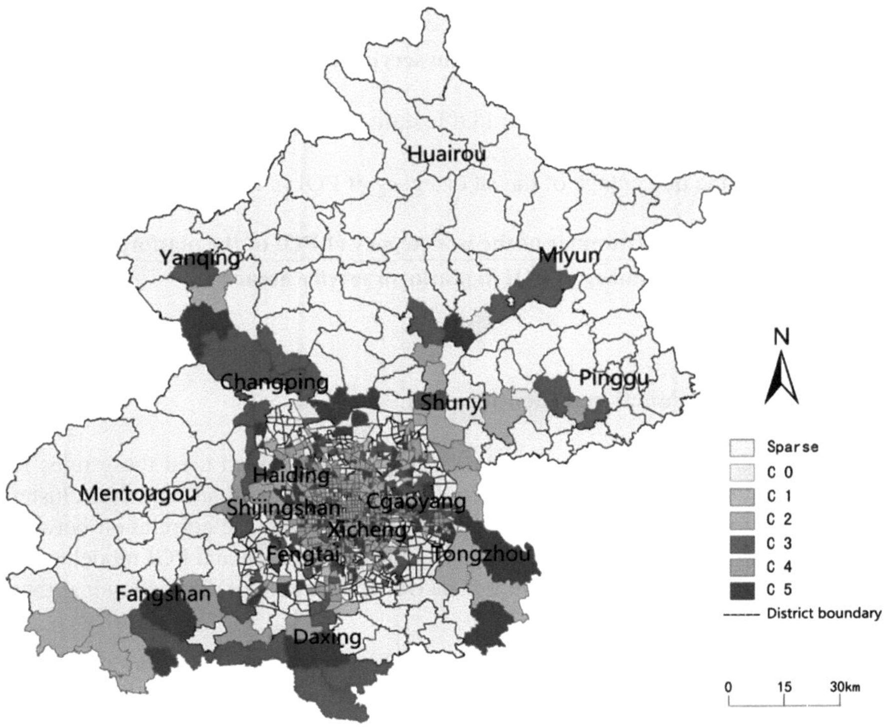

Fig. 10.8 Functional regions of the Beijing

10.4.2 Function Identification

10.4.2.1 Construction of POIs Model

According to the results of blind clustering of platforms in 10.4.1, a POIs model was built for each cluster (C0–C5), and the value of FD (Frequency Density) and RCR (Rank of Category) of each functional region were calculated, as shown in Table 10.2.

10.4.2.2 The Feature of Residents' Traffic Flows

The features of flows (pick-up and drop-off number) on weekdays and weekends in a week of clusters clustered by EM are shown as Fig. 10.9, 10.10, 10.11 and 10.12.

Table 10.2 Overall POI feature vector and ranking of functional regions by DZoF (the color of a cell indicates its value)

	C0		C1		C2		C3		C4		C5	
POI	FD	RCR	FD	RCR	FD	RCR	FD	RCR	FD	RCR	FD	RCR
Automobile service	−0.077	7	−0.025	6	0.073	9	0.03	19	0.021	18	−0.02	6
Vehicle sales	−0.075	6	0.034	2	−0.006	19	0.089	14	0.073	13	−0.063	12
Automobile maintenance	−0.005	3	0.032	3	0	18	0.119	9	0.084	12	−0.012	4
Motorcycle service	0.063	1	0.006	5	0.057	13	0.085	15	0.041	16	0.117	1
Catering service	−0.186	18	−0.109	13	0.142	1	0.149	7	0.205	5	−0.095	18
Shopping service	−0.173	16	−0.141	16	0.039	15	0.214	3	0.107	11	−0.051	10
Life service	−0.156	13	−0.157	18	0.099	5	0.216	2	0.114	10	−0.026	7
Sport and leisure service	−0.16	14	−0.114	14	0.124	2	0.095	10	0.307	1	−0.057	11
Health care service	−0.106	9	0.013	4	0.06	12	0.187	5	0.056	15	−0.004	3
Accommodation service	−0.183	17	−0.14	15	0.075	8	0.187	4	0.18	6	−0.034	8
Place of interest	−0.129	12	−0.076	11	0.042	14	−0.033	20	0.167	8	−0.075	15
Business and residential area	−0.073	5	−0.094	12	0.072	10	0.152	6	0.07	14	−0.018	5
Government agency and social organization	−0.124	11	−0.18	20	0.082	6	0.135	8	0.224	2	−0.11	20
Science, educational and cultural service	−0.202	19	−0.173	19	0.068	11	0.067	18	0.22	3	−0.095	17
Transport facility	−0.173	15	−0.076	10	0.111	4	0.089	13	0.17	7	−0.066	13
Finance and insurance service	−0.214	20	−0.057	8	0.114	3	0.094	11	0.216	4	−0.105	19
Corporation	−0.12	10	−0.144	17	0.075	7	0.069	17	0.117	9	0.017	2
Road affiliated facility	−0.008	4	−0.039	7	0.031	16	0.092	12	−0.036	20	−0.068	14
Address information	0.015	2	0.039	1	0.016	17	0.084	16	0.021	17	−0.079	16
Public facility	−0.102	8	−0.068	9	−0.048	20	0.214	1	−0.036	19	−0.044	9

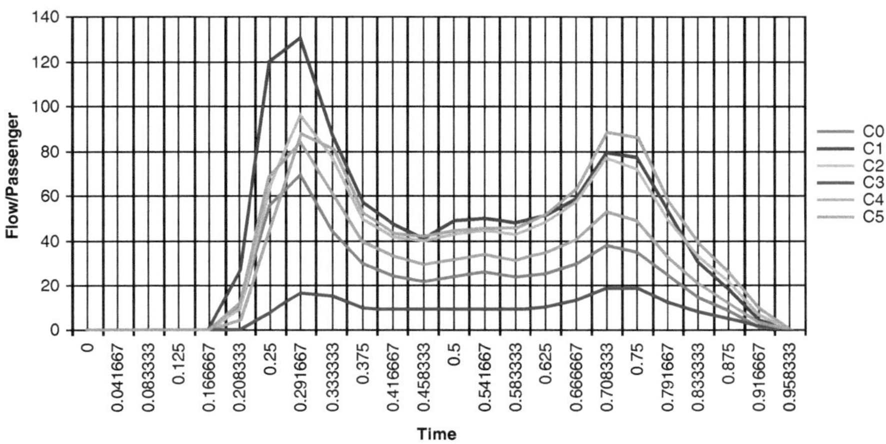

Fig. 10.9 The inflows on weekdays of clusters clustered by EM

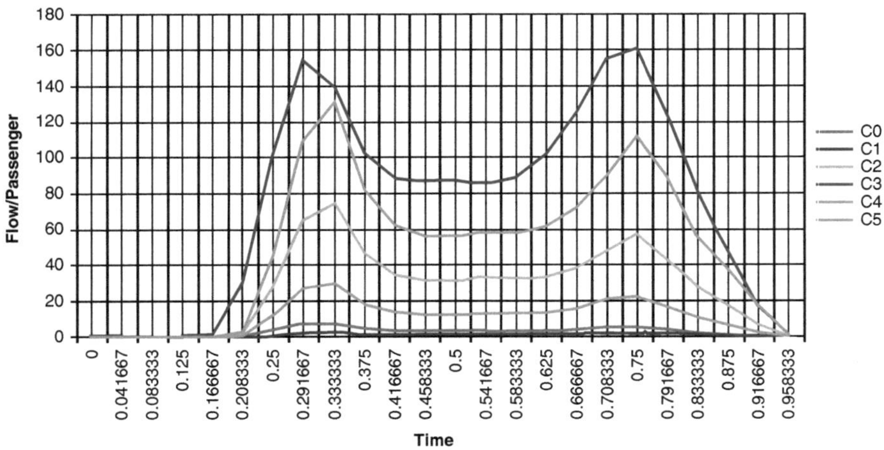

Fig. 10.10 The outflows on weekdays of clusters clustered by EM

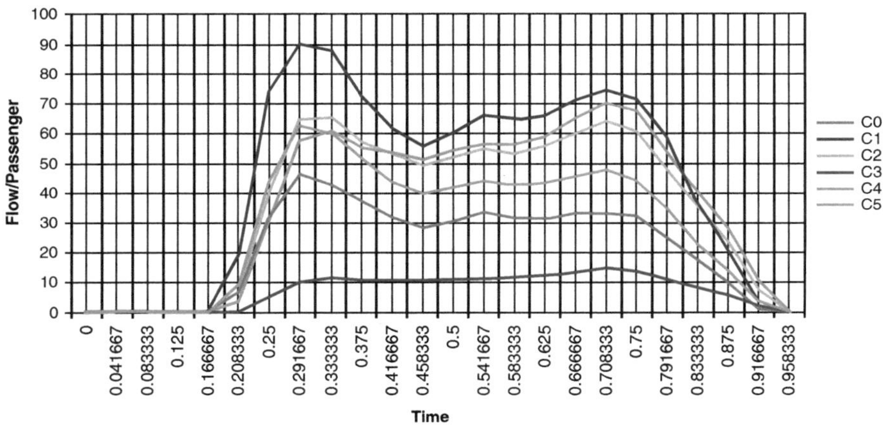

Fig. 10.11 The inflows on weekends of clusters clustered by EM

10.4.2.3 Identification of Results

Functions of clustering results of EM algorithm were identified as followed:

(a) Mature residential area (C0)

Regions of this kind have got widely distributed residential POI and a high proportion of serviced apartment (FD = −0.073), while services for residents' life like health care, accommodation and education service is distributed in equilibrium, which have shown the feature of POI distribution of a typical residential area.

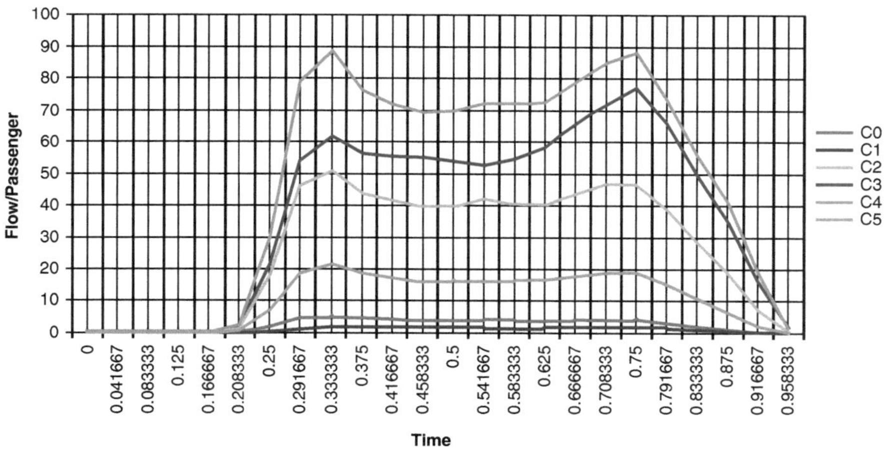

Fig. 10.12 The outflows on weekends of clusters clustered by EM

At the same time, through the analysis of flow data of the week, it was observed that flows of departure in weekdays of the region peaked in the early morning (7–8 o'clock in morning rush hours) and the peak of flows of return appeared in the evening (17–19 o'clock in afternoon rush hours), which have shown the trip mode of a typical residential area.

(b) Developing area (C1)

POIs of regions of this kind are mostly of motorcycle and car service category where 4S shops, motorcycle sales, automobile and motorcycle maintenance stations are relatively widespread and the surrounding infrastructure need to be improved.

(c) Scenic spots (C2)

Places of interest account for the highest proportion of POI in these regions. Compared with other kinds, regions of this kind have higher FD (Frequency Density) value (0.042) and are high-ranked in the external ranking of catering and accommodation service for tourists. And there is little difference between traffic flows of weekdays and weekends in these regions. Flows were relatively averaged in different hours of a day and there were usually higher flows in weekends than weekdays.

(d) Commercial and entertainment area (C3)

Regions of this kind have high FD value in catering service, shopping service and life service, respectively ranking second, first and first in all clusters. Meanwhile, the regions contain a high proportion of POI of catering service (with higher CR value. For example, clusters of catering service rank the 7th in CR value while clusters of shopping service rank the third). And form the above figures of traffic flows, it was observed that the peak of drop-off flows appeared in afternoon rush hours (17:00–19:00) on weekdays in these regions, which have shown that many people shopped and took part in recreational activities in these areas.

(e) Area of public management, science, education and culture (C4)

Regions of this kind have the highest percentage of POI of government agency and social association which have higher FD value (0.22) compared with other regions. POI of this kind account for 9.7 % of POI in these regions, ranking second in RCR value. And there are many POI of science, educational and cultural service in these regions where transport facility, sport and leisure service and accommodation service are high-ranked in the external ranking.

(f) New residential area (C5)

Regions of this kind have similar POI data structure with C0. While ranking according to the percentage that each kind of POI account for, residential service ranks the 5th, health care service ranks the third and life service ranks the 7th in these regions, which have shown the feature of POI distribution of a typical residential area.

On the other hand, form the data of traffic flows, it was observed that flows of departure in weekdays of the regions peaked in the early morning (7–8 o'clock in morning rush hours) and the peak of flows of return appeared in the evening (17–19 o'clock in afternoon rush hours). But compared with C0, these regions had smaller flows which were about 1/4 of flows of C0, which have shown that these regions don't have big traffic flows and still in a stage of developing.

(g) Unclassified area (sparse)

Some regions lack data of traffic flow for being covered with mountains, forests, rivers and others, so there are put in one category in this study.

According to the results of functional identification, the area and population of each functional area was collected, as shown in Table 10.3.

Table 10.3 Information of each cluster

Functional area	Number of TAZ	Area (km$^{2)}$	Population (people)
Unclassified (Spars)	357	11,061.13	2,561,345
Mature residential area (C0)	63	326.2918	572,609
Under developed area (C1)	25	82.10567	162,647
Scenic area (C2)	155	973.0869	1,878,303
Commercial and entertainment area (C3)	129	2068.882	1,911,066
Area of public management, science, education and culture (C4)	267	918.993	3,551,985
New residential area (C5)	122	974.2913	1,271,898

10.4.3 Examination of the Results of Identification

In order to test the accuracy of the results of DZoF model, the map of different functional areas in Beijing obtained from the study was compared with the map of land-use status quo in the overall urban planning of Beijing (2004–2020)[5] and Google map of the area. Some results of comparison are shown as Table 10.4.

Besides, we also compared results of the study with detailed land-use data of each TAZ in Beijing to test the overall accuracy of identification. After ranking 1118 TAZ according to the size of public land (including land of public facilities and municipal administration), selecting the top 50 TAZ and removing the TAZ without SCD information, there were a total of 44 TAZ left. 22 of them were identified as areas of public management, science, education and culture with an accuracy rate of 50 %. Using the same method, the residential land was analyzed and the accuracy rate was 58.06 %.The result is as shown in Table 10.5.

From the overall consideration of the highly-mixed land use status of Beijing and the contrastive analysis of the study, DZoF model has a certain degree of accuracy in effectively identifying main functional areas in Beijing.

10.5 Conclusion and Discussion

This study is based on 77,976,010 records of SCD of Beijing City in one week in April 2008 and 113,810 POIs of Beijing in 2010. By constructing a DZoF model, the study completed the identification of functional areas in Beijing and obtained seven kinds of functional areas such as public management and culture, scenic area, commercial and entertainment area, mature residential area, new residential area and unclassified area. The area of public management and culture covers 267 transportation analysis zones (TAZ) with a total area of 918.993 km^2. Business and entertainment area covers 129 TAZ with a total area of 2068.882 km^2. And scenic area covers 155 TAZ with a total area of 973.0869 km^2.

The results show that DZoF model has acceptable ability to recognize the characteristic of functional areas in Beijing. This study can help people understand the spatial structure of a complex city, assist city planners to carry out planning of different urban functional areas based on human mobility and POI data, and provide the guidance and reference to city planning and site selection for real estate development.

This research has potential contributions in three aspects: first, it studied the spatial structure of a city through human mobility based on massive SCD; second, it combined the traditional methods of urban studies and big data mining and identi-

[5] The data is from the urban planning institute of Beijing.

Table 10.4 Comparison and analysis between recognition results and the status quo of figure

Contrast area	A famous scenic place in Beijing: Shidu Scenic Zone
Contrast diagram	
Identification results	Region A (green) in the map of recognition results is a scenic area, which is in accord with region A in the map of land use status quo
Contrast area	Culture relic protection site in Beijing: the ruins of the Turnoff City (which is 1.5 km northwest of Badaling Guan City)
Contrast diagram	
Identification results	Region B (green) in the map of recognition results is a scenic area which corresponds to the location of the ruins of the Turnoff City
Contrast area	A delta in the watersheds of Yongding River in Beijing city which is mostly woodland

Contrast diagram	
Identification results	Region C (green) in the map of recognition results belongs to an unclassified area, according with region C in the map of land use status quo which represents the delta of Yongding River
Contrast area	Haidian District has many universities including Peking University, Tsinghua University and Renmin University of China as well as Chinese silicon valley, which is the region of science, education and culture in Beijing
Contrast diagram	

(continued)

Table 10.4 (continued)

Contrast area	A famous scenic place in Beijing: Shidu Scenic Zone
Identification results	Region D (purple) in the map of land use status quo is the region of science, education and culture, which is in accord with region D in the map of recognition results.
Contrast area	some parts of Dongcheng District
Contrast diagram	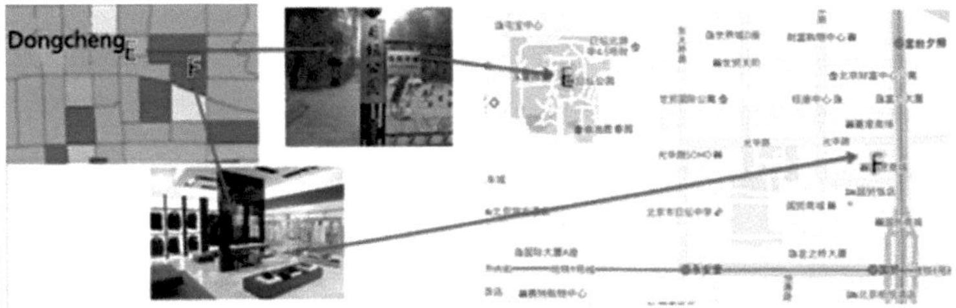
Identification results	In the map of recognition results, region E is a scenic area and region F is a commercial and entertainment area. And by comparing with Google map of the regions, region E is Ritan Park in reality and region F is actually a famous commercial and entertainment area in Beijing, which has a plenty of shopping centers like China World Mall, Traders Hotel, Jiali Mall and Wanda Plaza

Table 10.5 Comparison and analysis between recognition results and land use of TAZ

		Valid contrasting data	Identification results	Accuracy rate
Area of public management, science, education and culture	Number of TAZ	44	22	50.00 %
	Total area (m^2)	59,596,348	19,824,488	
Residential area	Number of TAZ	31	18	58.06 %
	Total area (m^2)	49,319,636	15,423,667	

fied the features of residents' trip behavior from existing survey data of residents' trip, and applied these features as rules to the identification of urban functional area; third, it constructed a DZoF (discovering zones of different functions) model and identified the function of different regions in the city. As a whole, based on data of POI and SCD, the research studied the dynamic spatial structure of city using data mining method, thus providing a new analysis and research methods for studies of metropolitan spatial structure.

In this study, there are still some deficiencies that need to be improved in further studies: (1) The ratio of travelling by public bus in Beijing is 28.2 % in 2010, while rail transit accounts for 11.5 %, taxi for 6.6 % and car for 34.2 %.[6] In future researches, data of rail transit and taxi data can be added into the study to complete the information of human mobility and get more accurate results. (2) In the real situation, mixed land use of commercial and residential services is widespread. The study ignored the possibility of mixed use and selected the function with the largest percentage to represent the region. Classification of mixed land use can be added into the future study.

References

Agrawal, R., Raloutsos, C, & Swami, A. (1993). *Efficient similarity search in sequence databases.* In Proceedings of the 4th international conference on foundations of data organization and algorithms. Berlin: Springer-Verlag, pp. 69–84.

Anas, A., Arnott, R., & Small, K. A. (1998). Urban spatial structure. *Journal of Economic Literature, 36*(3), 1426–1464.

Batty, M. (1990). Invisible cities. *Environment and Planning B: Planning and Design, 17*, 127–130.

Batty, M. (2009). *Cities as complex systems: Scaling, interactions, networks. Dynamics and urban morphologies*. Berlin: Springer.

Calabrese, F., & Ratti, C. (2006). Real time Rome. *Networks and Communication Studies, 20*, 247–258.

CH, J., & Hwang, C. (2010). *A time-geographic analysis of trip trajectories and land use characteristics in Seoul metropolitan area by using multidimensional sequence alignment and spatial analysis*. In: 2010 AAG annual meeting, Washington, DC.

[6] The data is from the 2011 report of transportation development of Beijing.

Gonzalez, M. C., Hidalgo, C. A., & Barabasi, A. L. (2008). Understanding individual human mobility patterns. *Nature, 453*, 779–782.

Goodchild, M. F., & Janelle, D. (1984). The city around the clock: Space-time patterns of urban ecological structure. *Environment and Planning A, 16*(6), 807–820.

Goodchild, M. F., Klinkenberg, B., & Janelle, D. G. (1993). A factorial model of aggregate spatio-temporal behavior: Application to the diurnal cycle. *Geographical Analysis, 5*, 277–294.

Hamilton, B. W. (1982). Wasteful commuting. *The Journal of Political Economy, 90*(5), 1035–1053.

Huang, Libin. (2006). *City rail transit sites influence evaluation research of regional development – in Shanghai Xujiahui rail hub*. Tongji: Tongji University.

Jiang, B., Yin, J., & Zhao, S. (2009). Characterizing human mobility patterns in a large street network. *Physical Review E, 80*(2), 1136–1146.

Jiang, S., Ferreira, J., & Gonzalez, M. (2012a). *Discovering urban spatial-temporal structure from human activity patterns*. UrbComp'12. Beijing, China, 12 Aug 2012.

Jiang, S., Ferreira, J., & Gonzalez, M. C. (2012b). Clustering daily patterns of human activities in the city. *Data Mining and Knowledge Discovery, 25*, 478–510.

Jiang Yuan, Zhang Zhaoyang, & Qiu Peiliang. (2005). Clustering algorithms used in data mining. *Journal of Electronics & Information Technology, 27*(4), 655–659 (In Chinese).

Korn, F., Jagadish, H. V., & Faloutos, C. (1997). *Efficiently supporting ad hoc queries in large datasets of time sequences*. In Proceedings of the ACM SIGMOD international conference on management of data. Tucson: ACM.

Lai, S.-K., & Han, H. (2009). *Complex: The new ideas of urban planning*. Beijing: China-Building Press (In Chinese).

Liu Zhilin, & Wang Maojun. (2011). Job accessibility and its impacts on commuting time of urban residents in Beijing: From a spatial mismatch perspective. *Acta Geographica Sinica, 66*(4), 457–467 (In Chinese).

Liu, Y., Wang, F. H., Xiao, Y., et al. (2012). Urban land uses and traffic 'source-sink areas': Evidence from GPS-enabled taxi data in Shanghai. *Landscape and Urban Planning, 106*, 73–87.

Long Ying, Yu Zhang, & Cui Chengyin. (2012). Identifying communing pattern of Beijing using bus smart card data. *Acta Geographica Sinica, 67*(10), 1–12 (In Chinese).

Lu, D., & Weng, Q. (2005). Urban land-use and land-cover mapping using the full spectral information of Landsat ETM+ data in Indianapolis, Indiana. *Photogrammetric Engineering & Remote Sensing, 71*(11), 1275–1284.

McMillen, D. P., & McDonald, J. F. (1997). A nonparametric analysis of employment density in a polycentric city. *Journal of Regional Science, 37*(4), 591–612.

Neuhaus, F. (2009). Urban diary-a tracking project. UCL working paper series. Paper 151. Available on line: http://discovery.ucl.ac.uk/19245/

Pan, G., Qi, G., Wu, Z., et al. (2013). Land-use classification using taxi GPS traces. *IEEE Transactions on Intelligent Transportation Systems (T-ITS), 14*(1), 113–123.

Pulliam, H. R. (1988). Sources, sinks, and population regulation. *American Naturalist, 132*, 652–661.

Qi, G., Li, X., Li, S., et al. (2011). *Measuring social functions of city regions from large-scale taxi Behaviors*. In The 9th IEEE international conference on pervasive computing and communications (PerCom'11), work in progress. Seattle, 21–25 Mar 2011.

Ratti, C., Pulselli, R. M., Williams, S., & Frenchman, D. (2006). Mobile landscapes: Using location data from cell phones for urban analysis. *Environment and Planning B: Planning and Design, 33*(5), 727–748.

Sun, J. B., Yuan, J., Wang, Y., et al. (2011). Exploring space-time structure of human mobility in urban space. *Physical A, 390*, 929–942.

Sun, L., Lee, D., & Erath, A. (2012). *Using smart card data to extract passenger's spatio-temporal density and train's trajectory of MRT system*. In ACM SIGKDD international workshop on urban computing. Beijing, 12 Aug 2012.

Sun Wenshuang, & Chen Lanxiang. (1994). *Multivariate statistical analysis*. Beijing: Higher Education Press.

Tan, P.-N., Steinbach, M., & Kumar, V. (2006). *Introduction to data mining* (pp. 2–7). Beijing: Post & Telecom Press.

Wang, D., & Chai, Y. (2009). The jobs-housing relationship and commuting in Beijing, China: The legacy of company. *Journal of Transport Geography, 17*, 30–38.

Witten, I. H., & Frank, E. (1999). *Data mining: Practical machine learning tools and techniques with Java implementations*. San Francisco: Morgan Kaufmann.

Xiao, J., Shen, Y., Ge, J., et al. (2006). Evaluating urban expansion and land use change in Shijiazhuang, China, by using GIS and remote sensing. *Landscape and Urban Planning, 75*(1–2), 69–80.

Yuan, J., Zheng, Y., & Xie, X. (2012). *Discovering regions of different functions in a city using human mobility and POIs*. In The 18th ACM SigKdd conference on knowledge discovery and data mining. Beijing, 12–16 Aug 2012.

Zhao Weifeng, Li Qingquan, & Li Bijun. (2011). Using urban POIs data extraction hierarchical landmarks. *Journal of Remote Sensing, 5*, 981–989 (In Chinese).

Part III
Planning Support and Its Future in Beijing

Chapter 11
An Applied Planning Support Framework Including Models, Quantitative Methods, and Software in Beijing, China

11.1 Introduction

For decades China has witnessed rapid urban growth, especially of its large cities. Urban planning plays an important role in leading a sustainable development pattern. Planning support systems (PSSs) to support urban plan compilation and evaluation have attracted extensive attention from scholars and decision makers. Most of the existing research on PSSs is related to system design, implementation and application as well as evaluation of a standalone system in one area, e.g. What If?, Community Viz and INDEX. The uptake of the developed PSSs is not advanced (Vonk 2005; Vonk et al. 2007). In addition, most of the planners in China have backgrounds in architecture and lack quantitative analysis skills. For this, the chapter describes the PSS framework we developed with an overview of urban plan and possible supporting models, existing software and quantitative methods. The PSS framework serves as an online query system of PSS toolkit for the Beijing Institute of City Planning (BICP). The result of the user evaluation confirms that the PSS framework serves well as an online query system for planning support within BICP. We hope this work can support various types of plans in China, based on a conceptual framework of the planning process, can promote the application of PSSs in practical urban planning.

The concept of a "planning support system" (PSS), initially proposed by Harris (1960), was considered to be the latest form of computer-aided planning system (Geertman and Stillwell 2004; Klosterman 1997). Several books on PSSs have been published in recent years (Brail and Klosterman 2001; Geertman and Stillwell 2003; 2009; Brail 2008). PSSs have been applied mainly in spatial plans (Geneletti 2008; Kammeier 1999), urban environment improvement plans (Edamura and Tsuchida 1999), industrial location choices (Kammeier 1999), and land use plans (Klosterman 1999). Single PSS implementations and applications are reviewed widely in the literature. Typical PSSs related to land use plan are listed in Table 11.1. Various approaches have been used in these PSSs, and numerous factors are input into PSSs

Y. Long, Z. Shen, *Geospatial Analysis to Support Urban Planning in Beijing*, GeoJournal Library 116, DOI 10.1007/978-3-319-19342-7_11

Table 11.1 An inventory of typical PSSs

Publication	PSS name	Approach (es)
Landis (1994), Landis and Zhang (1998a, b)	CUF/CUF-2	Rule-based land suitability analysis
Clark et al. (1997)	SLEUTH	Cellular automata
Wu (1998)	SimLand	Cellular automata, AHP
Shi and Yeh (1999)	N/A	Case-based reasoning
Klosterman (1999)	What if?	Rule-based land suitability analysis
Allen (2001)	INDEX	Rule-based land suitability analysis
Waddell (2002)	UrbanSim	Microsimulation, discrete choice models
Lautso (2002)	SPARTACUS (based on MEPLAN)	Input–output model, discrete choice models
Yeh and Qiao (2004)	KBPSS	Knowledge-based reasoning
Carmichael et al. (2004)	GB-QUEST	Rule-based land suitability analysis
Placeways, LLC	Community Viz	Rule-based land suitability analysis
Li and Liu (2008)	N/A	Cellular automata, multi-agent
Long et al. (2009)	BUDEM	Cellular automata, logistic regression

to predict land use patterns based on different scenarios. However, most of these PSSs focus on a single aspect of urban planning, and, to the best of our knowledge, there is no existing reported research proposing a framework for various kinds of plans. The most relevant research is Geertman and Stillwell (2004), which reviewed a basket of PSSs.

Research relating to PSSs in China has been carried out frequently since the concept of PSS was first introduced into China in 2003 by Liu (2003). Du and Li (2005) applied What if? to a Chinese urban master plan. Long (2007) focused on geospatial techniques for establishing PSSs. Li (2010) analyzed the current condition of and future prospects for PSSs. Li and Zhan (2011) developed a PSS named UPlan. Long et al. (2011) developed an urban containment PSS in Beijing. The development and application of a standalone PSS is still the emphasis for Chinese researchers. Now, researchers in China are beginning to focus on proposing general techniques or multi-cases related to PSSs. For instance, Niu (2012) published a PSS book in which he proposed over 40 GIS tools as PSSs, including six basic aspects: fundamental techniques, spatial overlay, 3D analysis, transportation network, spatial research and planning information management. A project conducted by the Chinese Academy of Urban Planning proposed a framework of digital techniques for urban planning and developed dozens of tools for urban master plans and detailed plans (Luo et al. 2009). In addition, decision support systems (DSS) or management information systems (MIS) are extensively applied in the Chinese urban management bureau or commission in the process of data management and issuing land permits. In contrast to the information techniques used in the field of urban management, PSS is not much used by agencies and institutes during compiling urban plans. In conclusion, there is a large body of literature in China in this field and some researchers have begun to combine various techniques to propose a framework of PSSs although most of them are still under development.

This chapter is organized as follows. Section 11.2 introduces various methods we have applied in establishing the PSS framework. Section 11.3 illustrates the framework we have developed as well as an online query system for an improved application of this PSS framework. We discuss the application of the framework in BICP and its potential contributions in Sect. 11.4. Last, we draw several conclusions and propose future research into the PSS framework.

11.2 Methods for Establishing the Framework

11.2.1 Requirement Analysis

We used two methods to conduct the requirement analysis for the PSS framework in BICP: two rounds of seminars and a survey. We held two seminars for planners to discuss the requirements of the PSS framework. One seminar was held at the very beginning of the PSS framework project, and the other was held when we were seeking feedback on the preliminary results of the PSS framework. In the first seminar, the 10 planners involved made 20 suggestions. Most of suggestions focused on the applicability of the PSS framework as well as more types of plans to be included. In the second seminar, the 20 planners involved gave us 30 comments and suggestions. Planning evaluation is also a very important aspect to include. The dataset required by each PSS is also necessary. In addition, at the very beginning of this project, we designed an online survey with six questions relating to the requirements for the proposed PSS framework. We conducted this in BICP and got 34 responses (from a total of about 300 planners in BICP). Figure 11.1 shows the questions and the statistical results. From 34 responses, 29 planners were interested in plan contents and compilation methods. Nineteen planners had never or seldom applied quantitative methods in their planning practice, methods such as GIS, statistical approaches, models and visualization techniques. Thirty planners saw opportunities for PSSs to support their work, and 30, 7 and 27, respectively, saw opportunities for existing condition analysis, plan scheme compilation and plan evaluation in their work. Overall, planners in BICP showed great interest in using PSSs in their work, although this is not yet common. Their involvement in seminars and the survey significantly improved the applicability of the framework.

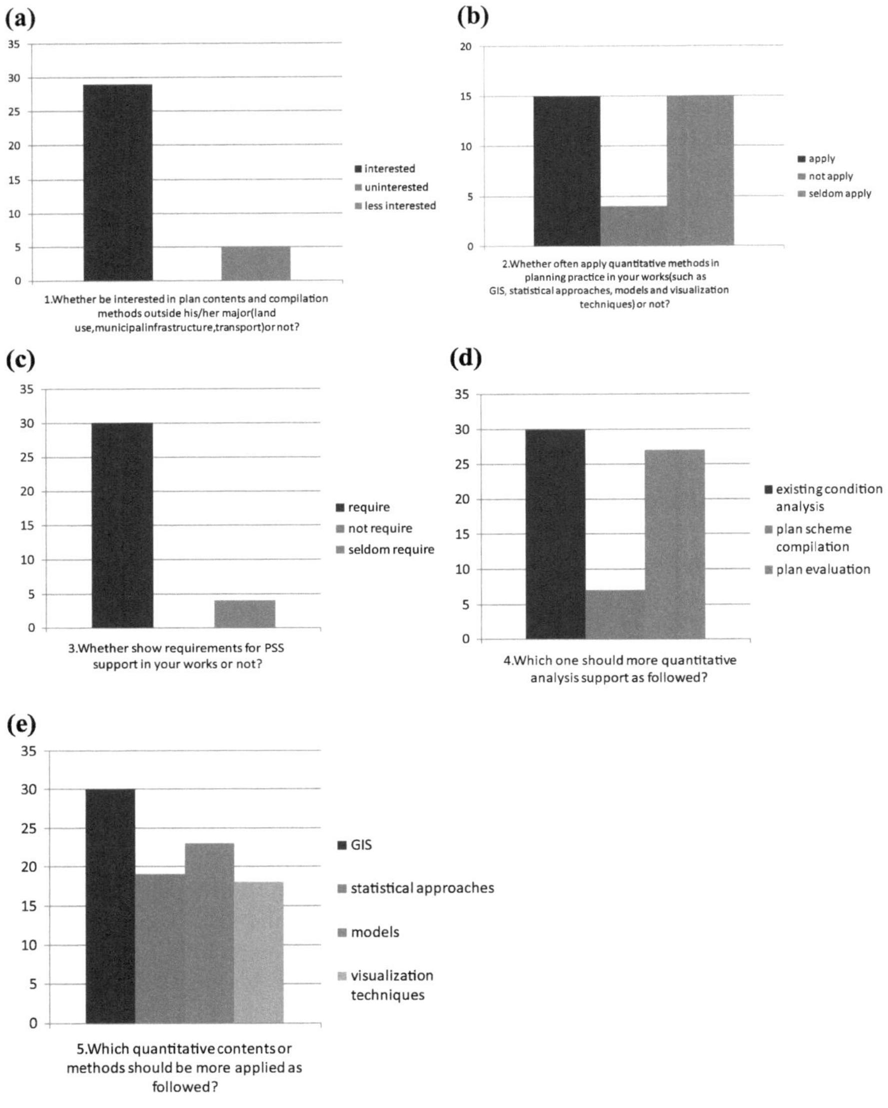

Fig. 11.1 The survey results regarding the requirements for the framework

11.2.2 Selecting the Form of the Framework

We designed the overall PSS framework based on the requirement analysis. First, we focused on both the urban plan compilation and evaluation. Second, we included both existing PSSs and PSSs under development by BICP in the PSS framework. Third, quantitative methods/theories, although not PSS entities, were included in the framework to broaden the horizons of BICP planners, who, we understand, mostly have backgrounds in architecture and lack quantitative analysis skills. Fourth, the PSS framework was designed as detailed as possible for planners' queries and applications. According to these principles, the structure of the PSS framework is in two dimensions (see Fig. 11.2), the vertical dimension is for each plan element, which is a part of a type of plan. The horizontal dimension is planning support tools and methods. Details of the framework format are as follows.

11.2.3 The Selection of Plan Elements

We based the selection of plan elements involved in the framework on existing urban planning laws, regulations and standards in China. The key reference was the City Planning Law of the People's Republic of China enacted in 2008, which defines the urban–rural planning system for plan compilation and plan evaluation. In Beijing, plan compilation was further divided into downtown plans, new city plans, town plans and rural plans, and each type was further divided into master plan and detailed plan levels. Therefore, plans involved in the framework are considered from the viewpoints included both plan compilation and evaluation, as shown in Table 11.2. In this stage, we focused more on plan compilation than evaluation in the framework in line with dominating historical and existing planning role of BICP, although Fig. 11.1 shows that more planners are interested in applying PSSs in plan evaluation. We will extend the framework for plan evaluation in a near future.

Each plan in Level 3 was further itemized into various plan elements, for which we proposed one or more PSSs in the framework. The plan elements were confirmed based on existing planning laws and guidelines, as well as recommendations

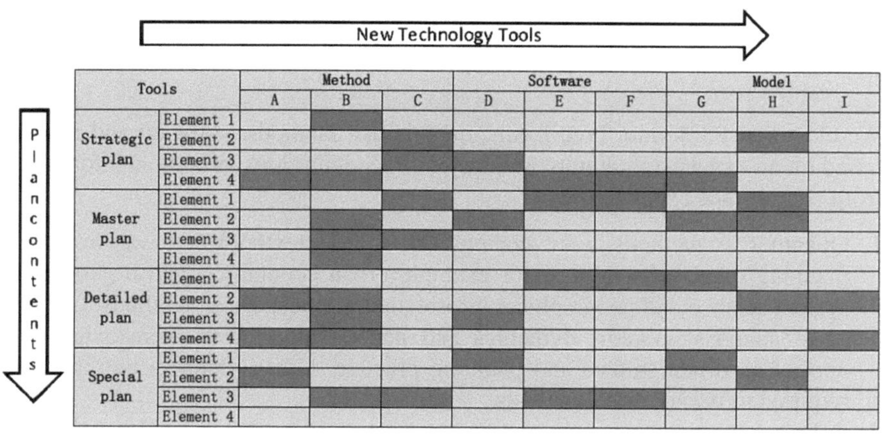

Fig. 11.2 The simplified PSS framework (Cells in *dark* means a PSS tool, a method, software or model, could be adopted to support the plan element)

Table 11.2 Plan types involved in the framework

Level 1	Level 2	Level 3-1	Level 3-2	Level 3-3
Part 1: Plan compilation				
	1 Strategic plan	1.1 Spatial development Research		
	2 Master plan	2.1 Downtown master plan	2.2 New city master plan	2.3 Town master plan
	3 Detailed Plan	3.1 Street level	3.2 Lot level	3.3 City design
	4 Municipal Topic	4.1 Water supply plan	4.2 Storm water drainage plan	……
	5 Transport Topic	5.1 Transport demand plan	5.2 Road network plan	……
	6 Special plan	6.1 Elementary education Facilities special plan	6.2 City fire equipment special plan	……
Part 2: Plan evaluation				
	1 Master plan evaluation	1.1 Urban master plan evaluation		

from BICP. For example, 1.1 Spatial development was divided into several plan elements, including landscape analysis, existing condition analysis, land use suitability analysis as well as population distribution analysis. All these elements should be addressed under spatial development within a strategic plan.

11.2.4 The Selection of PSS Types

We classified PSSs into three forms, quantitative methods, software and models, based on an extensive literature review into PSS definitions as well as information from face-to-face expert surveys.

1. Quantitative methods were documented in the textbooks of various urban planning-related disciplines, for example urban economics, urban geography, system science and geographical information science. These methods, like scenario analysis, systems dynamics and genetic algorithms, were extensively applied in urban studies and planning practice. Urban planners are generally required to master these methods.
2. Software in this chapter was defined as existing PSSs developed by developers outside BICP, like *What if*? and INDEX, commercial, shared, or free, which could support plan compilation and evaluation.

3. A model in our framework was defined as a tool specially developed to implement a function to support plan compilation or evaluation. Generally, models were all developed or will be developed by BICP, while existing models developed by third parties were excluded. Models in the framework were highlighted and will be regarded as the base for the next steps in various phases of PSS development.

11.2.5 Proposing PSSs for Plan Elements

Proposing appropriate PSSs for each plan element was the core procedure in establishing the PSS framework. Researchers with backgrounds ranging from urban planning, transport planning, municipal infrastructure planning to social planning were involved in this process. Literature review was the dominant approach used for proposing PSSs for each plan element. We also held several extra seminars for BICP planners to evaluate the proposed framework. More than ten urban planning experts were involved in developing this framework.

11.3 The New Framework

11.3.1 The Framework and Detailed Descriptions

We developed a comprehensive PSS framework for various types of urban plans in China. Using the methods for designing the framework, we proposed 128 methods, 59 software programs, and 58 models to be included in it. Table 11.3 shows a part of the framework. For an example of the plan element "urban growth boundaries" in spatial layout of master plan compilation, the full description for the element is "delimit the urban expansion and settle the boundary of built-up area". Various datasets including boundary and area of built-up area over the years, previous land use plans, DEM, socioeconomic status, municipal infrastructure, transport infrastructure, land use status, as well as constraining elements are necessary for establishing urban growth boundaries. Methods like cellular automata and trend analysis, and software like SWARM, REPAST, NETLOGO and ArcGIS (Spatial Analysis module) could be used in the process. Furthermore, we propose several existing or to-be-developed models, e.g. Beijing Urban Spatial Development Model (BUDEM), Urban Growth Control Model (UGCM) and Land Use Layout Analysis Model (LULAM), to support establishing urban growth boundaries.

We have detailed descriptions for each PSS in the framework, which readers can use to learn more about the PSS took it. In addition, several models have been developed in BICP.

Table 11.3 Part of the PSS framework proposed

Level 4	Planning element	Descriptions	Data	Method	Software	Model
Problem analysis	Topography and geomorphology	Analyze the topography and geomorphology, construct the digital elevation model, and compute the slave and aspect	DEM, RS		ArcGIS (3D Analyst Tools)	Basic topography model
	Current conditions	Analyze the current situations of natural resources, historical evolution, spatial layout, infrastructure and social and economic issues	Natural resources (ecological environment, land resource, water resources, etc.), engineering geological conditions, historical and cultural resources, land cover status, municipal infrastructure, transport infrastructure, population, industry		PSS tools of Chenghui, ArcGIS (Analysis Tools), Excel	Status comprehens model
	Land use suitability	According to the requirements of land cover, analyze the land cover suitability (usually divided into suitable, comparatively suitable and unsuitable levels), determine the constraining factors of exploitation, find out the optimal tray of land use and a sound plan scheme	Elevation, slope, existing land cover, existing land cover, municipal infrastructure, transport infrastructure, natural resources (water source, wet land, forest)	Grid algebra operation, multi-attribute evaluation, basic topography analysis model, grey system theory	ArcGIS (Spatial Analyst Tools)	Land use suitabil model

	Population spatial distribution	According to the population of each statistical unit, display and analyze the spatial distribution of population with a continuous surface of population density using spatial interpolation	Population and land use in towns and sub districts (total number of population and buildings)	Density core analysis, spatial interpolation, Monte Carlo	ArcGIS (Spatial Analyst Tools), GeoDA	Spatial distribut population predict
Forecast of development trend and scale	Population development	Analyze the scale of population in different historical stages and judge the development trend in the future	Demographic data over the years	Synthetic growth-rate method	SPSS, Excel	
	Urbanization development	Analyze the spatial distribution, expansion, direction and mechanism of urban construction land in different historical stages (e. g. location, accessibility and public policy)	Existing land cover, existing land use, DEM, municipal infrastructure, transport infrastructure over the years	Remote sensing interpretation, Logistic Regression model, principal component analysis, phase analysis, land use evolution analysis model	Erdas, Envi, ArcGIS (Spatial Analyst Tools.), SPSS (Logistic regression, correlation analysis, principal component analysis), GWR3X	Land use evolution model, beijing city development analysis

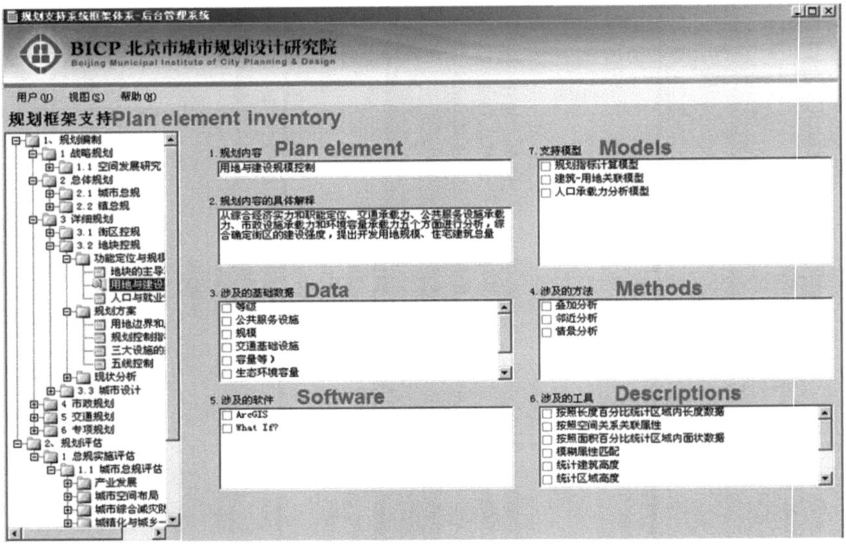

Fig. 11.3 The main graphical user interface of the online query system for the PSS framework

11.3.2 The Online Query System

In addition to a hard-copy version of the framework, we have created an online version of the framework, from which one could query related PSSs for a plan element, query the application fields of a specified PSS, and learn about a PSS with downloadable materials in PDF form. The main interface of the online query system is shown in Fig. 11.3. This browser-based system was developed using Asp.Net and C# on a Windows 2003 platform. The contents of the framework were stored in a Microsoft SQL Server 2005. The functions of the system included: (1) querying required PSSs/data/guidelines for a specified plan element; (2) querying the plan elements where a specified PSS could be applied; (3) downloading existing PSSs; (4) querying the person in charge of a PSS; (5) searching a plan element of PSS. This system was installed on the BICP intranet and could be easily accessed by planners in the institute via a browser.

11.4 Discussion

11.4.1 Application and User Evaluation of the Framework in BICP

The PSS framework has been available at BICP, a top official planning agency in China with more than 300 planners, for several months. It has attracted hundreds of application requests from planners. The applications of the framework include the following aspects.

First, we developers organized a large-scale training workshop for all planners in BICP so as to widely apply the framework in BICP. The contents of the workshop range from the development background, overall structure, user manual of the online system, as well as our further plans. Most of planners agreed that they have gained basic knowledge on the framework and its user manual.

Second, BICP planners regard the PSS framework as a knowledge base of PSS. In the framework, each PSS has been associated with a planner who is familiar with and experienced in its use. The users can contact the associated person to gain more knowledge in the usual way by Instant Messenger in BICP. Building on BICP's existing spatial databases, the framework can promote the application of new techniques in urban planning compilation and evaluation. Currently these techniques are not common in architecture-dominated official and private planning institutes in China, thus our framework has potential applications in those entities. The application of the framework could broaden the horizons of planners by introducing a large body of planning support techniques and then promoting their efficiency and increasingly scientific results.

Third, BICP planners regard this framework as a knowledge base of urban planning theories. New urban planners could complete practical tasks more effectively using information from the framework. Newcomers could get to know the detailed procedure of a specific job, like a detailed plan in a town, by checking the plan element rows of the framework. In addition, planners with various backgrounds like urban planning, transport planning and municipal infrastructure planning could learn more about unfamiliar plan fields. In current China, it is not easy for planners to familiarize themselves with different specialist areas. Assisted by the framework, this situation is expected to improve and new plan schemes are expected to be better than before because of more shared understanding among planners of various backgrounds. As a byproduct, planners in different fields could get familiar with plan contents of others by querying the framework, a process that was not easy prior to the launch of the framework.

Fourth, we drafted the development plan of PSSs in BICP using the new framework. We have decided to develop several fundamental urban models during 2011–2015, based on aggregating models in the framework. These models include the existing land analysis model, the urban spatial development model, the land use and transportation integrated model, the low carbon urban model, the urban planning implementation evaluation model, and the municipal facility evaluation model, as well as the storm water management model.

11.4.2 Potential Contributions

The contributions of our research in the regime of PSS are as follows. First, to our knowledge, this research is the first attempt to establish a comprehensive PSS framework including quantitative methods, models and software for various types of plans, rather than a standalone PSS. The framework is an integration of existing PSSs and those yet to be developed. Second, this framework is a form of urban planning knowledge base, in which users with different education background can share their knowledge both on urban planning theory and PSS toolkits. It will be a complementary digital infrastructure for the well-studied spatial database and is expected to promote the potential application of PSSs. It could also be used as training material for novices in this field. Third, the development plan for PSSs can be compiled based on the new framework, which provides a platform for both long-term goals and short-term development. Seven comprehensive models, e.g. urban spatial development model and existing condition evaluation model, have been proposed to develop by BICP during 2011–2015. Fourth, this research has passed the rigid review process by PSS processors and practitioners in China. The review reports said the work as a fundamental research has its potential extended application and could promote PSS development in the whole country.

11.5 Conclusions and Next Steps

In this chapter, we have proposed a PSS framework for various types of plans in China, e.g. master plan, detail plan, municipal infrastructure plan and transport plan. Based on an extensive literature review and several rounds of planner and decision maker surveys, the framework focuses on two aspects. On one hand, we itemized plan types (termed as "plan elements") into various steps for each type of plan, e.g. population forecasting and establishing urban growth boundaries (UGBs) in master plans. On the other hand, we listed related PSSs for each plan element. In our research, PSSs embody three forms, existing PSS software (e.g. What if? and INDEX), planning support models to be developed in the future or already developed, as well as quantitative methods (e.g. scenario analysis, systems analysis, and logistic regression). The two dimensional framework provides a full picture of PSS application in various types of plans. This framework has two forms of application, the hard copy and the online system. We have revised the established framework several times following discussions with and feedback from planners. This framework is a first systematic attempt to integrate existing planning support techniques and provides users/planners with a knowledge base in both planning procedures and PSS. It is being extensively used in BICP.

Finally, there are several actions that can enhance this study. First, more plan types are expected to be included in the framework, e.g. more special plans and planning evaluation, as highlighted in the latest City Planning Law of the People's Republic of China. Second, PSSs like existing methods and software can be continuously enriched by literature reviews and planner surveys. Third, the online query system can be further developed by linking existing data and PSSs to run the PSS directly in the online system. Fourth, BICP will develop new planning support models and specify them in the proposed framework.

References

Allen, E. (2001). INDEX: Software for community indicators. In R. K. Brail & R. E. Klosterman (Eds.), *Planning support systems: Integrating geographic information systems, models, and visualization tools*. Redlands: ESRI Press.

Brail, R. K. (Ed.). (2008). *Planning support systems for cities and regions*. Cambridge: Lincoln Institute of Land Policy.

Brail, R. K., & Klosterman, R. E. (Eds.). (2001). *Planning support systems: Integrating geographic information systems, models and visualization tools*. Redlands: ESRI Press.

Carmichael, J., Tansey, J., & Obinson, J. (2004). An integrated assessment modeling tool. *Global Environmental Change Part A, 14*, 171–183. doi:10.1016/j.gloenvcha.2003.12.002.

Clark, K. C., Hoppen, S., & Gaydos, L. (1997). A self-modifying cellular automaton model of historical urbanization in the San Francisco Bay area. *Environment and Planning B: Planning and Design, 24*, 247–261. doi:10.1068/b240247.

Du, N., & Li, Y. (2005). Planning support system (PSS) and its application to decision making for urban spatial development (in Chinese). *Engineering Journal of Wuhan University, 38*(1), 137–142.

Edamura, T., & Tsuchida, T. (1999). Planning support system for an urban environment improvement project. *Environment and Planning B: Planning & Design, 26*, 381–391. doi:10.1068/b260381.

Geertman, S., & Stillwell, J. (Eds.). (2003). *Planning support systems in practice, advances in spatial science*. Berlin: Springer.

Geertman, S., & Stillwell, J. (2004). Planning support systems: An inventory of current practice. *Computers, Environment and Urban Systems, 28*, 291–310. doi:10.1016/S0198-9715(03).

Geertman, S., & Stillwell, J. (Eds.). (2009). *Planning support systems best practice and new methods*. Berlin: Springer.

Geneletti, D. (2008). Incorporating biodiversity assets in spatial planning: Methodological proposal and development of a planning support system. *Landscape and Urban Planning, 84*, 252–265. doi:10.1016/j.landurbplan.2007.08.005.

Harris, B. (1960). Plan or projection: An examination of the use of models in planning. *Journal of the American Institute of Planners, 26*, 265–272. doi:10.1080/01944366008978425.

Kammeier, H. D. (1999). New tools for spatial analysis and planning as components of an incremental planning-support system. *Environment and Planning B: Planning & Design, 26*, 365–380. doi:10.1068/b260365.

Klosterman, R. E. (1997). Planning support systems: A new perspective on computer-aided planning. *Journal of Planning Education and Research, 17*, 45–54. doi:10.1177/0739456X9701700105.

Klosterman, R. E. (1999). The what if? Collaborative planning support system. *Environment and Planning B: Planning and Design, 26*, 393–408. doi:10.1068/b260393.

Landis, J. D. (1994). The California urban futures model: A new generation of metropolitan simulation models. *Environment and Planning B: Planning and Design, 21*, 399–420. doi:10.1068/b210399.

Landis, J. D., & Zhang, M. (1998a). The second generation of the California urban futures model. Part 1: Model logic and theory. *Environment and Planning B: Planning and Design, 25*, 657–666. doi:10.1068/b250657.

Landis, J. D., & Zhang, M. (1998b). The second generation of the California urban futures model. Part 2: Specification and calibration results of the land-use change submodel. *Environment and Planning B: Planning and Design, 25*, 795–824. doi:10.1068/b250795.

Lautso, K. (2002). The SPARTACUS system for defining and analysing sustainable urban land use and transport policies. In S. Geertman & J. Stillwell (Eds.), *Planning support systems in practice*. Heidelberg: Springer.

Li, Y. (2010). Planning support system: Current situation and reflection. *Urban Studies, 17*(5), 59–65.
Li, X., & Liu, X. (2008). Embedding sustainable development strategies in agent-based models for use as a planning tool. *International Journal of Geographical Information Science, 22*, 21–45. doi:10.1080/13658810701228686.
Li, S., & Zhan, Q. (2011). Application of UPlan planning support system (in Chinese). *Chinese Building, 9*, 90–91.
Liu, K. (2003). Planning-support system as an innovative blend of computer tools. *Foreign Urban Planning Construction, 18*(5), 15–20.
Long, Y. (2007). *Principles and applications of planning support system (in Chinese)*. Beijing, China: Chemical Industry Press.
Long, Y., Mao, Q., & Dang, A. (2009). Beijing urban development model: Urban growth analysis and simulation (in Chinese). *Tsinghua Science and Technology, 14*, 787–794. doi:10.1016/S1007-0214(09)70149-X.
Long, Y., Shen, Z., & Mao, Q. (2011). An urban containment planning support system for Beijing. *Computers Environment and Urban Systems, 35*, 297–307. doi:10.1016/j.compenvurbsys.2011.02.004.
Luo, J., Dang, A., & Mao, Q. (2009). Research on integration of the service-oriented digital city planning platform (in Chinese with English abstract). *Beijing City Planning & Construction Review, 2*, 113–116.
Niu, Q. (2012). *Application guide of urban planning GIS technology (in Chinese)*. Beijing: China Architecture & Building Press.
Shi, X., & Yeh, A. G. O. (1999). The integration of case-based systems and GIS in development control. *Environment and Planning B: Planning and Design, 26*, 345–364. doi:10.1068/b260345.
Vonk, G. (2005). Bottlenecks blocking widespread usage of planning support systems. *Environment and Planning A, 37*, 909–924. doi:10.1068/a3712.
Vonk, G., Geertman, S., & Schot, P. (2007). A SWOT analysis of planning support systems. *Environment and Planning A, 39*, 1699–1714. doi:10.1068/a38262.
Waddell, P. (2002). Modeling urban development for land use, transportation, and environmental planning. *Journal of the American Planning Association, 68*, 297–314. doi:10.1080/01944360208976274.
Wu, F. (1998). Simland: A prototype to simulate land conversion through the integrated GIS and CA with AHP-derived transition rules. *International Journal of Geographical Information Science, 12*, 63–82. doi:10.1080/136588198242012.
Yeh, A. G. O., & Qiao, J. (2004). Component-based approach in the development of a knowledge-based planning support system (KBPSS). Part 1: The architecture of KBPSS. *Environment and Planning B: Planning and Design, 31*, 517–537. doi:10.1068/b2721.

Chapter 12
The Planner Agents Framework for Supporting the Establishment of Land Use Patterns

12.1 Introduction

Urban master planning is a basic tool for the Chinese government to regulate urban development and promote sustainability. Land use patterns, also known as land use layout, is one of the key issues in compiling an urban master plan or a land use plan. In China, the government, urban planners and local residents are the involved agents participating in establishing the land use pattern. Particularly, government has a role on determining the overall goals of social, economic and environmental development under the constraints of local development conditions; planners play a role in negotiating with related agents, then establishing and evaluating land use patterns. Meanwhile, residents provide suggestions and feedback to the related agents. All these participants have varying requirements and preferences for land use patterns. For instance, government wants to improve social, economic and environmental development simultaneously, planners emphasize implementation of a specific planning concept or theory, and residents are concerned with parks or shopping centers having higher accessibility. In practical content, however, demands and inclinations frequently do not meet city development regulations, or even contradict them. Establishing land use patterns using traditional planning means depending largely on the planners, and these methods lack effective theories to reflect the planner's role in the process of establishing the land use pattern, and the communication and negotiation mechanisms among agents. This is likely to cause overly strengthen of planner's effect, and ignorance of or weak compliance with the requirements and preferences of other agents, especially residents, thereby reducing the plan's suitability as a result.

A Planning Support System (PSS), a computer-aided instrument specifically designed to support comprehensive tasks in urban planning, is mainly based on theories and technologies such as planning model, visualization, and GIS (Klosterman 1997; Brömmelstroet 2013). Compared with traditional planning

Y. Long, Z. Shen, *Geospatial Analysis to Support Urban Planning in Beijing*, GeoJournal Library 116, DOI 10.1007/978-3-319-19342-7_12

methods, PSS can help access and analyze spatial data, support the design process, evaluate urban planning schemes, and therefore improve public participation.

PSS has been widely discussed and applied in urban planning for decades (Geertman and Stillwell 2003; Geertman and Stillwell 2004; Conclelis 2005; Mao et al. 2006; Long 2007; Brail 2008; Vonk and Ligtenberg 2010; Long et al. 2010a, 2011a; Curtis 2011). Several research works are related to land use pattern establishment. For example, California Urban Futures (CUF) developed by Landis (1994) can replicate realistic urban growth patterns and the impacts of development policy at various levels of government, and allocate urban growth to sites based on development profitability. What if? which can efficiently indicate the influence of planning management, was proposed by Klosterman (1999), and this tool has been widely used in other studies such as growth management strategy evaluation and land use forecasting (Klosterman et al. 2006; McColl and Aggett 2007). INDEX, developed by Criterion Planner, can evaluate planning influence in multiple aspects, including environment, energy, transport and public finance (Allen 2001). iCity (Stevens et al. 2007), based on vector Cellular Automata (CA), is a novel model for urban growth simulation to aid spatial decision making for urban planners. Long et al. (2011b) developed an urban containment PSS in Beijing for automatically compiling the urban containment plan, which represented constraints on the land use pattern. Niu et al. (2008) formulated a land use plan based on Scenario Planning, Qin et al. (2010) simulated land use allocation based on CUF, and both used Goals Achievement Matrix (GAM) for scenario evaluation. All these studies, however, are not applications that are only aimed at supporting the establishment of the land use pattern, and they only can complete a portion of establishment tasks, such as urban land boundary formulation, planning evaluation, and constraints acquisition. Moreover, these applications do not deal with the establishment process from the planner perspective, which is closer to the real situation. Aspects such as model parameter acquisition, evaluation method, and considered planning impact factors (PIFs) should be improved to better support land use pattern establishment.

There are several studies related to Planner Agents (PAs). For example, Ligtenberg et al. (2001) defined actors as players (both individuals and groups) in the process of spatial planning. Actors would communicate, negotiate and decide upon the spatial organization of their environment, and simulate spatial change as a result of actor-based decision making. Agent iCity, developed by Jjumba and Dragicevic (2012), is an upgraded version of iCity, and can simulate the land use pattern by incorporating interactions of various stakeholders. Ligtenberg et al. (2009), based on their preliminary study (Ligtenberg et al. 2001), extended an existing approach with the principle of sharing knowledge among participating actors. Saarloos et al. (2005) defined agents as land use experts that initiate the development of plan proposals and communicate with each other over time, for drawing up proposals incrementally. The above, however, are mostly at a preliminary level and pay little attention to empirical studies. Additionally, communication is generally limited among planners, and coordination between the land use pattern and other types of plans (e.g., those for transport and public facilities) is not considered. To the best of our knowledge, in China, there is as yet no research related to Planner Agents.

In this chapter, we propose the PAs framework for establishing land use patterns based on existing PSS research. In this framework, we determine characteristics of and interactions among government, various planners and resident agents. We identify Planning Rules (PRs) for reflecting planner requirements and preferences through existing plan archives. The land use pattern can be established by the Spatial Planner Agent (SPA) combined with identified PRs, comprehensive constraints made by the Government Agent (GA), and special plans formulated by the Non-spatial Planner Agent (NPA). The established land use pattern can be evaluated by the SPA from multiple aspects, and the Resident Agent (RA) conducts a satisfaction evaluation of the established land use pattern. Finally, the Chief Planner Agent (CPA) determines the final ideal land use pattern. The framework of Planner Agents is proposed and described in detail in Sect. 12.2. The framework is tested in a small region of Beijing Metropolitan Area (BMA) in Sect. 12.3 to demonstrate its validity. At last, we conclude and propose the benefits and future research venues of this study in Sect. 12.4.

12.2 Framework and Methods

12.2.1 Basic Concepts

12.2.1.1 Planner Agents

According to the content of the work performed by urban planners in supporting the establishment of land use patterns, Planner Agents can be classified into three types: Non-spatial Planner Agent (NPA), Spatial Planner Agent (SPA) and Chief Planner Agent (CPA). The NPA is responsible for formulating special plans such as for transport, municipal public facilities and nature reserves, which correspond to data such as road network, public facilities and nature reserve patterns. The SPA is responsible for establishing and evaluating land use patterns. The SPA considers constraints of local development conditions, and communicates and coordinates with the NPA to confirm special plans that can support implementation of the established land use pattern. The CPA is responsible for determining the final land use pattern after a public participation process involving local residents.

12.2.1.2 Planning Rules

As we define them, Planning Rules (PRs) are criteria or guidelines of planner thinking and action when the planner establishes the land use pattern. The main content of PRs consists of the planner's considered planning impact factors (PIFs) and their weights. There are many PIFs for land use patterns, such as roads, rivers, parks and traffic noise. Different planners with varying demands and inclinations will consider different sets of PIFs, for which weights are usually different. As a case, planner A may believe parks and rivers are the most critical PIFs for a residential parcel pattern, whereas planner B only considers the park as a PIF, but just a normal one. In this case, the river is not a PIF for planner B, and the park weight of planner B is less than that of planner A. The planner's PRs reflect different requirements and preferences. For example, whether to consider the river and the determination of its weight for a residential parcel pattern reflects the demands and inclinations of a riverfront development strategy. In the following, requirements and preferences from planners are uniformly replaced by PRs.

PRs mainly consist of three aspects: parcel partitioning, land use type and development intensity determinations. In this chapter, we focus on how to determine land use type at present with primary consideration of planner requirements and preferences regarding parcel size, street scale, riverfront development, transit-oriented development, compact city, mixed use, and others.

12.2.2 The Framework of Planner Agents

The framework of Planner Agents for establishing the land use pattern contains eight steps. At first, the GA determines comprehensive constraints. Second, PRs are identified through existing plan archives (including existing special plans), questionnaire surveys, or other means. Third, the NPA formulates special plans (this formulation is not considered in this chapter, and formulated special plans are treated as exogenous variables). Fourth, the SPA establishes the land use pattern, combining the identified PRs, comprehensive constraints, and formulated special plans. Fifth, the SPA communicates and coordinates with the NPA to confirm formulated special plans can support implementation of the established land use pattern. If not, the SPA revises the established land use pattern or the NPA revises any special plans, until they meet the formulation requirements above (not considered in this chapter). Sixth, the SPA evaluates the established land use pattern from multiple aspects. Seventh, the RA does an evaluation of the established land use pattern (not considered in this chapter). And finally, the CPA coordinates the established land use patterns, and determines the final ideal one (Fig. 12.1).

12.2.3 Obtaining Comprehensive Constraints

The GA considers constraints of the planning laws, standards and physical geographic status of land use patterns when determining overall goals of social, economic and environmental development. These are called comprehensive constraints, and generally can be classified into two specific types: land use type and land use quota. Land use type constraints mean parcels are constrained to distribution as

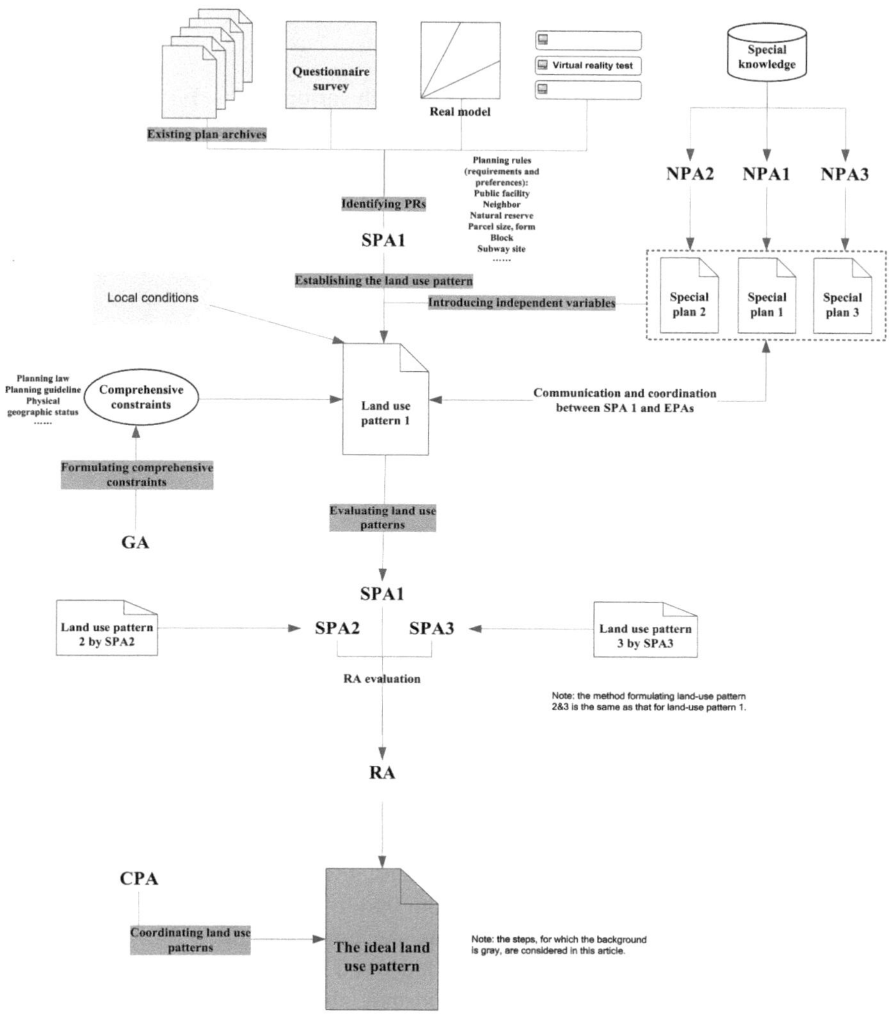

Fig. 12.1 The framework of planner agents

certain land use types, and they can be identified by uniform analysis zone (UAZ) method (Klosterman 1999; Long et al. 2006, 2010b). Land use quota constraints mean that the total areas of parcels of a certain land use type should be as similar as possible, but no more than the planned quota for an established land use pattern. This can be determined according to the specific objectives of city development; for example, the area of residential parcels should be no more than a certain quota according to the urban master plan.

12.2.4 Identifying Planning Rules

We identify PRs (PIFs and their weights) through various ways. Firstly, PRs can be identified through existing plan archives using multinomial logistic regression (MLR). In this process, the parcel is treated as a research unit, the parcel's planned land use type as a dependent variable, and the PIF as an independent variable to identify the weight of every PIF for every planned land use type. The detailed calculation method is as follows:

$$\begin{aligned} T &= \{t_k \mid k = 1,2,3,\ldots K\} \\ F &= \{f_i \mid i = 1,2,3,\ldots I\} \\ P &= \{p_n \mid n = 1,2,3,\ldots N\} \\ W &= \{w_{ik} \mid i \in [1,I], k \in [1,K]\} \\ P_{nk} &= \frac{e^{r_k + \sum_{i=1}^{I} w_{ik} \times f_i}}{1 + \sum_{k=1}^{K-1} e^{r_k + \sum_{i=1}^{I} w_{ik} \times f_i}} \end{aligned} \tag{12.1}$$

where t_k is the planned land use type, K is its number, f_i is the PIF, I is its number, p_n is the parcel, N is its total amount, w_{ik} is the weight of f_i for t_k, P_{nk} is the probability of p_n for t_k, and r_k is the corresponding constant term.

For existing plan archives, variables T (Residential R, Commercial C and Industrial M, and Others O), F (corresponds to special plans), P and P_{nk} (0 or 1) are known, so W can be calculated, and W and F constitute PRs.

PRs can also be identified through questionnaire surveys at the professional institutions. After discussion with survey respondents, PIFs can be confirmed by survey specialists, and PIF weights that ranked from 0 to 9 (insignificant to very important) are reflected in scoring by the respondents. Accumulating the information from a certain number of such questionnaires, the identified PRs become reasonable.

On another aspect, the PR identification may also be implemented using other methods, such as real models or virtual reality tests. For example, Hatna and Benenson (2007) identified the rules of city construction using building blocks, and Crompton (2012) calculated information content using LEGO® sets as a language.

In the following Beijing experiment, we identify PRs using existing plan drawings. Other methods would be tested in our further studies.

12.2.5 Establishing the Land Use Pattern

The NPA can formulate any special plans. Existing special plans are used to identify PRs, and formulated special plans are utilized to support establishment of the land use pattern in this work.

Based on identified PRs and existing special plans, as symbolized as T, F, P and W, we can calculate P_{nk}. Combined with comprehensive constraints, land-use pattern can be formulated. For T contains R, C, and M as an example, the detailed flow about how to allocate land use types, namely the establishment process, by the SPA is as follows:

- (1) Calculating P_r, P_c, and P_m of parcel *n*, combined with identified PRs, formulated special plans, and land use type constraints (if parcel *n* is constrained by land use type constraints, P_k is defaulted to be 0).
- (2) Comparing the size of P_r, P_c, and P_m of parcel n, to determine *n*'s suitable land use type *CompType*, for which the value is R, C or M, respectively.
- (3) According to land use quota constraints, comparing the size of P_r (then P_c and P_m) of all parcels, to determine which parcels are suitable to be distributed as R. The total area of suitable R parcels should be close to but no more than a certain planned quota. If parcel n is suitable to be distributed as R, then the comparative value of *CompTypeR* is R; otherwise, it is *NULL*.
- (4) For parcel *n*, if there is one of the variables *CompTypeR*, *CompTypeC* or *CompTypeM* for which a value exists (all other values are *NULL*), for example *CompTypeR*, *n*'s final distributed land use type *FinalType* is R.
- (5) For parcel *n*, if there are at least two of the variables *CompTypeR*, *CompTypeC* or *CompTypeM* for which values exist, there is a contradiction of the land use pattern. Then, the determination of *FinalType* is made according to size of the values of P_r, P_c, and P_m.
- (6) Calculating the distributed areas of R, C and M. If the distributed area of a certain land use type is smaller than the land use quota constraints, such as for R, then distribute the *FinalType* as R for the remaining parcels, for which the value of *CompType* is R, until the requirement quota is met.
- (7) After step (6), if the distributed area of a certain land use type is still smaller than land use quota constraint, then determine the *FinalType* for the remaining parcels randomly, until the quota is met.

12.2.6 Evaluating the Land Use Pattern

The SPA evaluates the land use pattern, taking multiple considerations into account. For example, the SPA analyzes spatial distribution using spatial clustering methods (Moran's I and others), evaluates urban forms using FRAGSTATS (Mcgariga and Marks 1994), and calculates potential transport energy consumption using the Urban Form-Transportation Energy Consumption-Environment MAS model (FEE-MAS) developed by Long (2011c). Various evaluation indexes are listed in Table 12.1, and a certain combination of indexes can be selected for evaluation, according to a specific objective.

Table 12.1 Evaluation indexes

Index A	Index B	Index C	Index D
Average area	Dimension index	Moran's I index	Shannon evenness index
Average center	Division index	Nearest neighbor distance	Smallest parcel area
Average perimeter	Edge density	Perimeter area ratio	Smallest parcel perimeter
Connectance index	Largest parcel area	Potential transport energy consumption	Total number of parcels
Contagion index	Largest parcel perimeter	Shannon diversity index	Total perimeter

12.2.7 Coordinating Land Use Patterns

When coordinating established land use patterns, the CPA first checks whether they contradict existing comprehensive constraints. Then, the CPA coordinates different elements according to results of the satisfaction survey completed by local residents. Thirdly, the CPA determines the final ideal scenario based on the coordination results. In the present stage, the CPA technically analyzes the evaluation results to determine the final ideal scenario, according to a comprehensive score. The detailed formula is as follows:

$$S = \sum_{n=1}^{N} \alpha_n \times p_n \tag{12.2}$$

where N is the number of elements considered in the evaluation, α_n is the weight of element n, p_n is the comparative score, and S a is the comprehensive score.

12.3 Beijing Experiment

In Sect. 12.3, we test the Planner Agents framework in Beijing to demonstrate its suitability. The PAs model was coded by employing Python language and Geoprocessing. When it comes to quantify the weight of PIFs, only the influence of accessibility (defined as the impact f here) is considered. The shortest Euclidean distance dist from the parcel to the PIF can be calculated using the Distance/Straight Line tool in ArcGIS. The impact f, determined by dist, is calculated according to Eq. 12.3, and we set $\beta = 0.001$ empirically.

$$f = e^{-\beta * dist} \tag{12.3}$$

The detailed information of Beijing experiment is in the following subsections.

12.3.1 Study Area

The Beijing Metropolitan Area (BMA, see Fig. 12.2a) has an area of 16,410 km^2. It has experienced rapid urbanization in terms of GDP and population growth since the Reform and Opening Policy of 1978, established by the Chinese central government. There are 16 districts under BMA jurisdiction, and four main districts under the jurisdiction of the Beijing Central Area (BCA). In 2010, the BCA urban area was 987.5 km^2, and areas of residential, commercial and industrial parcels were 194.6 km^2, 129.2 km^2 and 64.3 km^2, respectively.

In the Beijing experiment, there were four land use types, namely R, C, M and O. Existing plan archives included the BCA Detailed Controlling Plan (BCA-DCP) (Fig. 12.3) and special plans. To identify PRs, planned parcel samples were selected from the BCA-DCP. To identify three PRs by which we can establish three different land use pattern scenarios, three sets of planned parcel samples were chosen. These are distributed over the entire BCA, Regions A and B. On another aspect, Region C (Fig. 12.2a) in the Beijing Haidian District was chosen as an experimental area, in which the established land use pattern will be distributed. Because parcel division is not considered and the impact of the current land use pattern is not addressed in the Beijing experiment, Region C is treated as a "clearing space", in which parcel division is the same as that in the BCA-DCP, but has no current land use pattern. The boundaries of the aforementioned regions are shown in Fig. 12.2a. As shown in Fig. 12.2b and Table 12.2, there are 336 parcels in Region C, the largest one, with area 107.67 km^2 (41 % of the total area). The area of M is the smallest, about 0.4 % of that total.

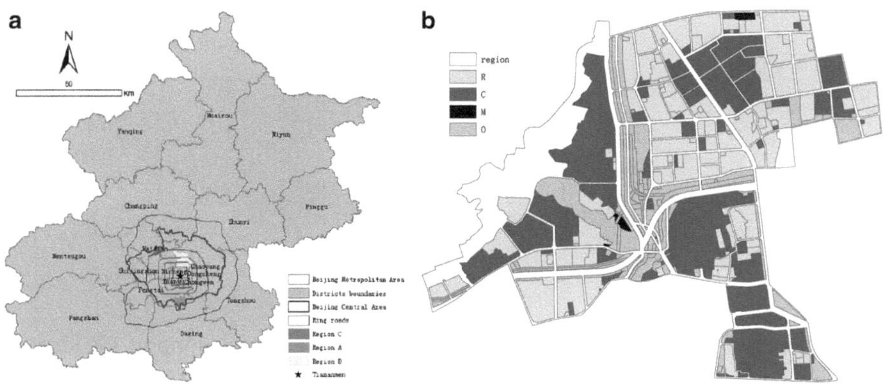

Fig. 12.2 (**a**) Beijing metropolitan area, (**b**) the planned land use pattern of Region C

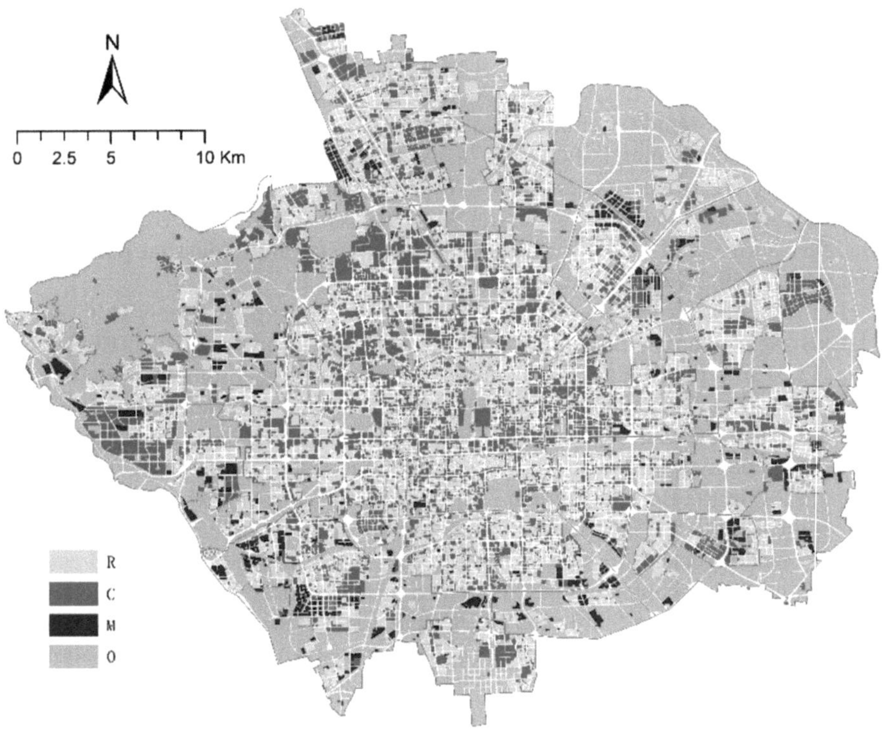

Fig. 12.3 BCA-DCP (the planned land use pattern of BCA)

Table 12.2 The planned land use pattern of Region C

Land use type	Parcel distribution		
	Number	Area (km^2)	Percentage
R	114	43.85	0.41
C	97	44.41	0.41
M	4	0.47	0.004
O	121	18.94	0.18
Total	336	107.67	1.00

Subject to availability, the existing special plan data, which support PR identification, and formulated special plan data by the NPA, which support establishment of the land use pattern, are the same, for which are all from the spatial database of Beijing Institute of City Planning. The fact that the planned parcel samples are from different regions means that there are differences among the three identified PRs, so the established land use patterns will vary, even using the same special plan data.

Table 12.3 shows the GIS spatial distribution of special plans in PIF format. The data of special plans covers the entire BMA. The impact f was calculated by Eq. 12.3.

12.3.2 Obtaining Comprehensive Constraints

The GA determines land use type constraints according to Beijing Limited Construction Zone Planning (Long et al. 2006; see Fig. 12.4). Land use quota constraints are determined according to the existing plan of Region C, such that areas of R, C and M parcels are no greater than 43.85 km^2, 44.41 km^2 and 0.47 km^2, respectively.

12.3.3 Identifying Planning Rules

Taking identification of PR 1 as an example, the planned parcel samples are distributed throughout the entire BCA, and are analyzed by multinomial logistic regression. In the BCA, there were 29,799 parcels included in the identification process. Among these, there were 9594 R parcels (32.2 % of the total number), 7516 C parcels (25.2 % of total), and 753 M parcels (2.5 % of total). Table 12.4 shows the identified parameters, i.e., the PIF weights. If the parameter is a positive number and closer to the particular factor, the parcel is more likely to be distributed as this land use type and vice versa. The −2 log likelihood decreases from 69,795.728 (intercept only) to 62,575.235 (final), and the significance of the likelihood ratio test is below 0.001, which indicates that the regression model is significant overall.

Identification of PR 2 and 3, namely the planned parcel samples in Regions A and B, is the same as that of PR 1, so the corresponding results and processes are omitted here.

12.3.4 Establishing Land Use Patterns

The SPA establishes the land use pattern using PR 1, 2, and 3 from existing plan archives. The established scenarios are shown in Fig. 12.5 and Table 12.5. The number of R and C parcels for scenarios A, B, and C varies, but these scenarios are relatively consistent. The middle and northeast parts of the region are predominantly R parcels, and C parcels are mainly distributed in the south. Scenarios A, B, and C are also relatively consistent with the existing plan (Fig. 12.2b). The difference is mainly the parcels along the river; for scenarios A, B, and C there are mainly C parcels, whereas there are mainly O parcels for the existing plan. This is partly because the existing plan is constrained more effectively by land use type constraints, which have kept these parcels as O types as a result of adopting no development strategy.

Table 12.3 Existing special plans (also formulated special plans)

No.	1	2	3	4
Data	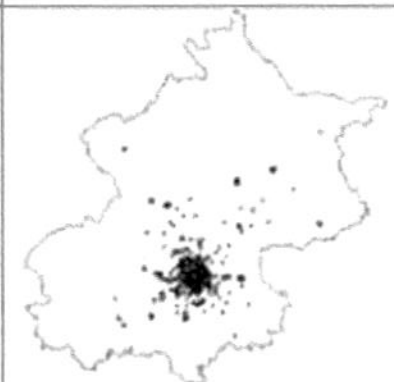	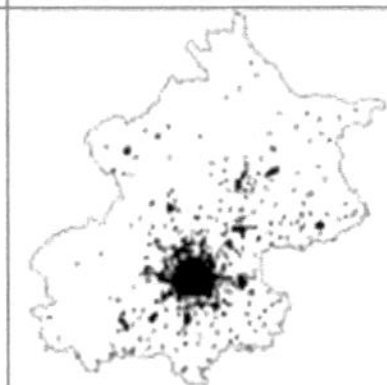	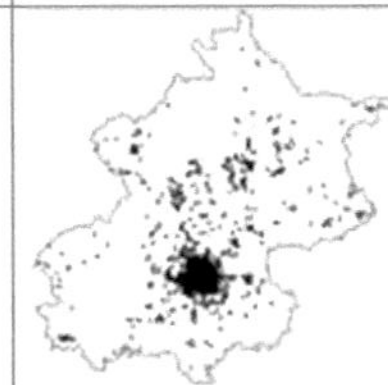	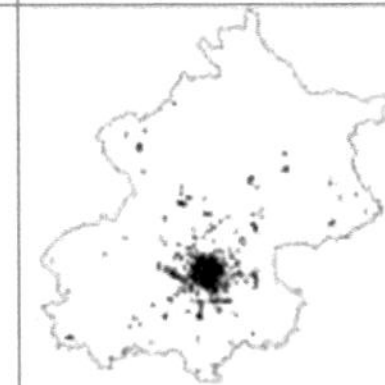
Name	C21 Markets	C22 Banks and insurers	C25 Hotels	C3 Recreational facilities
No.	5	6	7	8
Data	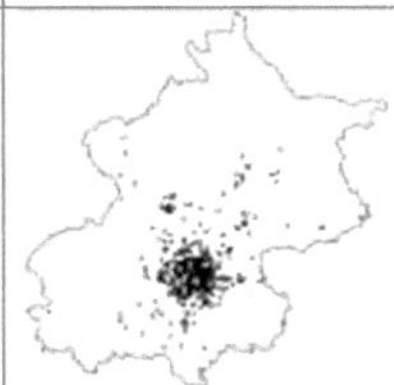	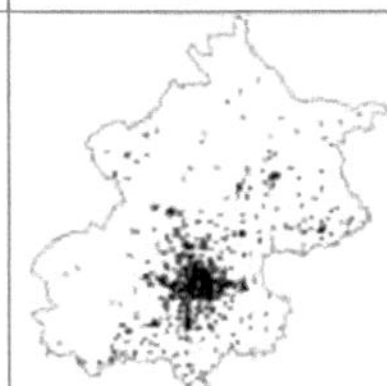	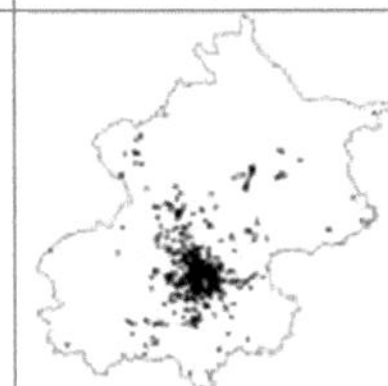	
Name	C4 Sports facilities	C5 Medical and health institutions	C6 Education and research institutions	CBD
No.	9	10	11	12
Data		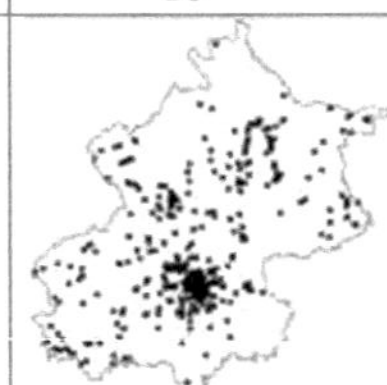	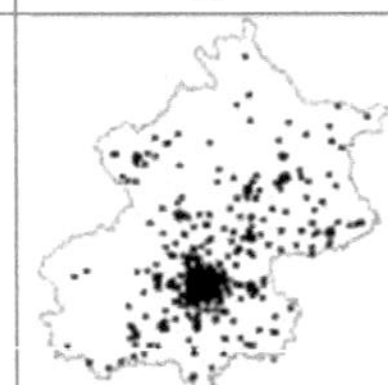	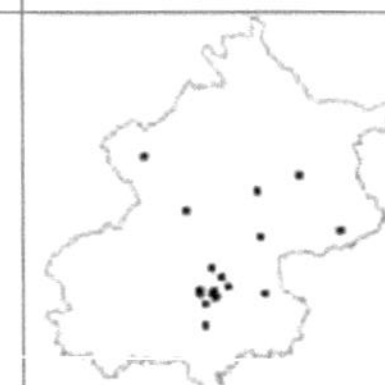
Name	EXIT Expressway exits	G Parks and attractions	GOV Government departments	HWST Highway stations

No.	13	14	15	16
Data			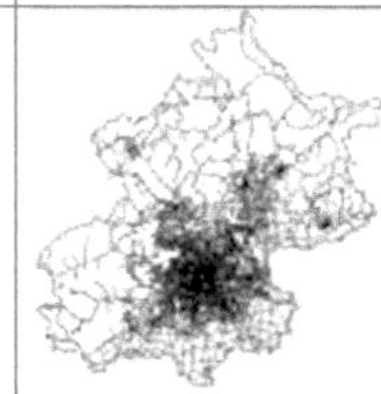	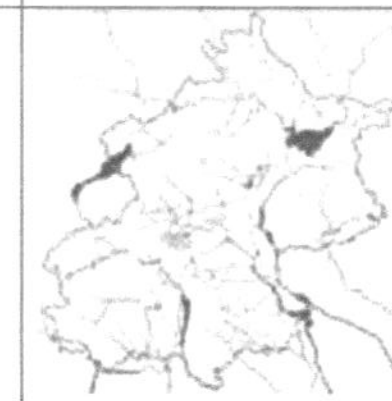
Name	NEWCTY New city centers	RAILST Rail stations	RD06 Road distribution in 2006	RVR Rivers
No.	17	18	19	20
Data			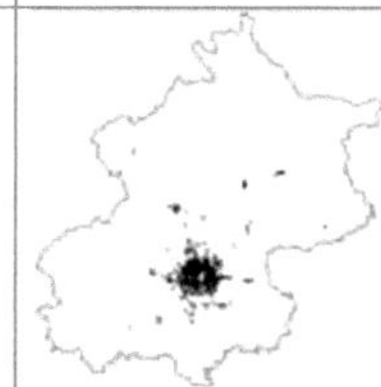	
Name	SUBST Subway stations	TAM Tiananmen	XZL Office buildings	YIZHG Yizhuang Development Zone
No.	21			
Data				
Name	ZGC Zhongguancun area			

Fig. 12.4 Land use type constraints of Region C

Table 12.4 Results of multinomial logistic regression

Parameter	Weight		
	R	C	M
Intercept	−.70203[a]	−2.24992[a]	−1.78990[a]
C21	.59824[a]	.10866	−1.50529[a]
C22	1.69092[a]	1.98993[a]	1.48453[a]
C25	0.27165[a]	.63531[a]	−1.50131[a]
C3	0.54465[a]	.53033[a]	.09401
C4	0.19670**	0.20072**	.34227
C5	1.01238[a]	0.71570[a]	−.37010
C6	0.59667[a]	0.83476[a]	.57046[a]
CBD	−3.13736[a]	−0.73107[a]	−7.74911[a]
Exit	−0.77072[a]	−0.81033[a]	.21059
G	0.06680	0.14353*	−.52322**
Gov	−0.22590[a]	0.11004	0.78724[a]
Hwst	−0.08708	−.28315**	−.95491*
Newcty	−8.33651**	−0.01048	−1.21120
Railst	−0.29179**	−.14296	0.79214[a]
Rd06	−2.09906[a]	−1.19993[a]	−1.10308**
Rvr	−0.26074[a]	−.71772[a]	−1.32691[a]
Subst	0.36312[a]	0.57882[a]	−.41520**
Tam	0.52299	1.24361[a]	−39.32950[a]
Xzl	0.31318[a]	0.52759[a]	1.24840[a]
Yizhg	−91.77109[a]	−101.64079[a]	33.57548**
Zgc	−1.49658[a]	0.16891	−23.24940[a]

Note: [a]p (significance) = 0.01; **p = 0.05; *p = 0.10

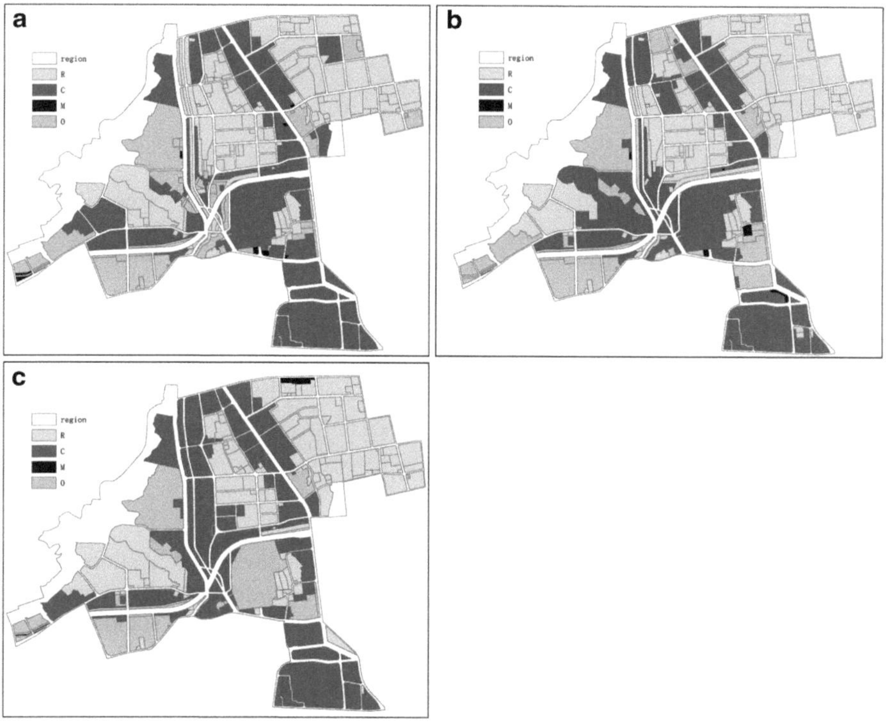

Fig. 12.5 Established land use patterns; (**a**), (**b**), and (**c**) represent scenarios using PR 1, 2, and 3, respectively

Table 12.5 Results of established land use patterns

Land use type	Parcel number (scenario A)	Parcel number (scenario B)	Parcel number (scenario C)
R	163	157	130
C	116	146	182
M	11	7	8
O	46	26	16
Total	336	336	336

12.3.5 Evaluating Land Use Patterns

In order to evaluate three established land use patterns schematically, we adopted three parameters, including Perimeter Area Ratio (PARA_MN), Euclidean Nearest Neighbor (ENN_MN), and Edge Density (ED). The evaluation (or calculation) process was conducted by FRAGSTATS software. In real situation, the requirements

Table 12.6 Evaluation results of land use patterns

Scenario	Land use type	PARA_MN	ENN_MN	ED	Land use type score	Comprehensive score
A	R	0.85	0.99	0.40	0.74	0.65
	C	0.71	0.97	0.23	0.63	
	M	0.60	0.13	0.99	0.57	
B	R	0.78	0.98	0.24	0.66	0.60
	C	0.79	1.00	0.07	0.61	
	M	0.65	0.00	0.99	0.54	
C	R	0.77	0.93	0.42	0.70	0.60
	C	0.91	1.00	0.36	0.75	
	M	0.00	0.11	1.00	0.37	
Existing plan	R	0.87	0.93	0.08	0.62	0.63
	C	0.65	0.93	0.28	0.62	
	M	1.00	0.97	0.00	0.65	

reflected to the index are different for cities have diverse development strategies. To trace the process, a premise is set as that the smaller the index above, the better the results will be to the city development. The range of index score is standardized to 0–1 using feature scaling, and the impact weights of index are all 0.33, the scores of index are calculated by formula (2), comprehensive score is the average score of land type score. The evaluation results are shown in Table 12.6. According to land use type score, the R pattern of scenario A, C pattern of scenario C, and M pattern of the existing plan (Fig. 12.4) are the best. The R pattern of the existing plan, C pattern of scenario B, and M pattern of scenario C are the worst. According to the comprehensive score, scenario A is best, followed by the existing plan; scenarios B and C are worst. According to comparison among scenarios A, B, C, the CPA determines that scenario A is the final ideal land use pattern. Given this comparison between scenarios A, B, C and the existing plan, it may be useful to investigate potential laws or problems that could either support or hinder the establishment of land use patterns using PAs.

12.4 Conclusion

In this chapter, we introduced two aspects of our work. First, the Planner Agents framework was proposed for supporting the establishment of land use patterns. PRs can be identified by several methods, such as the use of existing plan archives, questionnaire surveys, real models and virtual reality tests. Combined with identified

PRs, comprehensive constraints and formulated special plans, the land use pattern can be established. These land use pattern scenarios can be evaluated from multiple standpoints. Following this, the CPA coordinates these scenarios and determines the final, ideal one. Second, the Planner Agents framework was applied in a small region of Beijing, and PR identification was implemented through existing plan archives. Scenario evaluation and coordination of land use pattern scenarios were accomplished by calculating PARA_MN, ENN_MN, and ED indexes using FRAGSTATS.

As shown in the results, the framework of Planner Agents determines the characteristics of and interactions between government, planners and resident agents; it emphasizes the uniqueness and importance of planners, providing a useful framework that reasonably reflects the requirements and preferences of different agents. This approach was demonstrated to be more efficient and scientific than traditional planning methods. Our framework and corresponding methods use the parcel as the analysis unit, are based on existing traditional research on urban land coverage, and establish the entire city's spatial form scenario from the bottom up. Considering the rapid urbanization in China, the requirement for urban planning is much more stringent than before, so this chapter has potential practical application. For applicability to real world situations, the qualification of identified PRs and efficient public participation by the RA are crucial to establish a reasonable land use pattern. Although PRs can be identified quantitatively by diverse methods, it remains difficult to comprehensively reflect PR elements, especially the planner's subjective uncertainties. Moreover, participation by other agents is limited or excessive by many aspects of real society. The reasonability can be promoted by improvement of the Planner Agents framework, and by developing complementary social situations and technologies.

In the future, the Planner Agents framework could be improved in the following ways. First, communication and coordination between the SPA and NPA should be considered. This could be realized by the Form Scenario Analysis (FSA) approach proposed by Long et al. (2010b). Second, we aim to improve the methods for identifying PRs; for example, by considering PIFs more comprehensively and improving their data availability, identifying PRs using questionnaire surveys, real models, or virtual reality tests, and comparing effects of different methods. Third, public participation should be taken into account by introducing resident agents into our research, in this way, the evaluation of land use pattern scenarios can be facilitated based on the principle of resident utility maximization. Finally, we intend to extend application of the Planner Agents framework, e.g., by supporting the formulation of floor-area ratios in addition to the land use types used herein.

Acknowledgements We would like to thanks for the financial support from the Beyond the Frontiers Programme (2013–2014) of Radboud University Nijmegen.

References

Allen, E. (2001). INDEX: Software for community indicators. In R. K. Brail & R. E. Klosterman (Eds.), *Planning support systems: Integrating geographic information systems, models, and visualization tools* (pp. 229–261). Redlands: ESRI Press.
Brail, R. K. (2008). *Planning support systems for cities and regions*. Cambridge: Lincoln Institute of Land Policy.
Brömmelstroet, M. T. (2013). Performance of planning support systems: What is it, and how do we report on it? *Computers, Environment and Urban Systems, 41*, 299–308.
Couclelis, H. (2005). "Where has the future gone?" Rethinking the role of integrated land-use models in spatial planning. *Environment and planning A, 37*(8), 1353–1371.
Crompton, A. (2012). The entropy of LEGOÕ. *Environment and Planning B: Planning and Design, 39*, 174–182.
Curtis, C. (2011). Integrating land use with public transport: The use of a discursive accessibility tool to inform metropolitan spatial planning in Perth. *Transport Reviews, 31*(2), 179–197.
Geertman, S., & Stillwell, J. (2003). *Planning support systems in practice*. Heidelberg: Springer.
Geertman, S., & Stillwell, J. (2004). Planning support systems: An inventory of current practice. *Computers, Environment and Urban Systems, 28*(4), 291–310.
Hatna, E., & Benenson, I. (2007). Building a city in vitro: The experiment and the simulation model. *Environment and Planning B: Planning and Design, 34*(4), 687–707.
Jjumba, A., & Dragićević, S. (2012). High resolution urban land-use change modeling: Agent iCity approach. *Applied Spatial Analysis and Policy, 5*(4), 291–315.
Klosterman, R. E. (1997). Planning support systems: A new perspective on computer-aided planning. *Journal of Planning Education and Research, 17*(1), 45–54.
Klosterman, R. E. (1999). The what if? Collaborative planning support system. *Environment and Planning B, 26*, 393–408.
Klosterman, R. E., Siebert, L., Kim, J.-W., & Hoque, M. A. (2006). What if evaluation of growth management strategies for a declining region. *International Journal of Environmental Technology and Management, 6*(1), 79–95.
Landis, J. D. (1994). *The California urban futures model: A new generation of metropolitan simulation models*. Berkeley: University of California Transportation Center.
Ligtenberg, A., Bregt, A. K., & Van Lammeren, R. (2001). Multi-actor-based land use modelling: Spatial planning using agents. *Landscape and Urban Planning, 56*(1), 21–33.
Ligtenberg, A., Beulens, A., Kettenis, D., Bregt, A. K., & Wachowicz, M. (2009). Simulating knowledge sharing in spatial planning: An agent-based approach. *Environment and planning B, Planning & Design, 36*(4), 644–663.
Long, Y. (2007). *Planning support system theory and practice (in Chinese)*. Beijing: Chemical Industry Press.
Long, Y., He, Y., Liu, X., & Du, L. (2006). Planning of the controlled-construction area in Beijing: Establishing urban expansion boundary (in Chinese). *City Planning Review, 30*(12), 20–26.
Long, Y., Mao, Q., Shen, Z., & Dang, A. (2010a). Beijing urban spatial development model (in Chinese). *Urban and Regional Planning Study, 3*(2), 180–212.
Long, Y., Shen, Z., Mao, Q., & Dang, A. (2010b). Form scenario analysis using constrained cellular automata (in Chinese). *Acta Geographica Sinica, 65*(6), 643–655.
Long, Y., Mao, Q., Yang, D., & Wang, J. (2011a). A multi-agent model for urban form, transportation energy consumption and environmental impact integrated simulation (in Chinese). *Acta Geographica Sinica, 66*(8), 1033–1035.
Long, Y., Shen, Z., & Mao, Q. (2011b). An urban containment planning support system for Beijing. *Computers, Environment and Urban Systems, 35*(4), 297–307.
Long, Y., Shen, Z., Mao, Q., & Hu, Z. (2011c). Urban growth control planning support system: Approach, implementation and application (in Chinese). *City Planning Review, 35*(3), 62–71.

Mao, F., Wang, L., Zhou, W., & He, G. (2006). Urban–rural spatial planning support system based on GIS/RS technology (in Chinese). *Journal of Tsinghua University (Science and Technology), 46*(9), 1495–1498.

McColl, C., & Aggett, G. (2007). Land-use forecasting and hydrologic model integration for improved land-use decision support. *Journal of environmental management, 84*(4), 494–512.

McGarigal, K., & Marks, B. J. (1994). Spatial pattern analysis program for quantifying landscape structure. Dolores (CO): PO Box 606:67.

Niu, X., Song, X., & Gao, X. (2008). Land use scenarios: An approach for urban master plans formulation and evaluation (in Chinese). *Urban Planning Forum, 4*, 64–69.

Qin, X., Duan, X., & Yang, J. (2010). Scenario simulation of urban land-use allocation and scheme evaluation based on GIS, a case study of Taicang city. *Acta Geographica Sinica, 65*(9), 1121–1129.

Saarloos, D., Arentze, T., Borgers, A., & Timmermans, H. (2005). A multiagent model for alternative plan generation. *Environment and Planning B: Planning and Design, 32*(4), 505–522.

Stevens, D., Dragicevic, S., & Rothley, K. (2007). iCity: A GIS–CA modelling tool for urban planning and decision making. *Environmental Modelling & Software, 22*(6), 761–773.

Vonk, G., & Ligtenberg, A. (2010). Socio-technical PSS development to improve functionality and usability – Sketch planning using a Maptable. *Landscape and Urban Planning, 94*(3), 166–174.

Chapter 13
Big Models: From Beijing to the Whole China

13.1 A Golden Era of Big Models

Applied regional/urban models have attracted extensive attention from researchers in recent decades. Regional models are used for regional analysis and simulation at a macro-geographic level, such as for a collection of cities or a whole country, which generally involve a variety of spatial analysis approaches and statistical methods. On the contrary, urban models rely more on modeling and simulation approaches (Batty 2009). They are commonly used for understanding and predicting urban systems through abstracting and generalizing different components of a city. Urban models were firstly developed in the early 1950s and experienced several phases as they developed and evolved. Figure 13.1 presents the development line of urban models from static to dynamic models. The dynamic models further include top-down differential equation-based models and currently prevailing bottom-up models using cellular automata or agent-based approaches. The spatial unit of urban models is also in a transition from a larger geographical unit such as a large grid or a zone to a smaller unit such as a block, a parcel, or even a building (Hunt et al. 2005; Wegener 2004). Generally, these two types of models are utilized separately. According to existing research on applied regional/urban models, they are rarely used simultaneously or synthetically.

In practice, the existing applied regional/urban models can fall into two clusters based on their geographical scale and spatial unit. One is a fine-scaled model for a small area, e.g. part of a city or an entire city. The modeling spatial unit can be a parcel, a block, or a small cell. The other is a model for a large area, such as a region or an entire country. The modeling unit can be a county or a super cell. Because there is a general tradeoff between the spatial extent and the resolution due to the data paucity, it is hard to develop a model that can be applied to a large geographic extent but with a small spatial unit (see Fig. 13.2).

Y. Long, Z. Shen, *Geospatial Analysis to Support Urban Planning in Beijing*, GeoJournal Library 116, DOI 10.1007/978-3-319-19342-7_13

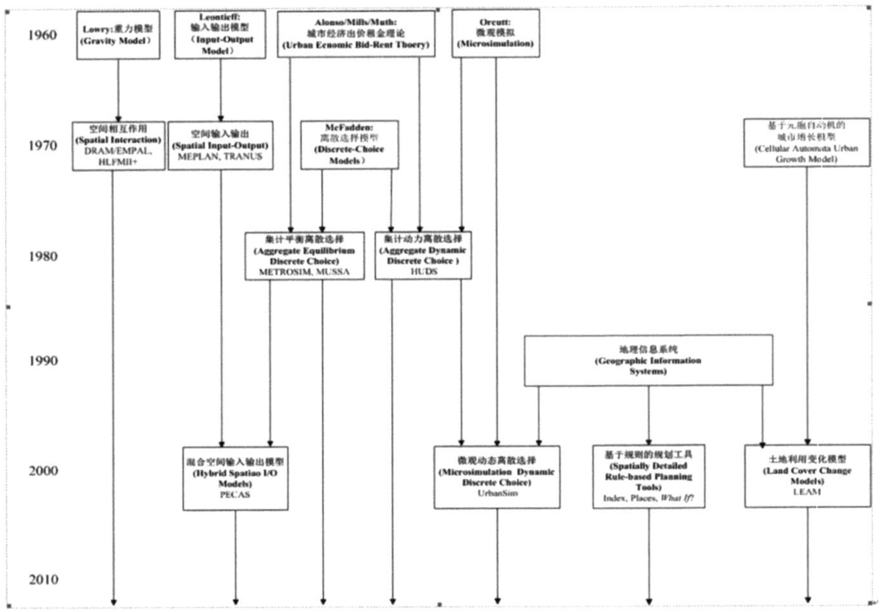

Fig. 13.1 The development line of Applied Urban Models (Adapted from Paul Waddell, Dynamic Microsimulation: UrbanSim, Webinar 5 of 8-part TMIP, Webinar series on land use forecasting methods)

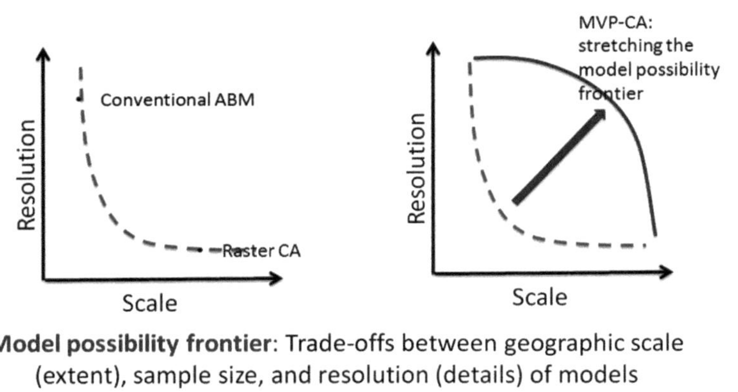

Fig. 13.2 Conventional models vs. "big models" (MVP-CA, a mega-vector-parcels cellular automata model, is our first big model for simulating urban expansion at the parcel level for all Chinese cities)

To the best of our knowledge according to extensive literature review, fine-scale applied urban models for a large area have been rare in academic research. As explained earlier, this stems largely from the lack of data and computation capacity limitation, which are particularly common in the case of China. In addition, collecting fine-scale data for feeding models in medium- and small-sized cities is often constrained by poorly developed digital infrastructures in developing countries. This condition, to some degree, has obstructed the progress of fine-scale urban simulation for a large area in developing countries in general and in China in particular. Overcoming data shortfalls has become the top priority for fine-scale urban simulation in developing countries, even in some developed countries, to support policy making.

In this chapter, we propose a term, namely "big model" for the fine-scale urban simulation model of a regional area with a large geographical scale. Big model is defined as data-driven regional analysis and urban simulation tools involving a variety of modeling approaches in this chapter as a new type of research paradigm for urban and regional studies, thus overcoming the trade-off between simulated scale and spatial unit. More importantly, as our ability to collect, store, and process data has increased remarkably in recent years since the digital revolution, big models would provide us with new opportunities for better understanding how cities work. There are four major reasons making the widespread use of big models happen. (1) Today, big data, such as mobile traces, public transport smartcard records, online check-ins/points-of-interest, and floating car trajectories, are becoming pervasively available. The spread of mobile technologies and computing has made generating, tracking, and recording individual data as partial representation of daily life, greatly supporting the analysis and modeling with rich datasets. Some scholars even advocate that data are models themselves (Batty 2012). (2) Open access to data has been improved significantly as there have been calls for governmental transparency and accountability. For instance, people can access the dataset inventory of planning permits from the official website of Beijing Planning Commission, land transaction records from Beijing Land Bureau, and housing projects from Beijing Housing and Construction Commission. Generally, these records are associated with detailed project-level information, including fine-scale physical characteristics and urban development status. Supported by online geocoding services, these records can be utilized in big models in the form of point datasets. Without painstaking efforts towards an "open government", no such things would have been possible in China. (3) Computational capacity has been largely improved for running big models by means of techniques like parallel computation and Hadoop. (4) For those bottom-up simulation methods adopted by big models, such as cellular automata, agent-based modeling, and network analysis, they have evolved and matured, allowing more sophisticate and powerful application of big models. Therefore, we argue that big models will mark a promising new era for the urban and regional study field.

The purpose of this chapter is to summarize the progress of our existing research on the application of big models in China. The next section elaborates the basic ideas and characteristics of big models. Section 13.3 reviews the methodology development and several case studies in applying big models in various urban and regional researches. In the end, we conclude with a summary of our findings and suggest directions for further research.

13.2 Big Models: A Novel Research Diagram for Urban and Regional Studies

Big models have the following characteristics. First, they need large-scale geographic data including so called "big data" or large-scale "open data" for initialization. The data may be collected at the individual observation level or based on small spatial units. Second, both the existing inter-city and intra-city analysis methods can be integrated in big models (see Fig. 13.3 for an illustration of a big model combining inter-city and intra-city approaches). Third, the geographic scale, or extent of big models is generally larger than that of conventional models but with similar spatial units of simulation. For instance, quality-of-life (QOL) studies can draw conclusions on a city using data at the block/parcel level. But with big models, the analysis of QOL can be conducted to a larger geographic scale, such as for a region or an entire country, and still maintain the same spatial resolution. Fourth, for the same geographical area, a big model can maintain at a higher spatial resolution when compared to a conventional model. A good example is that, in a national-scale population density research, the conventional models may only be applicable at the county or city level, whereas big models driven by fine-scale datasets make it possible to address the issue at the sub-district, block or parcel level, thus helping bring out more meaningful implication for urban spatial planning and policy-making.

Big models can be applied in the following avenues. First, urban dynamics from cities of all sizes can be investigated and examined using big models. Currently, most of applied urban models (AUMs) can only be adopted in large cities where

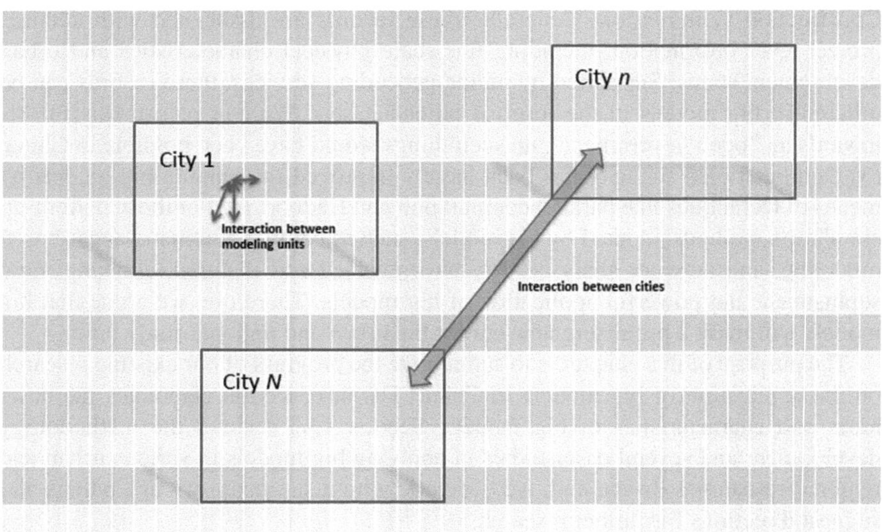

Fig. 13.3 An illustration of a big model integrating intra-and inter-city methodologies

data infrastructure and technical capacity are much better than those in middle- and small-sized cities in China. The introduction of big models could bridge the digital divide caused by data infrastructure. Second, focusing on individual data and fine-scale analyses and modeling, big models provide insightful solutions to various planning issues and contribute to a potential transition from a physical-concentration to a more collaborative and human-oriented planning process. Third, big models enable a variety of urban form and network indicators to be available and meaningful. These factors, combined with commonly adopted socio-economic aggregated indicators, can be adopted for inter-city analyses, which were particularly difficult previously due to lack of necessary road network and parcel geometries across many cities.

Simulating regional and urban dynamics using fine-scale and big models is advantageous as follows. (1) The adoption of local-level data with explicit geographical boundaries would be more appealing to local decision makers and citizens; (2) Land use regulations of spatial plan could be targeted directly at the fine spatial level. This would benefit those cities with limited capacity to analyze and forecast future development; (3) Such model can further be integrated with spatial interaction analysis (i.e. flows and networks).

13.3 Case Studies Using Big Models

Our efforts on the development and application of big models represent a first step towards a better understanding on cities using the emerging big data processing and analysis techniques in China. We outline our methodology development and research process of big models with several completed and ongoing research projects. As most of our case studies draw upon online data sources, the methodologies proposed in this chapter can be easily extended and applied to other cases.

13.3.1 Mapping Urban Built-Up Area for All Chinese Cities at the Parcel/Block Level

Urban built-up areas play a strong role in representing urban spatial development for planning decisions, management, and urban studies. They not only illustrate spatial patterns but also reveal socio-economic characteristics within the built-up areas, e.g., population aggregation, social interaction, energy consumption, and land use efficiency, thereby reflecting how a city evolves in a complex manner (Batty and Ferguson 2011). Conventional methods for delineating urban built-up area from the top down have been applied in major cities around the world on a large scale. However, such methods cannot be applied to most of cities in developing countries due to lack of high-resolution data (Long et al. 2013). Moreover, the research approach of the existing methods for fine-scale studies is conditioned by the presence of data and study context and hence varies from case to case. Against this backdrop, an automatic bottom-up approach was developed in this chapter. Built upon morphological and functional characteristics determined by street network as well as point of interests (POIs), the proposed approach creates a unified way to define fine-scale cities of all sizes.

The definitions and measurements of urban built-up areas have been varied. Urban built-up areas in the United States are defined as Urbanized Areas (UA) in a typical administrative model for spatial statistics. A UA comprises one or more "central places" areas and the adjacent densely settled surrounding "urban fringe" areas, with a total population of 50,000 or more (Morrill et al. 1999). A counterpart in Japan is called "Densely Inhabited District" (DID). DID is a district which has a population density of more than 4000 people per km^2. Urban Areas (UA) in UK are derived from entities-built areas, where certain real-estate densities are detected through satellite images (Hu et al. 2008). On the other hand, socio-economic factors are also adopted to describe the actual urban areas, e.g. labor force markets and commuter sheds are utilized to represent Metropolitan Area (MA) (Berry et al. 1969). Urban built-up areas can be utilized for different purposes with respect to population characteristics, economic status, and built environments attributes.

There are many ways of recording and mapping urban built-up areas. From the perspective of capturing morphological characteristics, an increasing attention has been focused on remote sensing images and street network. Remote sensing and night-time satellite imaging help us gauge urban activity and measure the extent and shape of built-up areas through capturing land cover information and interpreting light data (He et al. 2006). Apart from that, a number of indicators of street network have been introduced to describe the spatial layout of the built environment and predict their correlation with social effects. Examples are street intersection density (Masucci et al. 2012), fractal indices (Jiang and Yin 2014), integration, and accessibility. In terms of the functional characteristics, socio-economic statistics such as demographic densities (Rozenfeld et al. 2009), effective employment density (SGS Economics and Planning 2011, 2012), and infrastructures accessibilities (Hu et al. 2008) have emerged as a standard method of defining urban statistical areas (US Census Bureau 2014).

Nevertheless, these aforementioned approaches have some drawbacks. Firstly, such methods cannot be applied to most of cities in developing countries due to lacking necessary data and fine digital equipment. Moreover, these existing methods still require multiple steps according to unique conditions if achieving a fine-scaled result is expected. Furthermore, these existing approaches seem to isolate the spatial characteristics and the functional ones; therefore the real urban activities seem to be absent in snapping the urban areas by existing methods.

In light of this situation, this chapter employs an automated framework – "automatic identification and characterization of parcels (AICP)" – that was proposed by Long and Liu (2014) to delineate urban built-up areas at the parcel level, based on increasingly standardized roadway asset data from ordnance surveys and crowd-sourced point-of-interests (POIs) data. Roadway data are used to identify and describe parcel configuration, and POIs are processed to infer the intensity, function, and mixing of land use and human activities.

The working definition of a parcel is a geographical entity bounded by roads. Identifying land parcels and delineating road space are therefore dual problems. In other words, our approach begins with the delineation of road space, and individual parcels are formed as polygons bounded by roads. The delineation of road space and parcels is performed as follows: (1) All roadway data are merged as line features in a single data layer; (2) individual road segments are trimmed with a threshold of 200 m to remove hanging segments; (3) individual road segments are then extended on both ends for 20 m to connect adjacent but non-connected lines; (4) road space is generated as buffer zones around road networks. A varying threshold ranging between 2 and 30 m is adopted for different road types (e.g., surface condition, as well as different levels of roads); (5) parcels are delineated as the space left when road space is removed; and (6) a final step involves overlaying parcel polygons with administrative boundaries to determine whether individual parcels belong to a certain administrative unit.

We regard POI density as the ratio between the counts of POIs in/close to a parcel to the parcel area. We further standardized the density to range from 0 to 1 for better inter-city and intra-city density comparison using the following equation: standardized density = log(raw)/log(max), where raw and max correspond to density of individual parcels and the nation-wide maximum density value.[1] We also note that other measures (e.g. online check-ins and floor area ratio) can substitute POIs and approximate the intensity of human activities.

A vector cellular automata (VCA) model is adopted to identify urban parcels from all generated parcels. In this model, each parcel is assigned a value of 0 (urban) or 1 (non-urban). Initially, all parcels are assumed to be rural. To determine the actual status of each parcel, we should take into account not only the individual parcel's intrinsic attributes, such as population density, neighborhood attributes, and some other spatial variables, but also the status of neighboring parcels. The model stops at the iteration when the total area of simulated urban parcels reaches total urban land.

We applied this approach to map city boundaries for all Chinese cities and compared them with urban areas identified by GLOBCOVER, DMSP/OLS and population density. The simulation process and results highlight our proposed framework is more straightforward, time-saving and precise than conventional methods (see Fig. 13.4 for the results in typical cities).

The contribution of this work lies in three major aspects: data, methodology, and innovation. Firstly, the final product of this project is a database containing urban built-up area maps with detailed parcel features for 654 Chinese cities. Featured by fine-scale parcel information, the detailed road network and POIs datasets consolidated in this research can be applied to support a variety of planning and urban

[1] The unit is the POI count per km^2. For parcels with no POIs, we assume a minimum density of 1 POI per km^2.

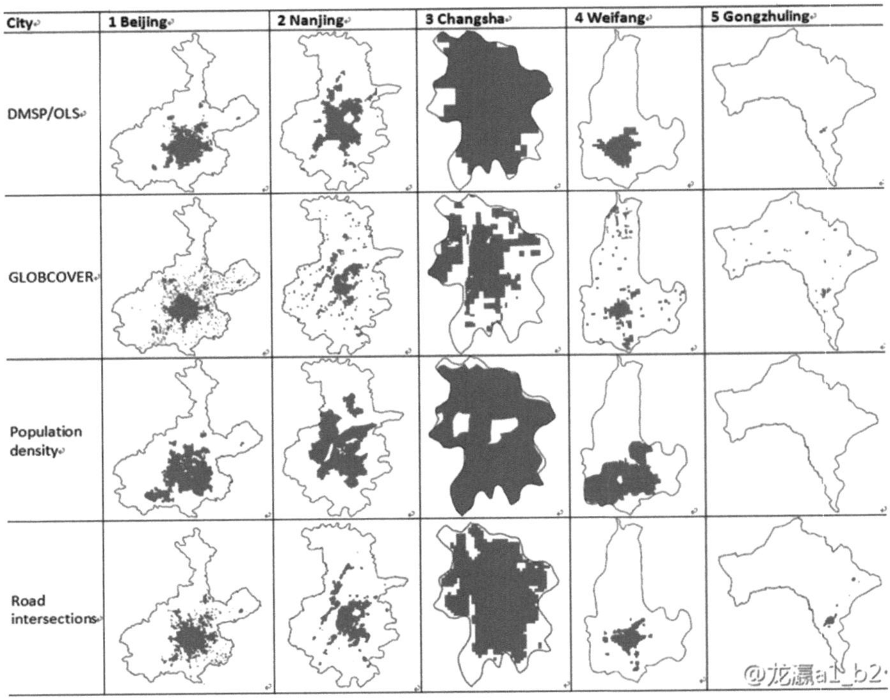

Fig. 13.4 Mapped urban areas in five typical Chinese cities by various methods

studies projects covering a wide range of geographic extent. Secondly, our research proposed a straightforward and consistent approach to identifying urban built-up areas across the country. Unlike previous methods that are somewhat laborious and subjective, our proposed methodology driven by VCA modeling is automatic, straightforward, and objective. The generated parcels can serve as basic spatial units for incorporating other high-resolution ubiquitous and spatially referenced data. In addition to the contribution of delineating urban built-up areas, this research also provides a robust framework for understanding complex urban system across cities from a bottom-up perspective.

13.3.2 Simulating Urban Expansion at Parcel Level for All Chinese Cities

China, as the largest developing country in the world, has experience rapid levels of urbanization in recent year since the introduction of Chinese Reform and Opening-up policies (Montgomery 2008; Liu et al. 2012). Featured by the history's largest flow of rural-to-urban migration and unprecedented economic growth, the urbanization process has shaped and transformed China from a rural to a more urban society. In light of this situation, increasing efforts on urban development assessment and management tools have been made in an attempt to promote a more sustainable development in China; among them are scenario-based urban simulation models (Zhang and Long 2013).

Large-scale simulation models are generally associated with large spatial units in space, like counties or super grids, sometimes reaching tens of square kilometers. Few applied urban models have the ability to pursue a large geographic scale extent with fine-level spatial units simultaneously due to data paucity and computation capacity limitation as discussed previously. Urban expansion simulation at a large geographic extent with a fine-scale (i.e. parcel scale) spatial unit could be promising for several reasons. Firstly, simulation and analysis at the parcel level would be more meaningful for local planners, decision makers, and residents to understand, administer, and monitor urban developments. Secondly, simulation modeling at the large geographic extent enables those administrative entities who have limited capacity to analyze and forecast the urban growth taking place within their boundaries by their own to have an insight on overall urban development scenario within the region and to gauge their growth and take action properly. Also, such simulation models make inter-city comparison possible.

In this section, we developed a mega-vector-parcels cellular automata model (MVP-CA) for simulating urban expansion in the parcel level for all 654 Chinese cities. Three modules, the macro module, the parcel generation module, and the vector CA module, were included in the MVP-CA, as shown in Fig. 13.5.

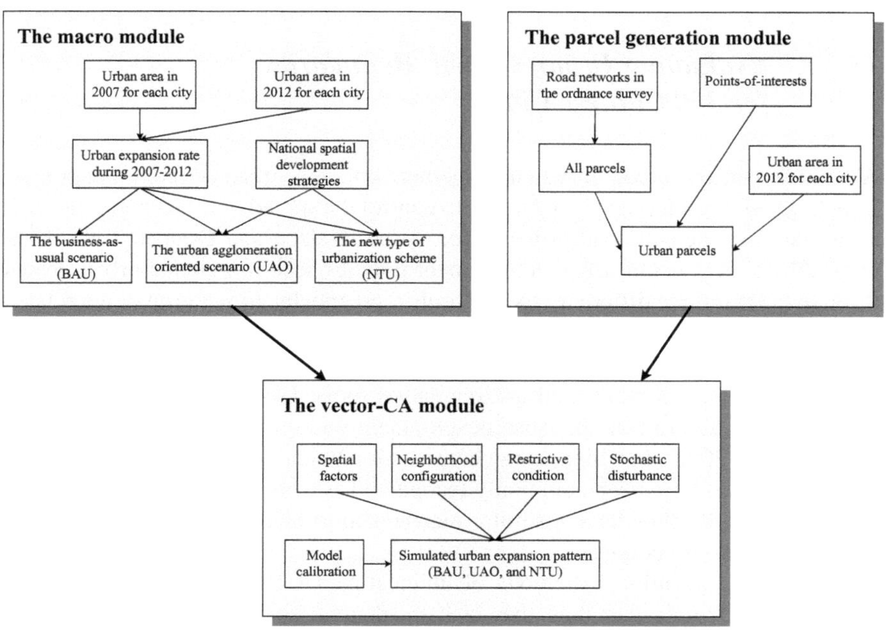

Fig. 13.5 The structure and flow diagram of MVP-CA

The macro module was responsible for setting urban expansion rate in the next five years for each city, taking into account historical urban expansion rate and national spatial development strategies. The parcel generation module was used for identifying existing urban parcels in 2012 using the framework of AICP (automatic identification and characterization of parcels) proposed by Long and Liu (2014). The vector CA module was applied for simulating urban expansion during 2012–2017. This module was examined using calibrated parameters abstracted from Beijing data. Three urban expansion scenarios – baseline, urban agglomeration, and new urban development- have been simulated during 2012–2017 by MVP-CA, respectively. The simulation results are shown in Fig. 13.6. We validated the simulation results by comparing the baseline scenario of Beijing with the results using a raster CA model BUDEM we developed previously.

As one of the first large-scale urban expansion models at the fine-scale for the whole China, our contributions of this chapter mainly lie in the following aspects. First, a vector-based cellular automata model was introduced for simulating urban expansion in a large geographical scale at the parcel level, which is rare in existing literature in the domain urban expansion modelling. Second, we proposed a solution for linking spatial development strategies with urban expansion via reflecting as the urban expansion speed of each city. This enables simulating macro policies in a very fine-scale through the channel of the MVP-CA model. Last, we simulated the near-future urban area for all Chinese cities in China, which, together with existing urban area, has already been shared online as an important data infrastructure for both practitioners and researchers.

13.3.3 Evaluating Urban Growth Boundaries for 300 Chinese Cities

Among the various urban growth management policies, urban containment policies have been widely adopted in an attempt to control the spread of urban areas, increase urban land use density, and protect open space (Nelson and Duncan 1995; Long et al. 2011). In general, urban containment policies seek to manage urban growth through at least three different types of tools – greenbelts, urban growth boundaries (UGBs), and urban service boundaries (USBs) (Pendall et al. 2002). UGB is one of the most widely discussed tools in the planning field. Through zoning, land development permits, and other land-use regulation tools, UGBs demarcate urban and rural uses and aim to contain urban development within the predefined boundaries (Pendall et al. 2002). In China, urban construction boundaries determined in master or detailed plans have been commonly recognized as Chinese/planned UGBs (Long et al. 2013), since they have a similar mechanism to UGBs in the U.S. as well as some other Western countries.

In China, conventional methods of delineating UGBs are based on planners' expertise and experiences; thus, they lack an adequate scientific basis and quantitative support. Consequently, the UGBs often fail to manage urban growth. According

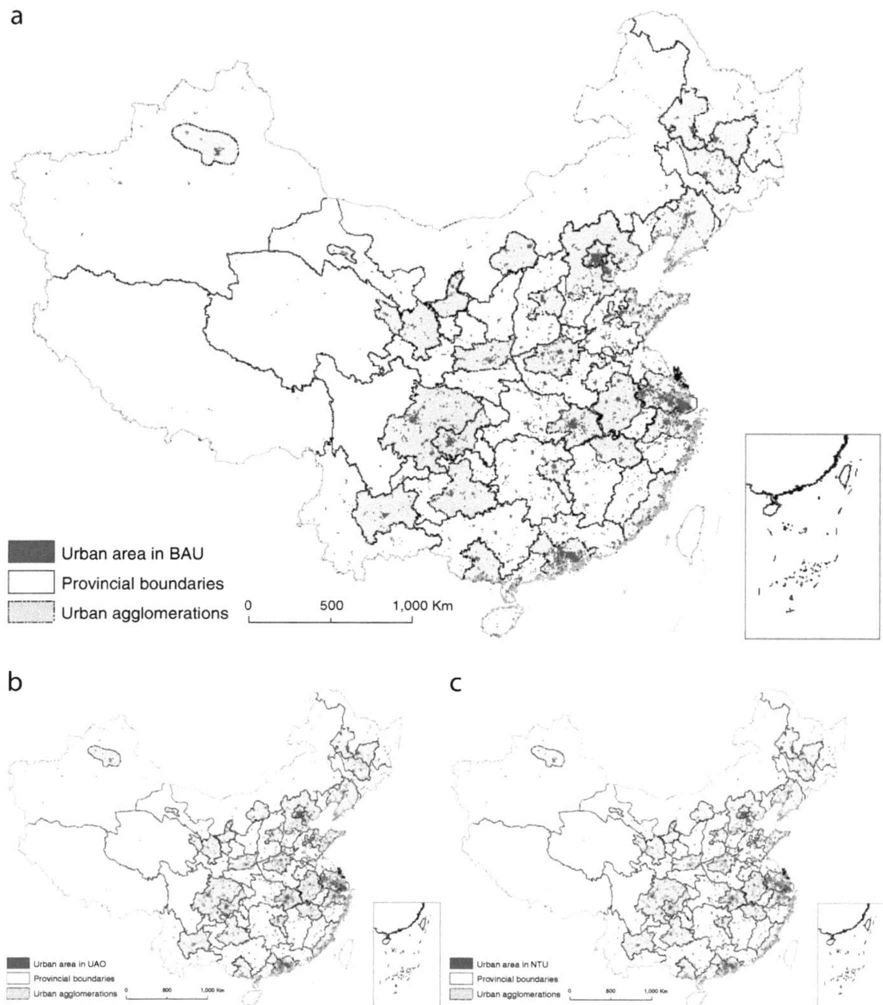

Fig. 13.6 Urban area of all Chinese cities (**a**), and urban expansion patterns of the entire China for three scenarios ((**a**) BAU, (**b**) UAO, (**c**) NTU)

to Han et al. (2009)'s study on the examination of the implementation of planned UGBs within the sixth ring road of Beijing using multi-temporal remote sensing images, more urban land developments were found outside than inside the UGBs during the previous two planning periods (1983–1993 and 1993–2005). Tian and Shen (2011) and Xu et al. (2009) also suggested that substantial urban development occurred outside of UCBs in Guangzhou and Shanghai in recent years. These findings were also supported by Long et al. (2012)'s research, which evaluated five master plans compiled and implemented in Beijing during 1958–2004. Though

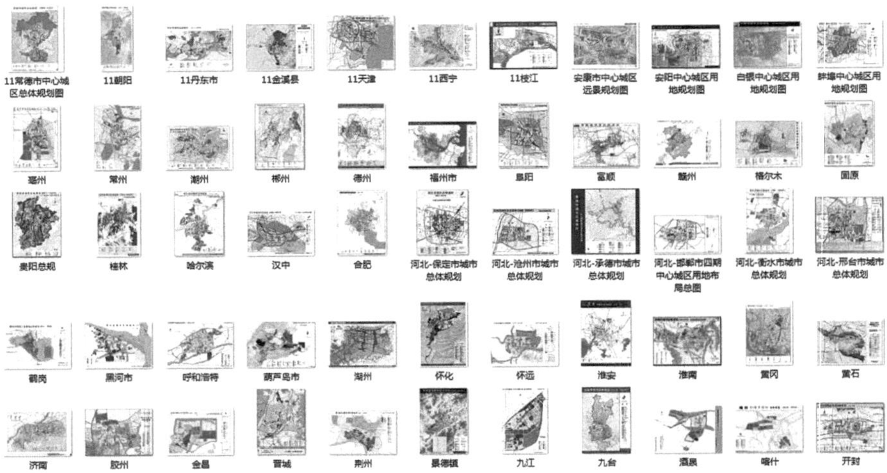

Fig. 13.7 The profile of raw figures for planned UGBs (partially shown)

considerable progress has been made in revealing and quantifying the extent of urbanization and/or evaluating the urban policies' effectiveness on managing urban growth, we have found that most of them have been focused on a single city or region, and little work on the city-level comparison of the performance of UGB's implementation has been done.

Driven by our proposed urban growth simulation model and other relevant big models studies, we launched effort to create a systematic approach to horizontally examine and evaluate the effectiveness of UGBs across cities and regions. We collected raw planning drawing maps on planned UGBs in over 300 Chinese cities (see Fig. 13.7 for a partial sample of cities) and digitalized the boundaries in GIS to facilitate spatial analysis and statistics on these planned UGBs. After that, the planned UGBs of a city were overlaid and compared with the actual extent of urban expansion in the past years since the plan was first implemented, and the ratio of legal development to all urban development can be directly calculated to facilitate city-level comparison. Furthermore, the ambitious degree of each city can be inferred by dividing the actual extent of urban expansion by the planned-to-be-development land area.

Compared with previous studies on big models to date, this research generalizes the planned UGBs across cities and regions and helps make sense of differing results of urban development. In addition, it can provide an insight of the overall trend of urban development in China and thus would be useful for planners to evaluate, monitor, and manage urban planning efforts.

Meanwhile, the digitalized UGBs can also be used to supplement the MVP-CA urban expansion model for all Chinese cities (see our first case study in this chapter) as an institutional constraint, thus accounting for the simulation results. In addition, the project may help identify some universal law of governing the pattern of planned UGBs among all Chinese cities.

13.3.4 Estimating Population Exposure to PM2.5

Chinese cities have for many years suffered from air pollution, which has been a major downside to rapid economic growth and increased urbanization. Currently, few studies of air pollution have been conducted to assess population exposure to PM2.5 over large geographical areas and time periods in China. The existing studies mainly focus on air pollution's effects on health and ecosystems or relevant monitoring methods and measurement, but less has been done on the link between urban spatial structure and air pollution exposure, not to mention their spatiotemporal pattern.

In this study, we collected daily PM 2.5 concentrations during April 08, 2013 and April 07, 2014 from 945 monitoring stations in 190 cities across China.[2] The air quality data were acquired from China National Environmental Monitoring Center (http://www.cnemc.cn). These datasets enable us to understand the PM 2.5 concentration of each station all year round, and can be used as a key input for our estimation. Considering the sparse distribution of monitoring stations across China, we further used Moderate Resolution Imaging Spectroradiometer (MODIS) Aerosol Optical Depth (AOD) retrievals to supplement the PM 2.5 estimates on a daily basis. Demographic statistics were drawn from China's 2010 census data. The spatial distribution of population density across China was determined by geocoding population density of each sub-district based on Google Map API. In total, there are 39,007 sub-districts[3] in China, and the average population density for all sub-districts is 977 persons per km^2. Population have been divided into three age groups (age 0–14, age 15–64, and 65 years and older), with an aim to differentiate the exposure estimates for different sensitive groups such as children and seniors. It is worth mentioning that this is the first time to use sub-district population density for estimating human exposure to air pollution in China, whereas former studies were conducted at the county level at best.

The population exposure estimation involves three major steps. (1) Interpolate the PM2.5 concentration site data into surface data using both ground station-level data and MODIS ADO: PM2.5 concentration data were obtained from all air quality monitoring stations across the country and supplemented with MODIS ADO data. Using numerous spatial interpolation methods, the station-level data can be interpolated into surface data. The outcome of this step is the average daily PM2.5 concentration over the entire area and over time. (2) Estimating population exposure to PM2.5 for each sub-district. Based on interpolated PM2.5 data, a daily PM2.5 concentration above the national standard of 75 mg/m^3 is considered to be unhealthy and thus defined as "exposed". In this way, the total exposed days all year round of each sub-district can be estimated. Further, the exposure intensity for each sub-

[2] There are 657 cities in mainland China as of the end of 2012.

[3] There are three forms of township-level administrative units in China, sub-districts (*jiedao*), towns (*zhen*), and township (*xiang*). Jiedaos are mainly in city area. *Jiedao*'s counterparts in the rural area are towns and townships. Hereafter in this chapter, we use the term sub-district for representing all types of township-level administrative units in China.

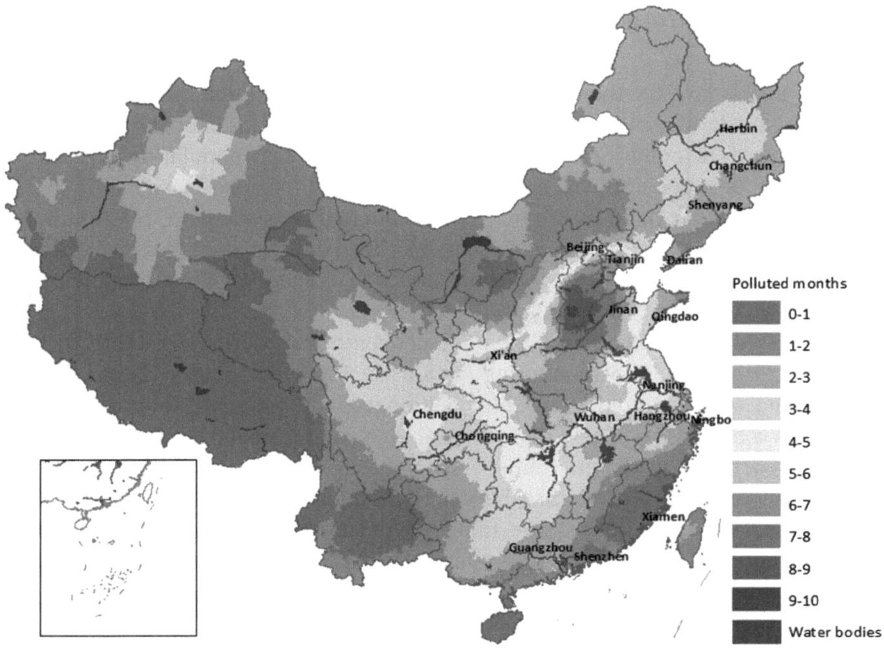

Fig. 13.8 The number of total exposed months for each sub-district in China

district can be calculated using the Equation: Exposure intensity = Population density * Exposed days. The greater exposed days or population density for a sub-district, the higher exposure intensity. This indicator reflects the strength of population exposure to PM2.5. The population density can be subject to specific sub-population groups for estimating the effects on members of sensitive groups. (3) Aggregating the estimated results spatiotemporally. To gain ideas on spatiotemporal pattern of population exposure to PM2.5, we can further aggregate the estimated results in both temporal and spatial dimensions. For the temporal dimension, the total number of exposed month can be calculated for each sub-district, thus presenting a big picture of population exposure to air quality over time. For the spatial dimension, the exposure of each city can be inferred by averaging the estimation results of all sub-districts in each city's administrative boundary.

The number of months subject to exposed condition across the entire country is presented in Fig. 13.8.

The daily exposure for each sub-district was further aggregated by each month. Table 13.1 displays the percentage of exposure days per month from April 2013 to March 2014.

The exposure intensities were obtained by multiplying population density for each sub-district with the estimated exposure days during the period. The final result is presented in Fig. 13.9. It is worth pointing out that the overall exposure intensity pattern generally coincides with the distribution of population density across the country.

Table 13.1 Exposed days in each month for each sub-district in China

Apr 2013	May 2013	Jun 2013	Jul 2013	Exposed percentage in each month
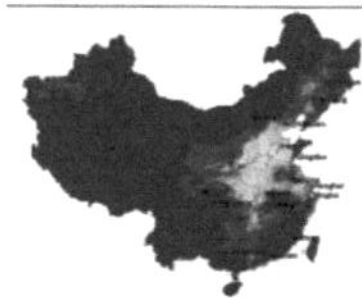	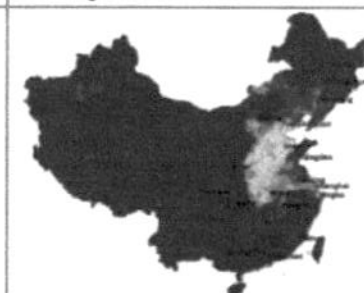		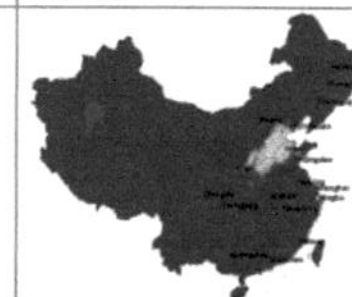	.0 - .1 .1 - .2 .2 - .3
Aug 2013	**Sep 2013**	**Oct 2013**	**Nov 2013**	
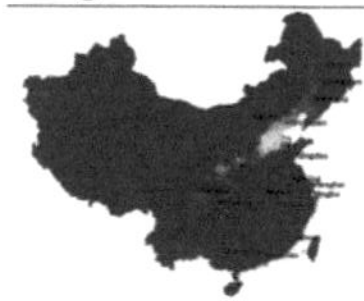	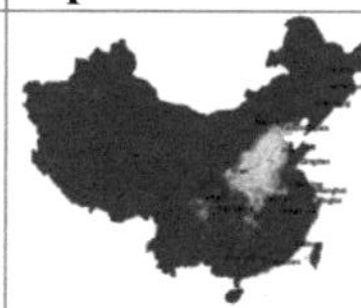	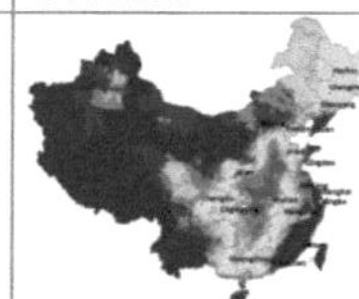	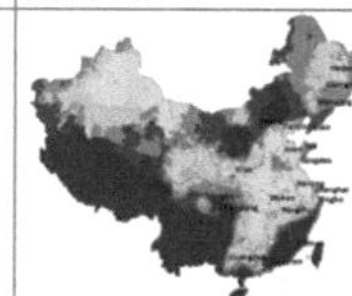	.3 - .4 .4 - .5 .5 - .6
Dec 2013	**Jan 2014**	**Feb 2014**	**Mar 2014**	
	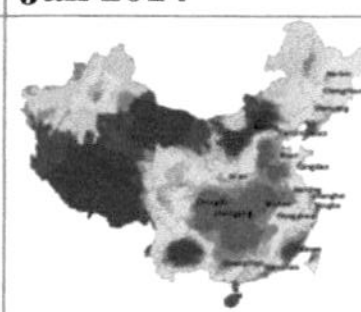	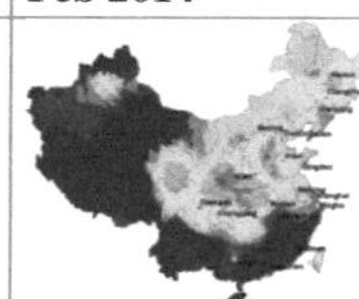	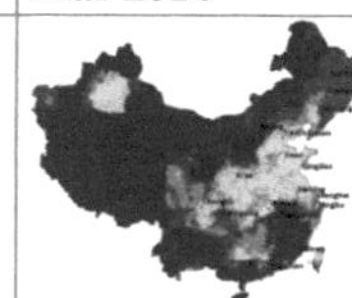	.6 - .7 .7 - .8 .8 - .9 .9 - 1.0

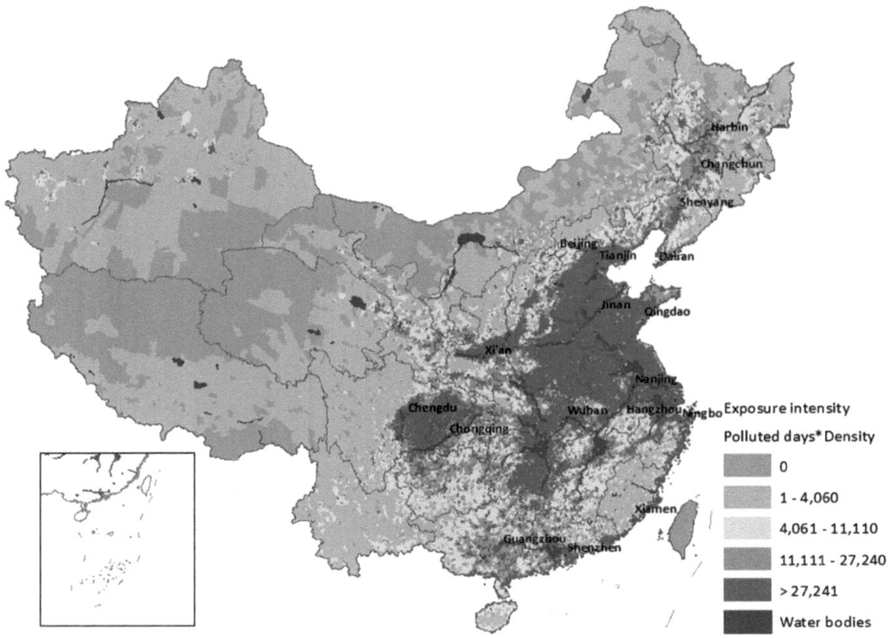

Fig. 13.9 Exposure intensity at the town level of China

13.4 Conclusions and Future Directions

This chapter has proposed the concept of big model as a novel research paradigm for regional analysis and urban studies. The concept, characteristics, and potential applications of big models have been elaborated. Meanwhile, we addressed several case studies to illustrate the progress of research and applications of big models, including mapping urban areas for all Chinese cities, performing parcel-level urban simulation, and several ongoing research projects. Most of these applications can be adopted across the whole country, and all of them are focusing on a fine-scale level, such as a parcel, a block, or a township (sub-district), which is quite different from the existing studies using conventional models. Believing that big models will mark a promising new era for the urban and regional studies in the era of new data environment, we hope our efforts on urban analytics and modeling in Beijing City will set new research agenda and inspire innovative ideas all over the country.

There are several avenues on big models that deserve further studies. First, it is necessary to combine both intra-city and inter-cities methods in big models. Existing case studies in this chapter mainly rely on bottom-up intra-city approaches. City level connections are essential to be included in big models in the next step. For instance, a spatial equilibrium module considering the city level input-output would be helpful for the current MVP-CA model via addressing the interaction between cities. Second, more general theory on big models can be identified through more in-depth case studies analysis.

Acknowledgement We thank Ms Yichun Tu for her editing the language of this chapter.

References

Batty, M. (2009). *Urban modeling. International encyclopedia of human geography*. Oxford: Elsevier.

Batty, M. (2012). Building a science of cities. *Cities, 29*, S9–S16.

Batty, M., & Ferguson, P. (2011). Defining city size. *Environment and Planning B: Planning & Design, 38*(5), 753–56.

Berry, B. J. L., Goheen, P. G., & Goldstein, H. (1969). *Metropolitan area definition: A re-evaluation of concept and statistical practice 28*. Washington, DC: US Bureau of the Census.

Han, H., Lai, S., Dang, A., Tan, Z., & Wu, C. (2009). Effectiveness of urban construction boundaries in Beijing: An assessment. *Journal of Zhejiang University SCIENCE A, 10*, 1285–1295.

He, C., Shi, P., Li, J., Chen, J., Pan, Y., Li, J., Li, Z., & Ichinose, T. (2006). Restoring urbanization process in China in the 1990s by using non-radiance calibrated DMSP/OLS nighttime light imagery and statistical data. *Chinese Science Bulletin, 51*(13), 1614–1620.

Hu, Y., Wu, Z., Xiong, W., & Pan, C. (2008). Study of identifying urban built-up area: Taking Wuhan as an example. *City Planning Review, 32*(4), 88–92.

Hunt, J. D., Kriger, D. S., & Miller, E. J. (2005). Current operational urban land-use–transport modelling frameworks: A review. *Transport Reviews, 25*(3), 329–376.

Jiang, B., & Yin, J. (2014). Ht-index for quantifying the fractal or scaling structure of geographic features. *Annals of the Association of American Geographers, 104*(3), 530–540.

Liu, Z., He, C., Zhang, Q., Huang, Q., & Yang, Y. (2012). Extracting the dynamics of urban expansion in China using DMSP-OLS nighttime light data from 1992 to 2008. *Landscape and Urban Planning, 106*(1), 62–72.

Long, Y., & Liu, X. (2014). Automated identification and characterization of parcels (AICP) with OpenStreetMap and points of interest. arXiv preprint. arXiv:1311.6165.

Long, Y., Shen, Z., & Mao, Q. (2011). An urban containment planning support system for Beijing. *Computers, Environment and Urban Systems, 35*(4), 297–307.

Long, Y., Gu, Y., & Han, H. (2012). Spatiotemporal heterogeneity of urban planning implementation effectiveness: Evidence from five master plans of Beijing. *Landscape and Urban Planning, 108*, 103–111.

Long, Y., Han, H., Lai, S.-K., & Mao, Q. (2013). Urban growth boundaries of the Beijing metropolitan area: Comparison of simulation and artwork. *Cities, 31*, 337–348.

Masucci, A. P, Stanilov, K., & Batty, M. (2012). Limited urban growth: London's street network dynamics since the 18th century. arXiv preprint. arXiv:1206.5298.

Montgomery, M. R. (2008). The urban transformation of the developing world. *Science, 319*(8), 761–764.

Morrill, R., Cromartie, J., & Hart, G. (1999). Metropolitan, urban, and rural commuting areas: Toward a better depiction of the United States settlement system. *Urban Geography, 20*(8), 727–748.

Nelson, A. C., & Duncan, J. B. (1995). *Growth management principles and practices*. Chicago/Washington, DC: Planners Press/American Planning Association.

Pendall, R., Martin, J., & Fulton, W. (2002). *Holding the line: Urban containment in the United States*. Washington, DC: The Brookings Institution Center on Urban and Metropolitan Policy.

Rozenfeld, H. D., Rybski, D., Gabaix, X., & Makse, H. A. (2009). The area and population of cities: New insights from a different perspective on cities (No. w15409). National Bureau of Economic Research.

SGS Economics & Planning. (2011). Melbourne's economy: A stunning success or captured by complacency?, presentation at State Library of Victoria by Terry Rawnsley.

SGS Economics & Planning. (2012). Melbourne metro: Move to more productive jobs, draft version 2.0, 3 February 2012.

Tian, L., & Shen, T. (2011). Evaluation of plan implementation in the transitional China: A case of Guangzhou city master plan. *Cities, 28*, 11–27.

United States Census Bureau. (2014). History: Metropolitan areas. U.S. Census Bureau. http://www.census.gov/history/www/programs/geography/metropolitan_areas.html. Accessed 31 March.

Wegener, M. (2004). Overview of land-use transport models. *Handbook of Transport Geography and Spatial Systems, 5*, 127–146.

Xu, Y., Shi, S., & Fan, Y. (2009). Methodology of Shanghai city master planning in new position. *Urban Planning Forum, 2*, 10–15.

Zhang, Y. P., & Long, Y. (2013). *Urban growth simulation using V-BUDEM: A vector-based Beijing urban development model*. Beijing: The conference of Spatial Planning and Sustainable Development.

MIX
Papier aus verantwortungsvollen Quellen
Paper from responsible sources
FSC® C105338

If you have any concerns about our products,
you can contact us on
ProductSafety@springernature.com

In case Publisher is established outside the EU,
the EU authorized representative is:
Springer Nature Customer Service Center GmbH
Europaplatz 3, 69115 Heidelberg, Germany

Printed by Libri Plureos GmbH
in Hamburg, Germany

MIX
Papier aus verantwortungsvollen Quellen
Paper from responsible sources
FSC® C105338

If you have any concerns about our products,
you can contact us on
ProductSafety@springernature.com

In case Publisher is established outside the EU,
the EU authorized representative is:
Springer Nature Customer Service Center GmbH
Europaplatz 3, 69115 Heidelberg, Germany

Printed by Libri Plureos GmbH
in Hamburg, Germany